Edited by
Richard C. Alkire, Philip N. Bartlett, and Marc T. Koper

Advances in Electrochemical Science and Engineering

Volume 18
Electrochemical Engineering: The Path from Discovery to Product

Advances in Electrochemical Science and Engineering

In collaboration with the International Society of Electrochemistry

Edited by
Richard C. Alkire, Philip N. Bartlett, and Marc T. Koper

Advances in Electrochemical Science and Engineering

Volume 18
Electrochemical Engineering: The Path from Discovery to Product

WILEY-VCH

The Editors

Prof. Richard C. Alkire
University of Illinois
Department of Chemical & Molecular Engineering
600 S. Mathews Avenue
Urbana, IL 61801
USA

Prof. Philip N. Bartlett
University of Southampton
Department of Chemistry
Highfield
SO17 1BJ Southampton
United Kingdom

Prof. Dr. Marc T. Koper
Leiden University
Leiden Institute of Chemistry
Einsteinweg 55
2333 CC Leiden
The Netherlands

Library of Congress Card No.: applied for

British Library Cataloguing-in-Publication Data
A catalogue record for this book is available from the British Library.

Bibliographic information published by the Deutsche Nationalbibliothek
The Deutsche Nationalbibliothek lists this publication in the Deutsche Nationalbibliografie; detailed bibliographic data are available on the Internet at <http://dnb.d-nb.de>.

Print ISBN: 978-3-527-34206-8
ePDF ISBN: 978-3-527-80718-5
ePub ISBN: 978-3-527-80720-8
oBook ISBN: 978-3-527-80721-5

Cover Design Schulz Grafik-Design, Fußgönheim, Germany
Typesetting SPi Global, Chennai, India
Printing and Binding C.O.S. Printers Pte Ltd, Singapore

Printed on acid-free paper

10 9 8 7 6 5 4 3 2 1

Contents

Series Preface

It is with sincere gratitude that we express appreciation to our co-editor and good friend, Professor Jacek Lipkowski, who has indicated his desire to bring more than two decades of editorial leadership to closure with his 2017 publication of Volume 17 of this series, entitled "Nanopatterned and Nanoparticle-Modified Electrodes." His deep technical knowledge and gracious personal manner have made it a genuine pleasure to work with him through the years.

With this volume, we are pleased to welcome Professor Marc Koper as co-editor. Professor Koper studied chemistry at the University of Utrecht (the Netherlands) and the Université Libre de Bruxelles and received his PhD from the University of Utrecht in 1994. After a postdoctoral stay at the University of Ulm (Germany), he first became research fellow and then Associate Professor at the Eindhoven University of Technology. In 2005, he moved to Leiden University (the Netherlands) where he is currently a professor of fundamental surface science. His research interests focus on electrocatalysis, electrochemical surface science, and theoretical and computational electrochemistry. Marc Koper is a member of the Royal Netherlands Academy of Arts and Sciences, fellow of the International Society of Electrochemistry, and has received awards from the Royal Society of Chemistry, the International Society of Electrochemistry, and The Electrochemical Society.

The purpose of this series is to provide high-quality advanced reviews of topics of both fundamental and practical importance for the experienced reader.

Preface

The path from scientific discovery to impact on society has many steps. Today, new science and engineering approaches are being developed to facilitate that task of moving atomistic-scale discoveries and understanding into well-engineered products and processes based on electrochemical phenomena. However, people working on one particular application often find it difficult to recognize highly relevant new methods that were developed for entirely different applications. That is, especially at the atomistic scale, routine science and engineering methodologies are still in an early phase of development for use in assessing emerging business opportunities.

Therefore, the focus for each chapter in this volume includes not only the overarching science and engineering roadblocks that were faced, but also the *reusable* approaches that were developed to inform the technological/business applications with the underlying atomistic science. Many such reusable methods could be relevant beyond their initial use. Bringing together such approaches for diverse applications that span the nano–bio–photo–micro landscape will facilitate recognition and cross-fertilization between seemingly different applications and will lessen the need to “reinvent the wheel” for each.

Alivisatos and Osowiecki, in their Introductory Perspective, emphasize the central importance of the value to society of new candidate technologies. Whether a proposed breakthrough is valuable enough depends in large part on its theoretical performance limits as compared to the current state of the art. Also important for electrochemical applications are selectivity and control of the desired reaction. They illustrate these points with the example of quantum dots, once unusual materials that are today produced at the ton scale and used in commercial display technologies. They comment on several electrochemical applications where considerations based on limits-selectivity-control could provide guideposts on the path from discovery to product.

Brushett describes a new paradigm for battery research: tight integration of discovery science, battery design, research prototyping, and manufacturing collaboration within a single highly interactive organization. This new paradigm, pursued at the Joint Center for Energy Storage Research (JCESR), seeks transformational change in transportation and the electric grid driven by next-generation, high-performance, low-cost electrochemical energy storage. Although JCESR focuses exclusively on “beyond lithium-ion batteries,” the overall systems approach is portable and can be applied to other applications in order to accelerate the pace of discovery and innovation, while reducing the time from conceptualization to commercialization. To this end, the chapter presents JCESR’s motivation, vision, mission, as well as outcomes and lessons learned in the development, execution, and refinement of this mode of operation.

Balsara and colleagues review the redox pathways that have been proposed for the cathode of a Li–S cell as it is charged and discharged. The use of various *in situ* spectral methods to identify the fingerprints of reaction intermediates is discussed. The electrode design required of such methods should guarantee unimpeded access to the species of interest and avoid transport bottlenecks. The basis for understanding the role of the electrolyte for achieving high specific energy is described. The general approach, based on in situ spectroelectrochemical methods, is reusable for other electrochemical systems where intermediate steps in the overall reaction are not yet resolved, such as in alkaline fuel cells and carbonate- or ether-based electrolytes.

Deligianni and coworkers describe the scientific and technological path from laboratory research to early industrial development of electrodeposition for inorganic solar cells. An overarching consideration is to design earth-abundant materials whose elements are amenable to massive-scale application. The chapter describes initial investigations on copper–indium–gallium–diselenide (CIGS) solar cells, before turning to fabrication approaches used for electrodeposited precursor materials, associated fundamentals of electrodeposition, and development of solution chemistries for copper-based earth-abundant electrodeposited kesterite precursor materials such as Cu_2ZnSn (Se,S). A comprehensive description of scale-up procedures is described from the laboratory scale of a rotating disk, to 15×15 cm glass plates, to 30×60 cm modules, and full-size 60–120 cm module, leading up to an industrial-scale production line for producing solar cells at the rate of $1\ m^2\ min^{-1}$.

Ohashi describes the perspectives in Japan that guided evolution of the thin film head technology within the hard disk drive business sector. The technical issues included understanding the properties of candidate magnetic materials and their suitability for use in thin film heads, such as stress, thermal decay, and noise emanating from domain walls of finite thickness. The business issues included recognizing trade-offs between different candidate technologies whose suitability depended on the application. The fluid state of new technologies and new business opportunities creates "The Innovator's Dilemma," which occurs when a new technology brings a value proposition that is different from any ever proposed by existing customers, that is, choosing between sustaining a proven path forward and investing in a potentially disruptive technology.

Orazem and colleagues address the problem of separating water from clay suspensions generated as a waste stream in beneficiation of phosphate ores. The sequential development of a continuous electrokinetic separation process was accomplished with experimental and computational methods that moved atomistic-scale discoveries and understanding toward a well-engineered process. The approach involved empiricism guided by understanding how solids content depended on applied electric field and elapsed time. While electron microscopy and surface analyses provided insight into the structure of the clay, the design of successive prototypes relied on intuition, insights gained from previous prototypes and informed engineering judgment.

Taylor and coworkers investigate pulse reverse-current electroplating and surface-finishing operations. While there are well-establish traditional paradigms for such processes, the authors report various research and development activities carried out with a balance between current fundamental understandings combined with a willingness to pursue non-conforming observations that lead them to paradigm-breaking conclusions. Examples include copper electrodeposition with pulse-reverse cathodic current and decreased used of chemical additives, deburring of non-passive metals with use of forward-only anodic pulses, deburring of passive metals with reverse-pulse anodic and decreased hydrogen evolution, and pulse reverse voltage electropolishing of niobium that uses low concentration acid devoid of HF.

Tsukuhara describes development of a trace moisture sensor that was developed via a series of investigations, spanning several decades, of puzzling phenomena and blind alleys in their pursuit. The tortuous path from discovery to product included non-intuitive results, failed hypotheses, unexpected phenomena, and the need to rethink past observations. All of these roadblocks served to seed fresh ideas guided by background knowledge of laser ultrasonics, surface acoustic waves, propagation of waves in elastic spheres, hydrogen interaction with Pd/Ni films, deposition of amorphous silica, and chemical interaction of water with silica glass, among others. The guiding strategies that emerged from this project include the following: find the thermodynamic limit, do not dismiss something that is apparently wrong, think about observations many times over and be prepared to change you previous opinions, test experiments with numerical simulation, and look broadly into cumulative knowledge available in other fields.

Williams reports on development of low-cost sensors based on gas-sensitive semiconducting oxides. A key concept illustrated in this case is to focus on what limits the path forward. For example, the central technical requirement involved correct control of the sensor and sample, while the economic barriers included the cost of calibration, maintenance, and the cost of an erroneous or unreliable reading. These constraints, along with the knowledge of fundamental science aspects of high sensitivity and low selectivity, led to innovative design of catalyst layers and T-programmed desorption routines. The scientific and engineering threads brought together in this work included understanding surface reactions on semiconducting oxide, sensor development, instrument development, and big data associated with the application. These illustrate how deep scientific understanding led to well-engineered products and markets for them, which, in turn, generated questions that further stimulated the quest for knowledge in fields that were never in view at the start.

In the long run, the reduction to routine use of such methods as described in this volume will provide the foundation for next-generation science and engineering at the molecular scale. In this spirit, we take inspiration from Carl Wagner,[1] who wrote in an earlier volume of this series:

> … molecular engineering may be important in the future development of industrial electrochemical processes.

June, 2018 *Richard Alkire*
Urbana

---

1 Wagner C. (1962). The scope of electrochemical engineering. In: *Advances in Electrochemistry and Electrochemical Engineering*, vol. **2** (ed. C.W. Tobias), 2.

1

Introductory Perspectives

A. Paul Alivisatos [1,2,3,4] *and Wojciech T. Osowiecki* [1,2]

[1] *Lawrence Berkeley National Laboratory, Materials Sciences Division, Berkeley, CA 94720, USA*
[2] *University of California, Department of Chemistry, D43 Hildebrand Hall, Berkeley, CA 94720, USA*
[3] *University of California, Department of Materials Science and Engineering, Berkeley, CA 94720, USA*
[4] *Kavli Energy NanoScience Institute, Berkeley, CA 94720, USA*

The path from a scientific discovery to a commercial product is a long one, with many twists and turns. It is easy to list many examples of potential breakthroughs that never crystallized into real-life solutions. Instead, this book takes the approach of finding the best examples of discoveries that either have already been able to deliver products or are on their way to accomplish this goal. Each chapter discusses a different field, as we believe that there exist certain themes that unite all these success stories.

In order to create a viable product, one has to be brutally honest about what is actually likely to be of genuine value to society, represented by real markets. Many projects and companies struggle if they cannot realize what is fundamentally distinctive about their technology and whether the proposed breakthrough is valuable enough. This conceptualization of what makes a technology distinctive almost necessarily has to be in reference to its theoretical performance limits as compared to the current state of the art.

This comparison between limits and current performance informs the researcher of not only how long the path forward is, but also whether the new technology has any chance of supplanting others or of creating a new market. Research is very complex and unpredictable, and often, we see a clear progression of discoveries that "tell a story" only after the fact. Nevertheless, the question regarding maximum theoretical limits should never leave our minds. Only then can we focus on putting our efforts into the most promising endeavors.

In the case of electrochemical transformations, the most important issues to address after the theoretical limits are selectivity and control of the desired reaction. For example, how do we make sure that the bonds that break and form are ones that we intended and that the desired products are obtained? Sometimes, the inspiration can come from nature and other scientific fields. Enzymes control reactions with an awe-inspiring degree of precision, forming exactly the compounds organisms need, and reminding us how sophisticated and optimized chemical environments can be. Although scientists did not have billions of years

Electrochemical Engineering: The Path from Discovery to Product,
First Edition. Edited by Richard C. Alkire, Philip N. Bartlett, and Marc T. Koper.

of evolution to perfect their processes, this book is also intended as a reminder to look beyond one's area of expertise for encouragement and fresh ideas.

So how can the ideas of theoretical limits and control be applied in electrochemical engineering? The challenges of the current world dictate opportunities for researchers, especially those interested in seeing their work incorporated in crucial technological innovations of the future. In the twenty-first century, supplying energy to the ever-growing global population in a sustainable manner is, without doubt, one of the most important such challenges.

Each year, the U.S. Energy Information Administration forecasts changes in energy generation for the next few decades and we expect to see significant shifts in the fuel mix; as market forces oblige coal plants to retire, they will be successively replaced by renewable sources and oil and gas (Figure 1.1). Among the renewable sources, solar and wind energy are particularly enticing to the electrochemical community, as they promise a growing supply of cheaper electrons, decoupled from environmentally costly fossil fuel combustion.

In this new economy, scientists should be encouraged to think of electrons as crucial and readily available chemical reagents. Just like fossil fuel industry turned oil into a ubiquitous precursor to many compounds and products, now electrons will be involved in crucial processes such as fuel generation and energy storage. Indeed, the ability to store energy in chemical bonds solves one of the greatest challenges for renewables, namely their transience. Energy storage brings opportunities for a distinctive set of technologies to emerge.

We believe that in order to use electrons properly, the issues, highlighted above, of (i) *theoretical efficiency limits* and (ii) *control* must be addressed. To start,

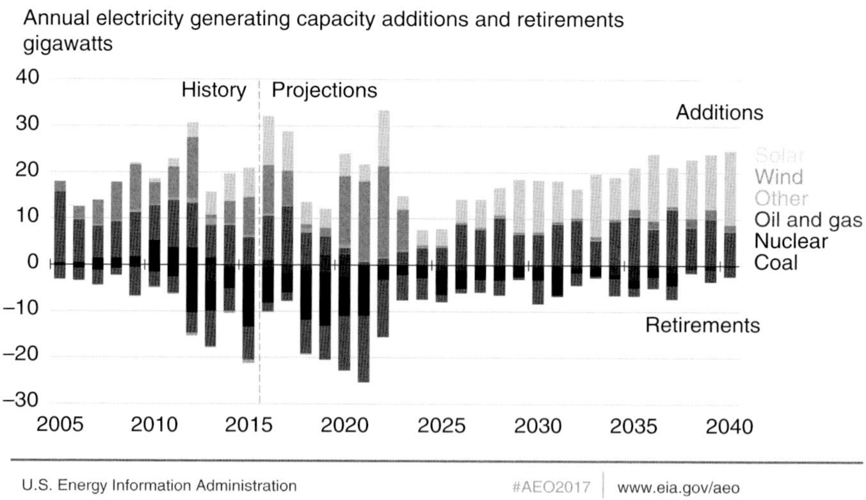

Figure 1.1 Annual electricity-generating capacity additions and retirements. Most of the wind capacity is expected to be built before the scheduled expiration of the production tax credit in 2023, although wind is likely to remain competitive without the credits. Substantial cost reductions and performance improvements strongly support continuous solar generation growth.

let us consider a field that the authors of this foreword are particularly familiar with: quantum dots (QDs). QDs are small semiconducting nanoparticles that possess bound, discrete electronic states [1]. The optoelectronic properties of these dots depend on the size and shape: the larger the particles are, the longer the wavelength of the emitted light is. After a long period of discovery and development, these once unusual materials are produced today at the ton scale and used in commercial display technologies. Their most distinctive edge comes from the color purity of their emission, which creates displays that realize a broader color gamut than previous technologies did.

For current commercial quantum dot display devices, such as televisions, the efficiency of light emission following absorption of a higher energy blue pump photon is crucial for the success of the product. Energy from every absorbed blue photon must be emitted as a green or red photon to form the full color image. The desired radiative rate competes with non-radiative processes that emit phonons instead of photons. Phonon emission not only decreases energy efficiency but also causes unwanted heating that may lead to device failure. This competition is measured by the quantum yield (QY): rate of radiative emissions divided by all rates. An ideal situation, where all energy is released as light, corresponds to the QY of 100%, also known as the unity QY.

As there is no *fundamental limitation* on particles achieving the unity QY (or at least 99.999%), scientists have been working for decades on perfecting synthetic recipes and treatments to get to this limit. Although initial QYs reported for CdS were below 1%, [2] it was very important to conceive of the then-unachievable maximum potential of QDs with much higher yields for biological and optical applications. The *selectivity and control* came with increased understanding of surface-related trap states, arising from insights spanning the fields of semiconductor surface chemistry, optics, electron microscopy, and theoretical modeling. A large community worked together in "constructive competition" to achieve relevant breakthroughs. Thanks to techniques such as particle shelling, [3] QYs were brought significantly above 90% [4]. With QD displays now successfully penetrating the market of TVs and electronic displays, scientists are already exploring new materials, such as perovskites [5, 6], to bring us even closer to the unity QY.

We believe that the same themes of finding maximum efficiency limits and reaching toward them by increasing control and understanding of the researched process apply to the fields presented in this book. Solar cell efficiencies are always compared to the famous Shockley–Queisser (SQ) limit. Although it concerns only a single p–n junction, the number of 33% has motivated researchers and engineers to search for continuous improvements [7]. Today, the market opportunity for a new single band gap technology is narrow because even if it exceeds the current 20% power efficiency of silicon and thin film technologies, it will likely at best be niche, while the ceiling of 33% is not that far off from the current technologies. Scientists may well be advised to look at technologies that naturally lend themselves to multi-gap configurations that have the potential to exceed the SQ limit. Likewise, by performing thermodynamic analysis, one can understand why scientists are excited about Li/air and Li/O_2 batteries. The latter can theoretically achieve 10 times higher energy density than present Li-ion batteries [8].

Currently, long-term stability issues are plaguing these devices but the promise set by the thermodynamic limit is worth years of research pursuing better control of charge/discharge cycles [9].

Another electrochemical field worth mentioning is catalysis, especially CO_2 reduction. From the perspective of thermodynamics, conversion of carbon dioxide to fuels such as ethanol should not be very energetically costly [10]. In reality, a large overpotential is needed to obtain an appreciable amount of reaction products [11]. Additionally, things get complicated due to lack of selectivity. Catalysis with Cu generates up to 16 different chemical species [10]. In electrochemistry, one can always adjust the voltage to speed up the reaction as long as it is kinetically controlled, but with poor control over the catalyst, there are too many electrons going into wrong places. We see here an analogy to exciting a semiconducting quantum dot significantly above the bandgap. High energy excitation increases the absorption of quantum dots and increases the rate of the electrochemical transformation. In both cases, however, the extra energy above a threshold opens up many new unwanted pathways, and in the quest for higher rates, it is easy to end up with unwanted processes. Is it possible to properly engineer the system to avoid them? Scientists are currently working on extending control over this reaction, and we hope to learn more about how to selectively form desired products.

We wish we could finish this foreword with a list of specific steps that scientists can take to guarantee improvement in the selectivity and control of solar cells, batteries, and catalysts. Instead, we have highlighted what we consider to be some of the important steps to consider while researching into improving these technologies. While the theoretical efficiency limits may never be achieved, the thermodynamic analyses will illuminate the most promising uses of electrons; the great challenge for the new generation of electrochemists is to conceive of entirely new approaches to guide reactions at the electrochemical interface by means that have not yet been tried. Creative nano-engineering seems as if it might be the key. Explicit thinking about the role of fluctuations and the creation of more structured sequences of local environments seem like they may lead to breakthroughs. We hope that this book will inspire scientists to consider their own research plans with equal measure of thought given to the limit of what may be possible and to the actuality of what is achieved in practice today. The path from discovery to product will surely prove fruitful if followed with both a realistic and ambitious mind.

References

1 Ashoori, R.C. (1996). Electrons in artificial atoms. *Nature* 379 (6564): 413–419.
2 Chestnoy, N., Harris, T.D., Hull, R., and Brus, L.E. (1986). Luminescence and photophysics of cadmium sulfide semiconductor clusters: the nature of the emitting electronic state. *J. Phys. Chem.* 90 (15): 3393–3399.
3 Peng, X., Schlamp, M.C., Kadavanich, A.V., and Alivisatos, A.P. (1997). Epitaxial growth of highly luminescent CdSe/CdS core/shell nanocrystals

with photostability and electronic accessibility. *J. Am. Chem. Soc.* 119 (30): 7019–7029.

4 Mcbride, J., Treadway, J., Feldman, L.C. et al. (2006). Structural basis for near unity quantum yield core/shell nanostructures. *Nano Lett.* 6 (7): 1496–1501.

5 Burschka, J., Pellet, N., Moon, S.J. et al. (2013). Sequential deposition as a route to high-performance perovskite-sensitized solar cells. *Nature* 499 (7458): 316–319.

6 Protesescu, L., Yakunin, S., Bodnarchuk, M.I. et al. (2015). Nanocrystals of cesium lead halide perovskites ($CsPbX_3$, X = Cl, Br, and I): novel optoelectronic materials showing bright emission with wide color gamut. *Nano Lett.* 15 (6): 3692–3696.

7 Rühle, S. (2016). Tabulated values of the Shockley–Queisser limit for single junction solar cells. *Sol. Energy* 130: 139–147.

8 Zu, C.-X. and Li, H. (2011). Thermodynamic analysis on energy densities of batteries. *Energy Environ. Sci.* 4 (8): 2614.

9 Chu, S., Cui, Y., and Liu, N. (2016). The path towards sustainable energy. *Nat. Mater.* 16 (1): 16–22.

10 Kuhl, K.P., Cave, E.R., Abram, D.N., and Jaramillo, T.F. (2012). New insights into the electrochemical reduction of carbon dioxide on metallic copper surfaces. *Energy Environ. Sci.* 5 (5): 7050.

11 Manthiram, K., Beberwyck, B.J., and Alivisatos, A.P. (2014). Enhanced electrochemical methanation of carbon dioxide with a dispersible nanoscale copper catalyst. *J. Am. Chem. Soc.* 136 (38): 13319–13325.

2

The Joint Center for Energy Storage Research: A New Paradigm of Research, Development, and Demonstration

Thomas J. Carney[1,2], *Devin S. Hodge*[1], *Lynn Trahey*[1], *and Fikile R. Brushett*[1,3]

[1] *Joint Center for Energy Storage Research, Argonne National Laboratory, 9700 South Cass Avenue, Building 200, Lemont, IL 60439, USA*
[2] *Massachusetts Institute of Technology, Department of Materials Science and Engineering, 77 Massachusetts Avenue, Cambridge, MA 02139, USA*
[3] *Massachusetts Institute of Technology, Department of Chemical Engineering, 77 Massachusetts Avenue, Cambridge, MA 02139-4307, USA*

Established in 2012, the Joint Center for Energy Storage Research (JCESR) is an Energy Innovation Hub managed by the United States Department of Energy's Office of Basic Science with a mission of developing the next generation of high-performance, low-cost electrochemical energy storage technologies for transportation and the electric grid. In pursuit of this transformative vision, JCESR introduced a new paradigm for battery research, integrating discovery science, battery design, and prototyping in a single, interactive organization, to accelerate discovery and innovation, to reduce the time from conceptualization to commercialization, and ultimately, to bridge the dreaded "valley of death" between research and industry. While JCSER exclusively focuses on energy storage, this organizational paradigm is malleable and can be applied to numerous technological challenges of societal importance. Thus, while JCESR's scientific accomplishments are described in the peer-reviewed literature, there is considerable value in disseminating JCESR's strategic approaches to promoting breakthrough energy science. This chapter seeks to provide insight into JCESR's mission and organizational structure, as well as to highlight important tools used to effectively connect research activities across the spectrum, from fundamental discovery science to cell design and prototyping intended to enable commercial deployment.

2.1 Background and Motivation

A grand challenge of the twenty-first century will be the evolution of the electrical power system (grid) to meet emerging energy demands while balancing environmental stewardship and cost-effectiveness. In the United States, 82% of the total energy consumed is derived from fossil fuel sources (i.e. oil, coal, and natural gas), dominated by use for electricity and transportation [1]. However, in

Electrochemical Engineering: The Path from Discovery to Product,
First Edition. Edited by Richard C. Alkire, Philip N. Bartlett, and Marc T. Koper.

the future, this dependence will not be feasible [2], as rising population and continuing economic growth in the developing world are projected to double global energy consumption by 2050 [3]. Moreover, the continued and increasing generation of anthropogenic carbon dioxide (CO_2) from fossil fuel combustion will likely have negative implications for the global climate [4]. Thus, a tremendous need exists for scientific and technological advances to address these challenges, sparking worldwide investment in low carbon/carbon neutral power generation, carbon capture and storage, and system-wide energy efficiency [5, 6].

Decarbonization of electricity generation will require the widespread integration of renewable, non-dispatchable energy sources (e.g. solar photovoltaic (PV), wind). However, the uncontrollable intermittency of these generation sources often leads to mismatches in electricity supply and demand, and thus only 15% of the US grid electricity is produced by renewables [7]. Energy storage technologies can smooth and meter the delivery of electricity from these variable resources as well as offset congestion issues within transmission and distribution infrastructure, thus deferring costly investments. Increased energy storage assets can also provide a range of high value services including grid stabilization and resiliency through backup power, introducing new revenue streams for a range of stakeholders [8].

2.2 Lithium-ion Batteries: Current State of the Art

Electrochemical energy storage, specifically rechargeable batteries, is poised to enable widespread penetration of electric vehicles to replace fossil fuels with electricity, reduce carbon emissions, and improve efficiency. Moreover, batteries can advance the grid by replacing its just-in-time delivery system with local inventories of stored energy that can be built up or drawn down to buffer time gaps in variable supply (i.e. intermittent renewable sources) and variable demand (i.e. customer needs). Both of these sectors require energy storage solutions with performance and cost metrics beyond the trajectory of current leading technology, spurring the formation of JCESR as an Energy Innovation Hub with a bold vision of creating next-generation battery technologies with the potential to transform the transportation sector and the electric grid in the same way lithium-ion (Li-ion) batteries revolutionized personal electronics.

Li-ion batteries represent the current state-of-the-art in energy storage, ubiquitous in personal electronics, and emergent in transportation and stationary applications. Conceptualized in the early 1970s, and commercialized in the early 1990s [9], Li-ion batteries store and release energy by shuttling lithium cations through an organic electrolyte, accompanied by electrons through an external circuit, between a positive electrode, typically a lithiated transition metal oxide or phosphate, and a negative electrode, typically graphite [10]. Over the past 25 years, Li-ion has become the predominant technology, due to its high energy density, good cycle life, and high charge/discharge efficiency, and has rapidly developed driven by scale, materials advancements, and manufacturing improvements. Indeed, over the past decade, Li-ion batteries have improved their energy density by 5% per year and reduced their cost by 8% per year [11]. A vibrant research topic, with ongoing contributions from academe, national laboratory, and industrial researchers, continued incremental improvements in energy density, lifetime, and cost can be expected for the foreseeable future. Yet, successive improvements in the technology are not anticipated to yield the affordable, high energy density storage required for transformational change in transportation or the grid.

2.3 Beyond Li-Ion Batteries

In contrast, the beyond Li-ion battery space is much larger, richer, and far less explored than the current Li-ion battery space. Unlike the single concept of intercalation at a negative and positive electrode in commercial Li-ion batteries, beyond Li-ion batteries embrace a wealth of novel concepts, including multiply charged working ions in place of singly charged lithium, high energy covalent chemical reactions at the electrodes in place of intercalation, and fluidized active materials in place of crystalline electrodes. These three energy storage concepts: multivalent intercalation, chemical transformation, and nonaqueous flow, detailed later, are the primary research directions that JCESR pursues. They are fundamentally different from the intercalation concept of commercial Li-ion batteries, have the potential of enabling transformative performance enhancements and cost reductions, as well as providing significantly greater design and operational flexibility and opportunity than commercial Li-ion batteries.

Despite this broad design space, at JCESR's inception, few researchers were exploring these opportunities as compared to conventional Li-ion battery research. Indeed, with a clear market and established value chains, incremental improvements in performance and cost of Li-ion batteries can produce high returns. There is significant risk in attempting to create an entirely new technology as this requires the discovery, development, and application of at least three new materials (one each for the negative electrode, electrolyte/membrane, and positive electrode), which must work harmoniously for the device to function. This represents a daunting challenge, beyond the resources of most research groups and small-to-medium sized companies, and thus necessitates a new approach to battery research. To this end, JCESR has assembled a broad multi-institutional team with expertise and resources to explore the possibilities of the beyond Li-ion battery space, to reduce scientific and technological risk encouraging other stakeholders to enter the field, and ultimately to create next-generation battery systems.

2.4 JCESR Legacies and a New Paradigm for Research

In order to accelerate research beyond Li-ion, JCESR developed a new paradigm for research combining discovery science, battery design, and rapid prototyping into a single organization and quantifying its successes through three legacies. Practically, JCESR focuses on novel electrochemical energy storage mechanisms and new device architectures. Researchers across JCESR seek to understand and control the underlying materials and phenomena that govern charge storage and then harness this knowledge to reliably engineer electrochemical cells and systems. To measure the success of JCESR, the organization set forth three quantifiable legacies:

1. *A library of fundamental science of the materials and phenomena of energy storage at atomic and molecular levels.* This library will be freely available through the literature and open source software with the goal of informing, inspiring, and accelerating the work of the broader battery community as "a rising tide lifts all boats." Traditionally, battery research operates by trial and error such that if a new material is developed and shows improved performance, it is adopted and published, while if the new material fails, it is discarded and often not reported. Thus, JCESR will also report candidates examined for improved energy storage, which failed techno-economic (TE) metrics to inform future researchers throughout the community.
2. *Two research prototype batteries one for transportation and one for the grid that, when scaled for manufacturing, achieved five times the energy density and one fifth the cost of current battery technologies at JCESR's inception (2012).* These two prototypes will differ in their format, materials chemistry, and operating conditions, but will be based on the same body of fundamental knowledge generated from the first legacy. Detailed performance metrics are noted below in Table 2.1. Although five years is a short time to achieve this legacy, JCESR has deliberately chosen not to diminish its vision or replace its mission with less aggressive outcomes that carry lower risk but are not transformative.
3. *A new paradigm for battery research and development that integrates discovery science, battery design, research prototyping, and manufacturing collaboration in a single highly interactive organization.* In the traditional battery community, each of these four functions is typically carried out by a separate research team, often in a different location by experts with differing skills, motivations,

Table 2.1 Goal performance metrics for JCESR prototypes when projected to battery packs, as written in the JCESR proposal.

Transportation	Grid
100 \$/kWh	100 \$/kWh
400 Wh kg^{-1} and 400 Wh l^{-1}	95% round-trip efficiency at C/5
800 W kg^{-1} and 800 W l^{-1}	
1000 cycles 80% DoD C/5	7000 cycles at C/5
15-year calendar life	20-year calendar life
EUCAR 2	Safety equivalent to natural gas turbine
Metrics based on a 350-mile range EV	Metrics based on 5 hours of storage

and focus. In JCESR, these four functions are pursued simultaneously and in close cooperation to exploit the dynamic interaction and inspiration that each function draws from the others. For example, early stage research prototyping can reveal battery design issues and discovery science challenges that can be passed rapidly to the appropriate functional team for analysis and solution, thus enriching the scope and expediting the progress of all areas. Such interactions are expected to quicken the pace of discovery and innovation and reduce the time from conceptualization to commercialization.

JCESR's new paradigm for accelerating the development of these energy storage technologies combines discovery science, battery design, research prototyping, and manufacturing consulting in a single, highly interactive, operationally nimble organization (Figure 2.1). Traditionally, in the battery community, fundamental materials discovery and cell engineering occurred separately with slow and periodic communication between them. Figure 2.1 shows the end-to-end innovation pipeline, from fundamental discovery science to manufacturing collaboration, integrated under "one roof" within the JCESR organization. The left portion of the diagram depicts the three transformative storage concepts that JCESR is pursuing, namely, multivalent intercalation, chemical transformation, and nonaqueous redox flow. The multivalent intercalation and chemical transformation concepts focus on transportation applications whereas nonaqueous redox flow focuses on energy-intensive grid applications. JCESR is pursuing these three beyond Li-ion scientific thrusts (Figure 2.2) because they have the theoretical potential to meet the legacy goals. Their active species and cell architecture deviate substantially from conventional Li-ion requiring the breadth of knowledge and expertise researchers at JCESR possess.

Multivalent intercalation utilizes a divalent cation (e.g., Mg^{2+}) instead of a monovalent cation (Li^+), as the intercalant at the battery positive electrode in an attempt to double the charge storage capacity and overall energy as compared to Li-ion [12, 13]. In addition, common multivalent intercalation cations deposit more uniformly than metallic Li during electrochemical cycling, opening the possibility of a metallic negative electrode to further increase the energy density of the system [14, 15]. However, discovering a stable electrode–electrolyte system that supports the plating/stripping at the negative electrode and intercalation at a high-voltage positive electrode remains elusive [16–20].

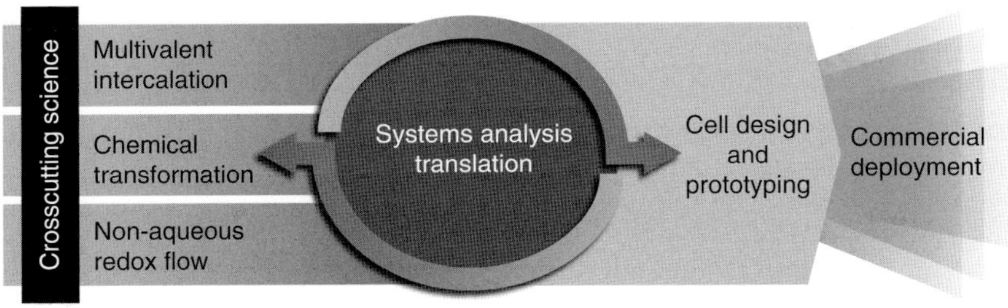

Figure 2.1 JCESR's new paradigm for battery research and development.

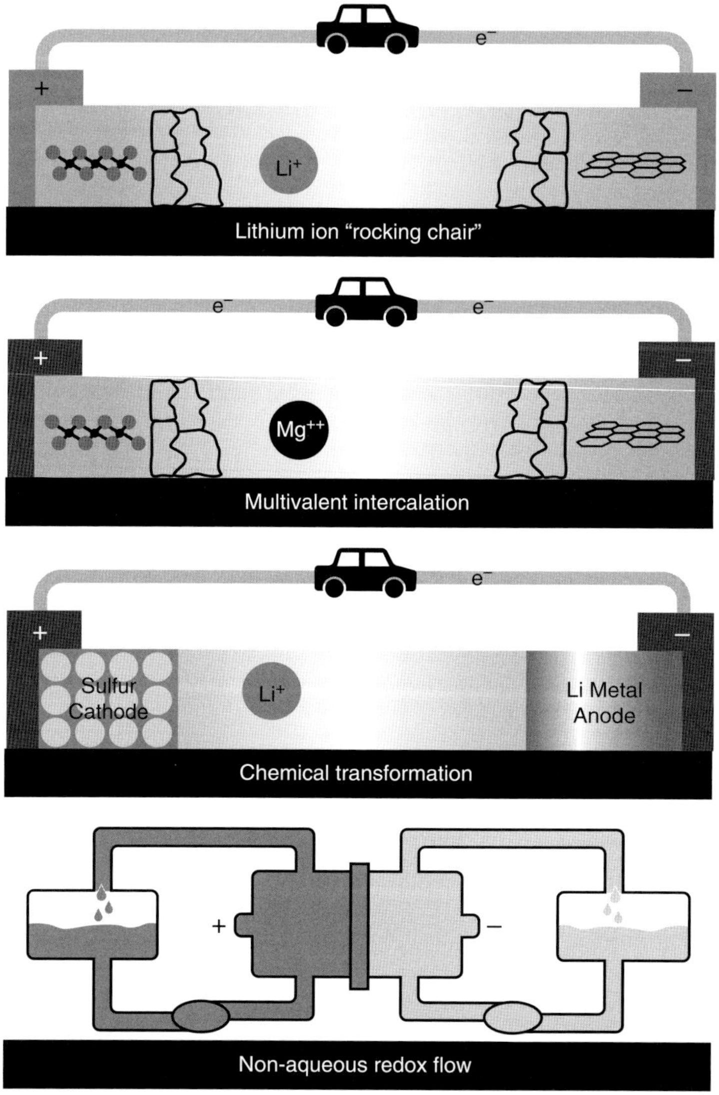

Figure 2.2 JCESR is pursuing three scientific thrusts beyond lithium-ion: multivalent intercalation, chemical transformation, and nonaqueous redox flow.

Chemical transformation harnesses a negative electrode and a positive electrode that rely on a phase change or alloying for storage such as the complete stripping of the metal-negative electrode solid (e.g. Li and Mg) into ions or the conversion between different solid structures through a liquid intermediate at the positive electrode (e.g. sulfur and oxygen). The benefits of these transformations are that more energy per reaction can be accessed and that the materials are inexpensive. However, the side reactions and significant volumetric expansion present in these systems contribute to rapid capacity loss and poor lifetimes [21–26]. Moreover, the knowledge of fundamental reaction mechanisms for these systems is limited and requires significant advancements before a commercially viable prototype cell [27–35].

Nonaqueous redox flow leverages active species that are either dissolved or suspended in organic electrolytes, stored in external tanks, and pumped to a power converting electrochemical stack where the active species undergo reduction or oxidation to alternately charge and discharge the battery. As compared to an enclosed architecture, a flow architecture offers several advantages including decoupled power and energy scaling and long operational lifetimes. However, flow battery energy density is inherently lower than that of enclosed batteries due to the additional electrolyte volumes required to solubilize the active materials. Moreover, expensive ion-selective membranes (e.g. Nafion) are required to prevent positive and negative electrolytes from mixing in the reactor.

As will be discussed later, connecting the three science concept thrusts, providing support, guidance, and technical expertise, are the Crosscutting Science thrust and the Systems Analysis and Translation (SAT) thrust. Crosscutting Science houses tools, such as computing and analytical expertise, that can predict and evaluate material properties as well as perform highly controlled experiments on battery chemistries. The SAT thrust integrates battery design and manufacturing needs by setting cell performance goals that will translate to high functioning battery packs and metrics, thereby providing feedback and guidance to JCESR researchers.

2.5 The JCESR Team

JCESR brings together a diverse team of scientists, engineers, and manufacturers, who would not work closely together in traditional R&D frameworks, to develop transformative energy storage technologies. The organization consists of 15 partner institutions and 5 funded collaborator institutions (Figure 2.3), together comprising more than 180 researchers, including students, postdoctoral associates, early career researchers, senior scientists, and engineers from leading universities (10 partners), national laboratories (5 partners), and industry (5 partners). Each group brings unique perspective, expertise, and tools to bear on this multifaceted project.

To further promote public–private partnerships, JCESR has also created an affiliate program that brings together more than 100 small and large businesses, non-profits, universities, and national laboratories. Many of these affiliates are involved in electrical energy storage, ranging from chemical and material manufacturers to battery system integrators and testers. As new technologies are developed and needs identified, affiliates have the opportunity to engage with the research to drive innovation. To this end, JCESR hosts an annual in-person workshop to allow affiliates to engage with each other and with representatives from JCESR staff, management, and specific research projects.

Managing a complex project with a range of stakeholders requires experienced leadership with a clear organizational hierarchy (Figure 2.4). Operationally, Dr. George Crabtree, the Director of JCESR, and a small team locally manage

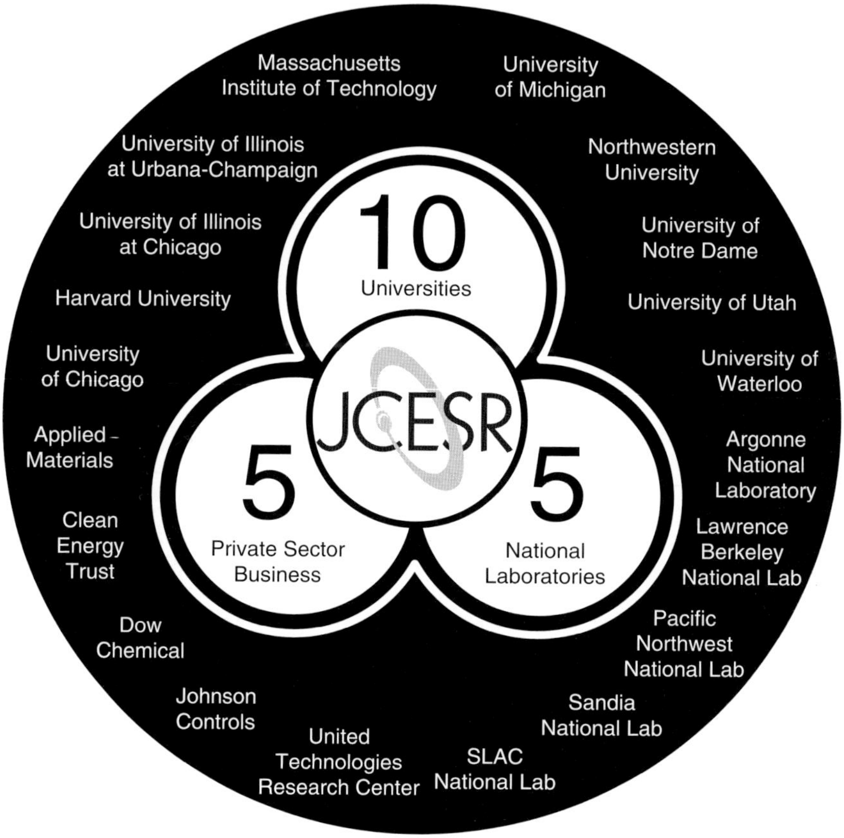

Figure 2.3 The JCESR team consists of 5 national laboratories, 5 industrial partners, and 10 universities (5 partners and 5 collaborators).

JCESR at Argonne National Laboratory. As JCESR Director, Dr. Crabtree directs the overall strategy and goals of the research program and the operational plan, acts as liaison to the executives of JCESR partner organizations, and represents JCESR with external constituencies and advisory committees. The governance committee (GC), chaired by the Argonne Laboratory Director, serves as an advisory board to which the JCESR Director reports. The JCESR executive committee (ExCom) comprises the Director, Deputy Director of Research and Development, Deputy Director of Operations, Research Integration Officer, and the two leads of transportation and grid arcs. The Deputy Director of Research and Development (Dr. Venkat Srinivasan) facilitates and integrates JCESR's research activities across the organization and reports the programs' progress to the ExCom. The Deputy Director for Operations (Mr. Devin Hodge) is responsible for effectively managing all operational aspects of JCESR: safety, budget, internal and external communications, report writing and delivery, outreach, intellectual property, and procurement. The Research Integration Officer, Dr. Lynn Trahey, leads scientific integration efforts across JCESR. The Chief Science Officer, Dr. Nenad Markovic, provides scientific guidance, advice,

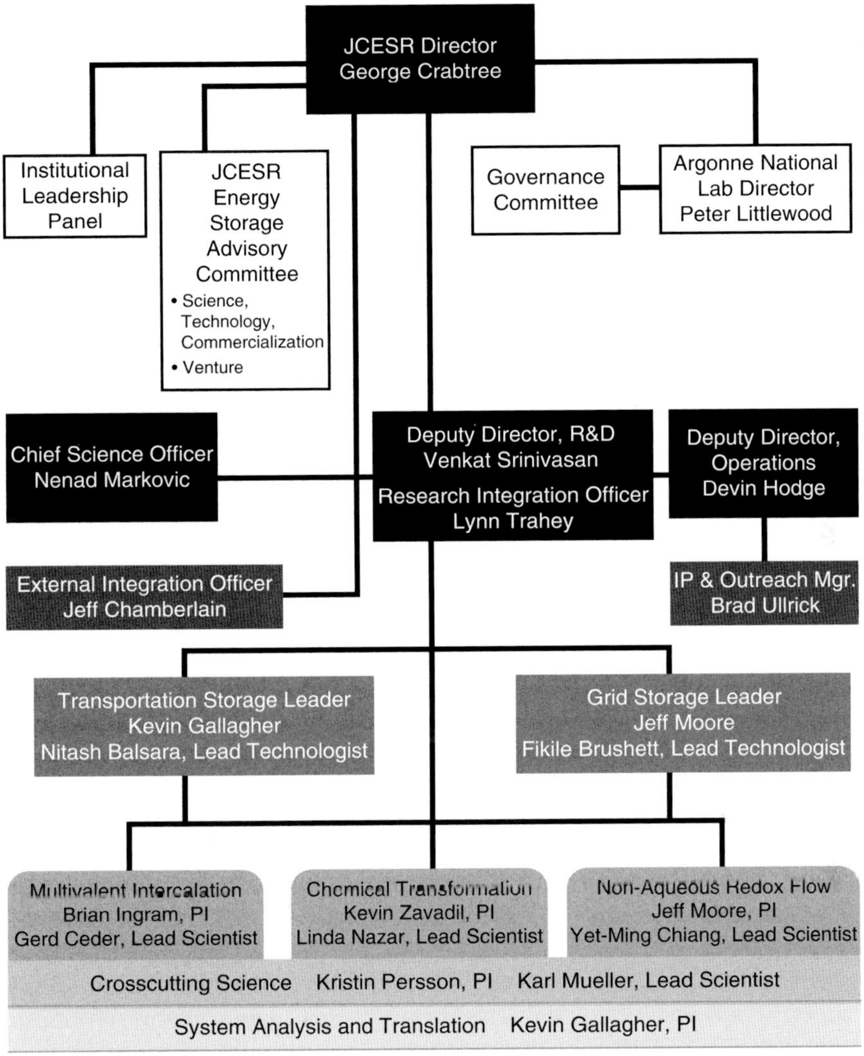

Figure 2.4 Organizational chart of JCESR from July 2015 to April 2016.

and support to the JCESR Director and reports on novel characterization tools and emerging research trends that could benefit JCESR. The JCESR Director, the Deputies, Research Integration Officer, Chief Science Officer, Thrust PIs, and designated Lead Scientists and Lead Technologists comprise the JCESR Directorate. The JCESR Directorate is the decision-making body, evaluating discovery science of materials and phenomena, integrating the findings into innovative battery designs, and creating research prototypes that will lead to next-generation batteries.

The JCESR Institutional Leadership Panel (ILP) consists of senior management (e.g., on the level of Associate Laboratory Director) from each JCESR partner and has institutional oversight for the partners' personnel and facilities identified in the JCESR program plan. The ILP provides recommendations and guidance to the JCESR Director and ExCom on matters related to Hub management, institutional resources (e.g., personnel and facilities), and internal communications. The ILP also ensures that the day-to-day needs of their home institutions are addressed to support JCESR and communicates issues or risks to the broader group.

The JCESR Energy Storage Advisory Committee (ESAC) consists of members representing the science, technology, and commercialization aspects of JCESR's mission. The ESAC advises the JCESR Director on all elements of the program from scientific, technical, legal, and managerial perspectives. In the early stages of the project, the JCESR ESAC focused primarily on evaluations of management processes and prototype definitions but, over time, focused more on scientific strategy, which became more valuable as the project matured. Throughout the project, the JCESR ExCom continuously improved its processes for obtaining external advice toward refining its scope and ensuring that strategic decisions were in line with JCESR legacies.

2.6 JCESR Operational Tools

An organization the size of JCESR requires flexible yet well-established operational tools that ensure robust communication, progress, and conflict resolution. To enable the monitoring and controlling of the mission-driven JCESR research, the JCESR Directorate defines the project work scope, the work breakdown structure (WBS) dictionary, and the companion project schedule and milestones. The ExCom, utilizing the WBS dictionary deliverables, establishes primary milestones that become the unifying goals for the Directorate to build the project tasks and lower level milestones from the bottom-up, outlining a critical path through the five-year project life. This is periodically refined over the project life to reflect major changes in scope/research direction. To accelerate near-term discoveries, JCESR utilizes a hybrid approach to project management that incorporates both agile and traditional project management principles. This is accomplished by the Thrust PIs developing one- to six-month duration "Sprints" (Figure 2.5) that enumerate the scientific and technological challenges that must be overcome to converge the selected prototype research directions to the proof-of-concept stage. A Sprint investigates a specific scientific

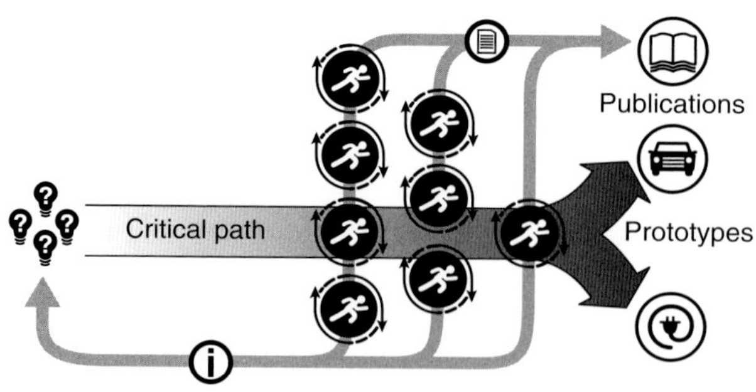

Figure 2.5 The JCESR sprint process answers a specific scientific question to provide guidance to management in choosing the direction of scientific thrusts.

question with a small team of 5–15 members that focuses nearly exclusively on that question. The Sprint team works on that challenge, meeting weekly or often daily to coordinate their progress and refine their research. The results of these Sprints are documented and disseminated, often altering the course of subsequent research and fine-tuning the critical path toward prototype development. Sprints also give early career researchers the opportunity to lead multidisciplinary, multi-institutional teams advancing their own professional development. Sprints represent a new paradigm in research (Legacy 3) in order to advance JCESR targets (Legacy 2) as well as generate new fundamental knowledge (Legacy 1). To date, there have been more than 22 JCESR sprints.

2.7 Intellectual Property Management

A critical operational challenge for JCESR is to establish intellectual trust and enable healthy collaboration among partner institutions. Accordingly, JCESR created the Intellectual Property Management Council (IPMC) that consists of representatives (technology transfer or legal personnel) from each member institution and is led by JCESR's Intellectual Property and Business Development Manager, who is fully dedicated to managing the JCESR intellectual property (IP). The council meets regularly and manages issues relating to IP within the JCESR. Notably, the council created a framework for information sharing through a novel and member-agreed-upon IP management plan and an umbrella nondisclosure agreement (NDA). As all JCESR members are signatories to the IP management plan and the NDA, JCESR scientists can discuss new results and ideas freely, shortening the innovation timeline. This policy reduces the barriers to collaboration and allows for effective and efficient information sharing among the researchers, regardless of institutional affiliation. Evidence of this success is documented by the more than 200 multi-institutional JCESR publications.

2.8 Communication Tools

JCESR's large size and geographically distant member institutions necessitate consistent and flexible communication tools to ensure that all members are apprised of current research results. Ideally, all researchers and their equipment would be located under one roof, where they could effortlessly discuss, plan, conduct, and redirect their research. While a single brick and mortar facility is unfeasible for JCESR, the establishment of the "JCESR Without Walls" strategy has continuously improved communication and increased interactions across institutional boundaries. The established JCESR virtual collaboration tools and in-person collaboration forums are summarized in Table 2.2.

2.9 JCESR Change Decision Process

JCESR developed a decision-making process that solicits feedback from stakeholders within the JCESR network to ensure that milestones are met and the scientific and engineering challenges are addressed. JCESR recognizes that strategic

Table 2.2 JCESR collaboration mechanisms.

Tool	Purpose	Collaborative benefits
Cloud file sharing	Provide a central and virtual repository for scientific document management	Fosters a sharing and peer-review forum
Secure video conferencing	Share operations, management, and scientific findings and ideas	Allows for face-to-face interactions without the travel costs
This week in JCESR newsletter	Weekly email to collaborators to communicate upcoming dates, relevant timely information, data requests, staffing changes, webinar links, and other announcements of interest	Routine dissemination of timely information to all JCESR personnel
Bi-weekly webinars	Share scientific findings and ideas	Supports the collaboration effort
Weekly thrust and executive committee web conferences	Assess progress and make JCESR-wide decisions	Ensures integration within and across thrusts
Rotating quarterly directorate meetings	Review the state of the research and discuss JCESR operations and management	Challenges thinking, ensures thrust integration, and shares ideas
Annual all-hands meeting	Present and exchange new data and cutting edge ideas to promote scientific discovery through information sharing across research thrusts	Fosters greater collaboration through intentional information sharing and face-to-face interactions

changes regarding the direction of certain scientific thrusts or research priorities are difficult to make as complex dependencies exist across institutional and project/thrust boundaries. The generic steps involved in executing the Change Decision Process include the following:

1. JCESR Directorate members propose a change that involves significant change to scope, cost, or schedule.
2. A structured debate of the pro/con(s) of the change occurs at a gathering (real or virtual) of the JCESR Directorate.
3. The Thrust PIs meet separately (sequestered) to come to a consensus decision/suggestion to recommend to the Directorate. If a unanimous decision is not reached, a dissenting opinion may be documented for consideration by the JCESR ExCom.

4. The ExCom, led by the JCESR Director, confirms the change, requests additional information, or overrides the Thrust PIs recommendation.
5. The ExCom communicates the change decision through the ILP and the GC.
6. JCESR Business Operations implements the formal Change Control Process.
7. The ExCom and the affected Thrust PI(s) drive change in the thrust(s).

In the event that implementation of these larger strategic changes involves the termination of individuals and/or research teams, a phased plan to terminate funding for these project efforts is developed to ensure that current research efforts are completed and/or publications in process are submitted. This phased termination of funds must include consideration for finalization of graduate student research and post-doctoral fellow placement elsewhere in JCESR, to the extent practical, and every attempt is made to place the displaced research personnel in newly emphasized research areas.

2.10 Safety in JCESR

Safety of researchers, research tools, and the environment is of paramount importance to the Department of Energy and JCESR management and weighs heavily into public perception of the quality of JCESR research. Safety cultures, however, can vary widely between institutions, which presents challenges if people, procedures, and samples are to move freely among members. The JCESR Operational Group is tasked with investigating researcher safety and institutional safety cultures to ensure work across JCESR is performed in a safe, healthful, and environmentally responsible manner. The group conducts field visits to member institutions to ensure the member's safety protocols meet JCESR requirements. During these engagements, JCESR Environment, Safety, and Health (ESH) staff meet with researchers as well as the member's own ESH team to develop and support the six functional JCESR ESH programs as follows:

- *Safety Culture*: Promotion of ESH awareness among all JCESR researchers.
- *Hazard Identification*: Identification of battery research hazards through the completion of a web-based Hazard and Training Assessment (HATA).
- *Training*: Declaration by researchers that they received training to identify and control the battery research hazards they identified in the HATA.
- *Preparedness and Risk Reduction*: Assessment of laboratory spaces with emphasis on emergency response preparedness and risk reduction.
- *Accident Reporting*: Reporting and investigating incidents.
- *Lessons Learned*: Sharing lessons learned with JCESR partners.

Central to the JCESR ESH program is the web-based HATA questionnaire. The HATA asks researchers to self-identify: (i) the health and physical hazards they may be exposed to while conducting research, (ii) the work practices they may perform, and (iii) the training they have received from their host institution for the hazards they identified. Each researcher's HATA response is reviewed by the JCESR Safety Lead, who replies with an individual report summarizing the researcher's responses and provides them with suggestions to improve safety and health practices.

2.11 Battery Technology Readiness Level

Battery technology readiness levels (BTRL) were introduced to track the progress of the various projects across JCESR and to make strategic decisions regarding future research directions (Figure 2.6). JCESR adapted the concept of BTRL, from technology readiness levels metrics found in various fields, to describe the progression of battery technology development from discovery science through conceptual design, testing, prototyping, and commercial product. The BTRL concept and charts were developed with input from collaborators including Johnson Controls, NASA's Glenn Research Center, and the Army's Tank Automotive Research Development and Engineering Center. The BTRL diagram codifies stages of development beginning with basic science breakthroughs in understanding of materials and phenomena (BTRL-1). It then continues through identification and synthesis of promising classes of new materials (BTRL-2). Next, the materials are examined for compatibility and verified in half-cells consisting of single electrode–electrolyte combinations (BTRL-3). Subsequently, the materials are placed in full cell systems comprising two electrodes and an electrolyte and measured against specific performance metrics generated from JCESR TE models (BTRL-4). Finally, the technology is translated outside of JCESR to an industrial firm or consortium that will launch prototyping of larger systems as models for manufacturing and commercialization (BTRL-5). JCESR's

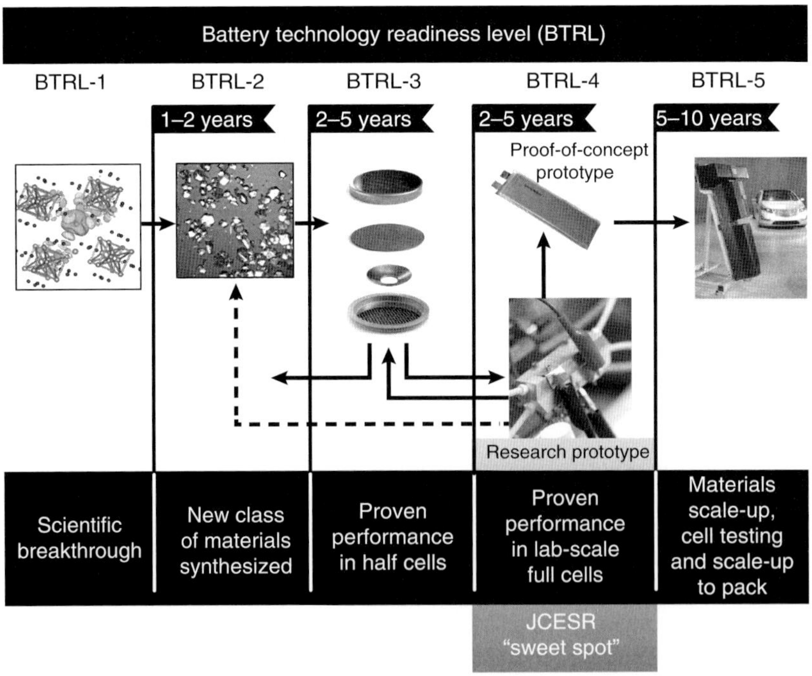

Figure 2.6 All activity in JCESR can be placed into a battery technology readiness level. The results from BTRL-4 are fed into techno-economic models as well as new directions for fundamental science (BTRL-1).

target for prototyping is captured as BTRL-4, proof of concept prototypes that demonstrate the potential to achieve factors of five improvements in performance and reduction in cost that will drive transformative changes in transportation and the electricity grid. Laboratory characterization of JCESR's prototypes provides materials and system-level performance data that are fed back into TE models as well as new directions for fundamental science (BTRL-1).

2.12 JCESR Deliverables

JCESR developed performance metrics to track project progress, provide updates to stakeholders, and allow for redistribution of resources when critical. As a research organization pushing the boundaries of scientific knowledge, progress is not anticipated to be linear, as might be expected for a development project. However, it is important to develop metrics of success to quantify and demonstrate progress. JCESR developed success metrics that spanned the program from basic science (e.g. genomic calculations) through more applied research (e.g. research prototypes developed). In addition, mission support metrics (i.e. Regional Outreach Events) were also developed that validate JCESR's new paradigm for research. Moreover, the majority of JCESR's publications were multi-institutional, highlighting enhanced connectivity among researchers within JCESR. Goals are evaluated on an annual basis and reported quarterly to the program sponsor. JCESR also followed a plan for tracking and managing the JCESR publication progress. As draft publications manuscripts are sent to JCESR management, they are reviewed for a proper acknowledgement statement and affiliation. Upon receipt, the manuscript is placed in a shared box folder, which can be accessed for review by all JCESR personnel for at least one week. Personnel have the ability to comment on the manuscripts, thereby serving as an internal peer review process and to avoid conflicts of authorship and research overlap.

Publications: The JCESR Director solicited input from the various institutions performing research, to set forth a goal of 100 publications per year with a quarter of these being multi-institutional papers. As of October 2017, there have been 327 publications of which 200 have been multi-institutional.

Invention Disclosures: The DOE's Energy Frontier Research Centers (also managed by the Basic Energy Science program) look to achieve a patent application rate of 5% of the "published" output. As JCESR does not have control over whether an institution decides to file a patent application and does not have a budget for patent prosecution, JCESR decided that invention disclosures are a better metric to track. As of October 2017, there have been 58 invention disclosures.

Genomic Calculations: The Electrolyte Genome, which utilizes high-throughput calculations to serve as a discovery engine, are discussed in the next section of this chapter. JCESR targeted approximately 5000 calculations per year for the first four years of operations. As of October 2017, the Electrolyte Genome has currently finished over 24 314 first-principles calculations of molecules.

TE Models: TE models are described in more detail in the next section of this chapter. JCESR set a goal, when looking at both developing new models and refining existing models, of two new or updated models per year. As of October 2017, JCESR has developed four independent TE models for the scientific thrusts.

Research Prototypes: JCESR consulted with the teams that are focused on both grid and transportation prototypes. Based on these consultations, JCESR targeted three to seven research prototypes per year. A documented outcome of each of the prototype sprints is the completion of a research prototype. As of October 2017, JCESR has developed seven research prototypes to meet JCESR's legacy goals.

Regional Outreach Events: JCESR holds public events around the country to bring the broader community (industry, academia, utilities, and government) together to share updates and ideas. The events inform JCESR of the emerging needs of different regions of the United States. As of October 2017, JCESR has held nine regional events ranging from microgrids in Texas to agricultural energy storage opportunities in Mississippi.

2.13 Scientific Tools in JCESR

JCESR is a novel, large research organization that seeks to accelerate the traditional timeline from discovery to prototype and accordingly has had to develop flexible, modular tools suitable for large-scale research. These tools are housed in the SAT and cross-cutting thrusts, as shown in Figure 2.7, a redrawn and simplified schematic of the JCESR organization.

At the inception of JCESR, techno-economic modeling, the Electrolyte Genome, and the Electrochemical Discovery Laboratory were envisioned to advance and expand capabilities across the JCESR science to manufacturing

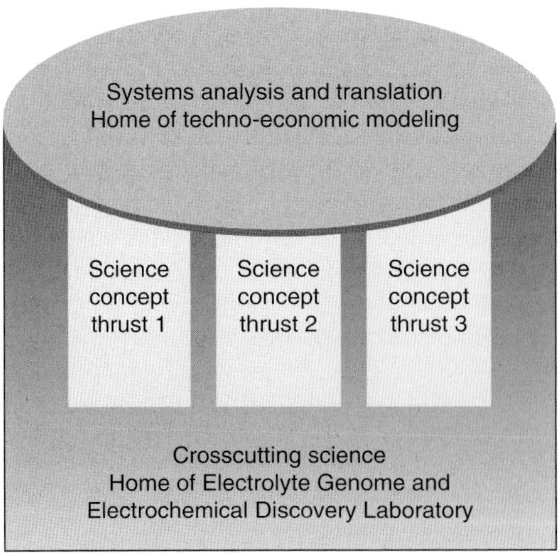

Figure 2.7 JCESR science engineering initiatives drawn to show the ownership and integration of scientific tools.

spectrum. These tools are called upon by the three scientific thrust leaders, whose goal is to translate transportation and grid-level targets into subsystem goals and requirements.

2.14 Techno-economic Modeling

JCESR uses techno-economic modeling to make strategic decisions on scientific thrusts: starting a new research direction, pivoting, or ceasing certain research directions. Techno-economic modeling utilizes scientific laws, technical specifications, and economic models to predict the future cost of an emerging technology. It is an analysis of the performance (techno) and cost (economic) of hypothetical battery packs. A battery is commonly a collection (a "pack") of electrochemical cells, where each cell consists of a negative electrode, positive electrode, and electrolyte, along with a required amount of "inactive," supporting material. Research scientists break down battery problems and discover new materials by working at the level of the cell because cell–level performance translates to performance of a larger battery pack while exhausting fewer materials. Meaningful quantification and metric parameterization from the models is enabled through industrial partnerships such as with Dow Chemical for chemical pricing and with United Technologies Research Center (UTRC) for cell engineering and system design. In JCESR, techno-economic models are further refined through close interaction and collaboration with industrial partners. The results of these models establish feasible, economic materials property windows for the discovery science teams examining new electrodes and electrolytes [36]. This is especially applicable for those in prototype development who have to rapidly iterate and reject research pathways that do not lead to economically viable JCESR deliverables. JCESR's fundamental TE modeling efforts aim to quantify and compare the performance and economic potential of a diverse set of established and conceptual batteries for transportation and the grid. This modeling framework is used across JCESR in three ways: (i) to "back translate" for system-level performance and price goals to materials and component-level targets, (ii) to "forward-evaluate" the challenges, cost, and ultimate performance of different technology approaches, and (iii) to chart progress and allocate limited resources effectively.

2.14.1 Techno-economic Modeling of a Metal–Air System for Transportation Applications

An early JCESR TE study investigated the feasibility and challenges of integrating a lithium–air (Li–air) battery into a transportation system [37]. A Li–air battery is technically a lithium–oxygen ($Li–O_2$) battery that comprises lithium metal as the negative electrode and oxygen gas as the active material at the positive electrode [38]. This battery couple has a high OCV (3.0 V) and very high specific energy (3505 Wh kg^{-1}) as compared to Li-ion (387 Wh kg^{-1}) when only the reactants are considered [39]. Accordingly, the $Li–O_2$ couple is the subject of intense research efforts around the world [40–42]. The study aimed to uncover how much capacity could be realized when the full system is considered. Closed and open system designs were considered. In an open design, filtered air from outside the vehicle provides the oxygen gas. In a closed design, oxygen gas is supplied from a pressurized tank stored in the vehicle. The Li–air TE model also aimed to pinpoint the pivotal research areas and materials needs for technology advancement. JCESR partnered with General Motors and Opel on the study.

To complete the model, several scientific advances were assumed, such as the identification of an electrolyte that enables reversible reduction and oxidation chemistry at both the positive and the negative electrodes. Another assumption was the identification of an air filter (for the open system) that would block nitrogen, water, and carbon dioxide from the air stream, allowing only oxygen gas into the battery chamber. Although small amounts of water have been shown to catalyze oxygen reduction, water and nitrogen are highly detrimental to the efficient cycling of the lithium metal negative electrode.

The Li–air study concluded that the open system has more barriers to construction and lower feasibility in transportation applications than the closed battery system (Figure 2.8). However, the closed system still suffers from the high

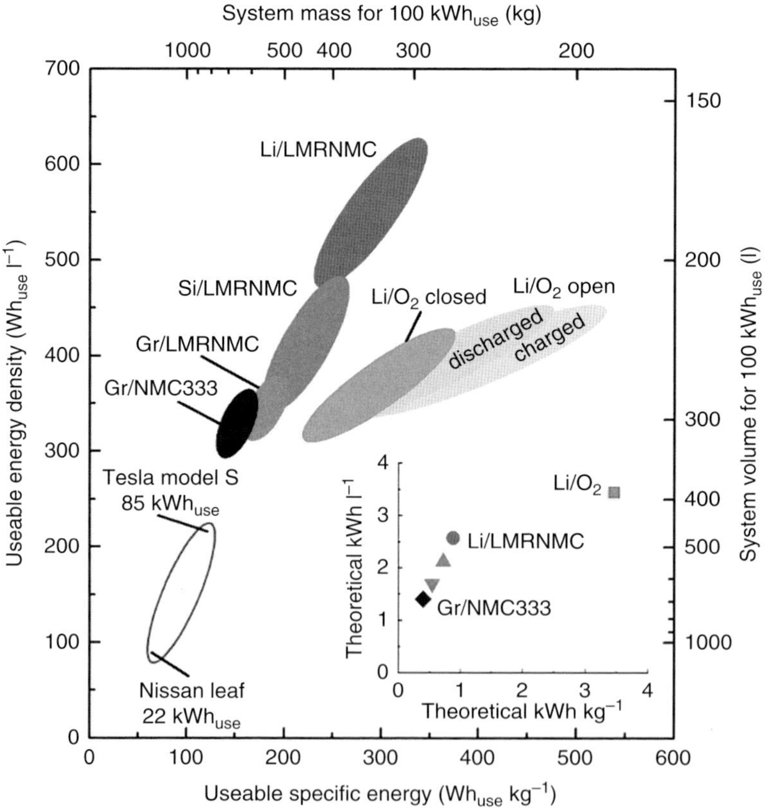

Figure 2.8 Calculated systems-level energy density and specific energy for both Li-ion and beyond Li-ion systems. Source: Gallagher et al. 2014 [37]. Reproduced with permission of Royal Society of Chemistry.

weight burden of supplied gaseous oxygen. Furthermore, the assumptions state that lithium metal is the negative electrode and that it cycles efficiently. If this was possible, lithium-metal-negative electrodes in Li-ion batteries (replacing the conventional graphite negative electrodes) paired with advanced positive electrodes would offer nearly the same performance at lower cost and risk. This analysis revealed the importance of system-level considerations and identifies the reversible lithium-metal-negative electrode as the common, critical high-risk technology needed for batteries to reach long-term automotive targets. Thus, a critical pathway in need of increased investment was identified. Ultimately, JCESR moved resources away from Li–air research and toward Li-based systems that do not rely on gaseous intake.

In the years since this pivot, research on Li–O_2 systems in the wider research community has progressed substantially. Taking into account JCESR's TE modeling of the system, researchers have aimed to make the concept function in open air without the need for rigorous filtering. Seeing how JCESR would need to narrow focus down to one transportation prototype in a five-year time frame, a different transportation concept took priority (the lithium–sulfur (Li–S) system). However, JCESR is nimble and attuned to scientific developments worldwide and, as such, would work on the Li–O_2 system if given a longer time frame to incorporate recent breakthroughs.

2.14.2 Techno-economic Modeling of Flow Batteries for Grid Storage Applications

TE modeling was also instrumental in guiding the redox flow scientific thrust and provided numerous design spaces for JCESR researchers to explore. While the majority of flow batteries are based on aqueous chemistries, transitioning from aqueous to nonaqueous electrolytes offers the opportunity to increase cell voltages, via the wider electrochemical stability windows of nonaqueous electrolytes, and to leverage new electrochemical couples that are incompatible with aqueous electrolytes due to their low solubility, electrochemical instability, or redox potential outside the stability window. Taken together, these advantages lead to higher energy densities, smaller system footprints, and lower costs of energy, provided suitable redox couples, electrolyte formulations, and selective membranes are identified and developed [43–46].

JCESR aimed to survey the broad flow battery design space using a TE model to determine the most promising design space for nonaqueous and aqueous redox flow batteries. It was determined that 170 g $(\text{mol}\cdot e^-)^{-1}$ or 154 mAh g^{-1} is the floor value that must be surpassed if redox organic molecules are to meet JCESR cost targets (assuming a neat electrolyte and open-circuit voltage of 2 V). This limits the number of molecular structures of negative and positive electrolyte active materials that may be possible. Studies of other molecular structures that do not meet the defined threshold would then be discouraged in JCESR unless researchers could show a clear connection to fundamental understanding.

Pathways toward developing aqueous and nonaqueous flow batteries were examined [47–51]. Both aqueous and nonaqueous flow batteries are promising technology platforms capable of achieving the low costs required for widespread implementation. Nonaqueous electrolytes should be no greater than

\$5 kg^{-1} and aqueous electrolyte should be less than \$0.1 kg^{-1}. Nonaqueous systems enable higher cell voltages (>3 V) than their aqueous counterparts (1.5 V) but also require higher active material solubility (5 M compared to 2 M) to offset higher electrolyte costs. Nonaqueous systems also benefit from incorporating cation-based counter ions to reduce salt cost in the electrolyte (Figure 2.9). For both battery types, a key enabling development will be the discovery of molecules that are long lived, provide large cell voltages, and are low cost.

These techno-economic studies also motivated JCESR researchers to investigate the economic and performance trade-off between the size of redox active molecules and the membrane used in the cell [52]. JCESR invented, demonstrated, and developed the concept of redox-active macromolecules, including polymers (RAPs) [53–55] and colloids (RACs) [56, 57] as charge storage media, thereby opening up new horizons of active molecules for a new generation of redox flow batteries [56, 58, 59]. In a similar manner, JCESR invented, demonstrated, and developed a new separator concept within the class of polymers of intrinsic microporosity (PIM) that uses tunable pore size to block crossover of small molecules between liquid electrolytes in redox flow batteries. This concept, introducing ultra-low cost size selectivity of small molecules, applies to both Li–S and macromolecular redox flow batteries and is unique in the field [32, 60].

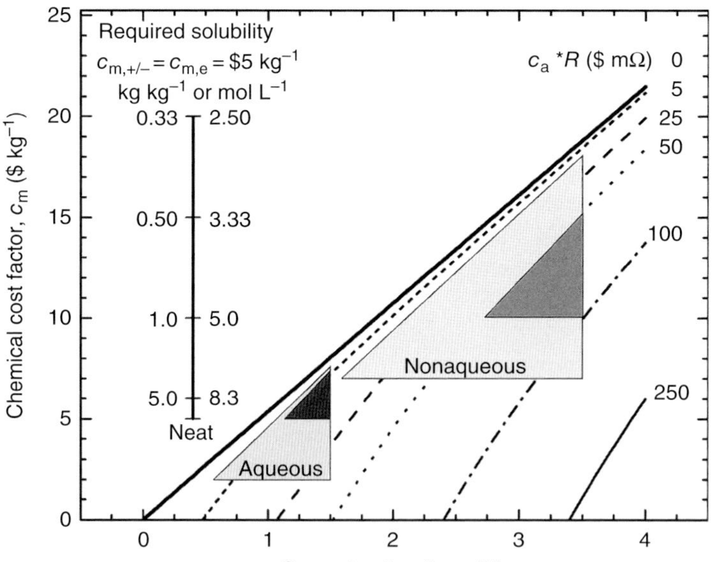

Figure 2.9 Chart showing modeled "design space" for aqueous and nonaqueous flow batteries that may meet cost targets. Source: Darling et al. 2014 [48]. http://pubs.rsc.org/en/Content/ArticleLanding/2014/EE/c4ee02158d#!divAbstract. Licensed under CC BY 3.0.

2.15 The Electrochemical Discovery Laboratory

JCESR founded the Electrochemical Discovery Laboratory (EDL) to answer critical scientific questions across JCESR that require specialized equipment and technical expertise of career electrochemists. The EDL is located at Argonne National Laboratory and staffed by a small group of Argonne employees (Figure 2.10). All JCESR scientists can access the facility and work closely with EDL staff. At the EDL, scientists can synthesize model materials, perform in situ analysis of wet and dry interfaces under controlled conditions, and study electrochemical trends in activity, selectivity, and stability. The EDL specializes in constructing model systems based on energy storage devices, extracting fundamental knowledge from the model system, and then translating that information back to the device to further prototype development. Knowledge gained at the EDL on highly defined systems is used to bridge scientific gaps in each of the thrusts.

2.15.1 The Effect of Trace Water on Beyond Li-ion Devices

Understanding the fundamental reactions that govern energy storage mechanisms is essential to commercializing beyond Li-ion technology because degradation mechanisms hinder operational lifetime. Small amounts of impurities play a disproportionate role in electrochemistry. Accordingly, the EDL has acquired expertise to produce high-purity solvents for electrochemical systems. In the $Li{-}O_2$ system, an initial system of study for the JCESR chemical transformation thrust, small amounts of water were found to enhance the kinetics of the discharge reactions. The EDL purified various organic solvents used in $Li{-}O_2$ down to a water content below <1 ppm. Water was then added back into the electrolyte in a controlled manner to study its effect on the system using Raman spectrometry, Fourier transform infrared spectroscopy (FTIR), rotating ring disk electrode studies on well-defined gold surfaces, and atomic force microscopy. The EDL found that water, at parts per million levels, catalyzed the conversion of lithium superoxide, a discharge intermediate, to lithium peroxide, the discharge product [61]. The EDL has also reported on the function of trace water in other

Figure 2.10 Picture of the Electrochemical Discovery Laboratory showing ultra-high vacuum, multi-module synthesis, and characterization system. Source: ANL/JCESR.

systems such as Mg^{2+} desolvation at hydrated positive electrode interfaces [62] and passivation of magnesium metal negative electrodes [63], applicable to the multivalent intercalation thrust.

2.15.2 Stability of Redox Active Molecules

The stability of redox-active molecules represents a maximum lifetime for any redox flow technology that incorporates them and thus the EDL's ability to measure molecular decay served as an important benchmarking tool for the non-aqueous redox flow thrust. JCESR researchers observed that during regular cyclic voltammetry studies, which measure the reversibility of reduction and oxidation reactions, the redox-active molecules under examination appeared stable and reversible in most electrolytes. However, during bulk electrolysis studies, where the concentration in solution changes drastically over time, performance degradation was evident in all electrolyte studies [64]. To understand these results, in situ FTIR was performed at the EDL to monitor structural changes to the redox active molecule as a function of applied potential. Crucially, these results were then combined with density functional theory calculations to determine the structure of active species in solution and the potential-dependent interactions as a function of electrolyte constituents [65]. Results indicated that the nature of the salt cation plays a key role in the stability of the active molecule and that the coordination complex formed between them dissociates upon electrochemical oxidation. This result led researchers to better design redox active materials and participatory electrolyte components for redox flow batteries.

2.16 Electrolyte Genome

Moving beyond Li-ion batteries requires the discovery of new working ions, electrodes, and electrolytes and thus JCESR developed the Electrolyte Genome to rapidly screen new material candidates. Traditional research relies on intuition and experience of the researcher in selecting and evaluating new materials. The Electrolyte Genome uses high-throughput computing to accelerate the traditional research process by identifying new candidates for battery materials, predict their performance, and select the most promising ones for experimental validation. Just as the goal of the Human Genome Project was to speed the course of understanding disease and evolution to advance medicine, the goal of the Electrolyte Genome is to speed the course of electrolyte discovery for new battery systems. As of 2016, large parts of the Electrolyte Genome were released to the public for electrolyte searches for various applications.

Electrolytes are a crucial element in batteries as they are the medium that carries ions between the positive and negative electrodes. An electrolyte is made up of solvents and salts and can be a liquid, gel, or solid, as long as it is electronically insulating and ionically conducting. Liquid is the most common phase owing to the high mobility of ions in most liquids. A problem with developing new electrolytes for new battery couples is the sheer number of possibilities to try experimentally. There are dozens of common solvent choices and hundreds more are possibilities. Solvents can also be mixed together in different ratios further expanding the design space. Combining with innumerable different salts and salt combinations, make for an impossible matrix of experimental possibilities. The Electrolyte Genome is intended to reduce this complexity and provide researchers with fewer, smarter options for laboratory experimentation.

The Electrolyte Genome works by first populating with several thousand candidate molecules defined by an automated molecule generator [66, 67]. The properties of the molecules determined via quantum chemical prediction (density functional theory electronic structure calculations), including redox potentials, solubilities, and structural stability, are screened and down selected for synthesis, characterization, and validation (Figure 2.11). Results of those experiments help define structure–activity relationships of molecules, providing insight into design principles. This information is also fed back into the Genome to further improve the predictive capabilities [68].

2.16.1 Screening of Redox Active Molecules for Redox Flow

Seeking to quickly move from discovery to prototype, JCESR enlisted the Electrolyte Genome to perform an initial screening of positive and negative electrolytes for redox flow batteries. Specifically, the Electrolyte Genome was used to narrow the field of possible redox-active organic materials from a starting set of

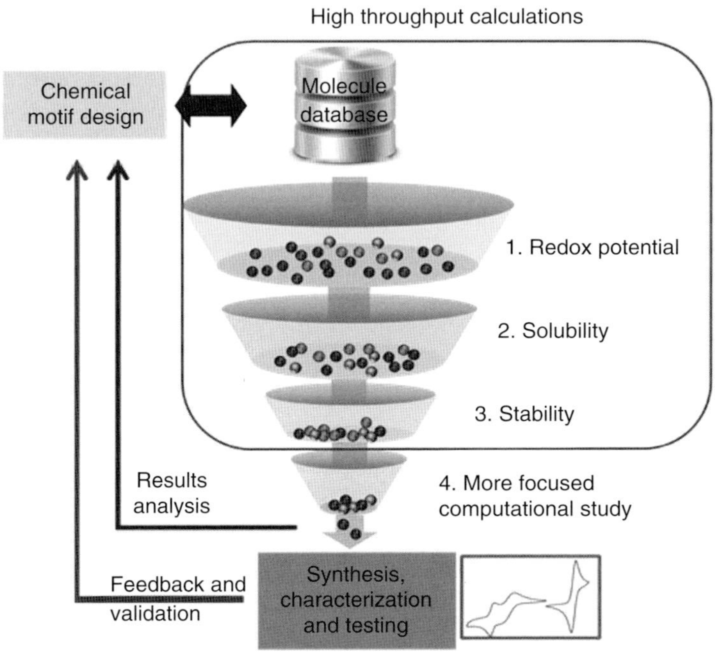

Figure 2.11 Depiction of how the Electrolyte Genome is utilized to down-select redox-active materials for flow batteries and the feedback loop that improves future predictions. Source: Cheng et al. 2015 [66]. Reproduced with permission of Royal Society of Chemistry.

4178 molecules [69]. These molecules included combinations of 23 possible base groups, nitrogen, oxygen, and sulfur heteroatoms, saturated and unsaturated bonds, and 16 possible functional groups, both electron-withdrawing and electron-donating substituents. The researchers initially focused on the reduction and oxidation potential as these are critical in developing an economically feasible prototype. With the knowledge gained from this study, promising molecules were identified, synthesized, tested to verify reduction and oxidation potentials, and then assessed for solvation and stability.

The study concluded that the base groups and functional groups have strong effects on redox potentials. Unsaturated base groups (containing double bonds) lead to higher reduction potentials and lower oxidation potentials because π-bonding in the ring structure facilitates electron addition and removal. More generally, electron-withdrawing groups led to higher reduction potentials and electron-donating groups led to lower oxidation potentials. The results indicated that molecules with nitro and trichloro groups should be investigated for possible positive electrolytes, while amine groups should be investigated for negative electrolyte materials. This guidance was disseminated to experimentalists throughout JCESR, thereby decreasing their time spent making materials with redox potentials and molecular weights that align with Legacy 2 goals.

2.16.2 Examination of Multivalent Intercalation Materials

The technical knowledge gleaned from screening redox-active molecules was also applied in the search for new multivalent intercalation materials. Embodied as the Materials Project [70], this tool helps researchers identify electrode materials that will intercalate and deintercalate Mg^{2+} cations with appreciable mobility and capacity at desired potentials vs a magnesium-metal-negative electrode [71]. JCESR modified and used the Materials Project to screen novel multivalent-positive electrodes, reducing thousands of possibilities into dozens that were then made and tested. A particular success was the identification of specific spinel compounds, which upon experimental testing, showed evidence of the elusive Mg^{2+} intercalation behavior at high potentials [72–74].

A breakthrough will come when an electrolyte is discovered that functions well at both the positive and negative electrode of the Mg-ion battery. To meet the TE goals of JCESR, the negative electrode of these cells must be magnesium metal and the positive electrode must operate at a high potential (~3 V) vs Mg^0/Mg^{2+}. Electrolytes that work for magnesium plating and stripping have been identified, yet often contain constituents, like the chloride anion (Cl^-), which corrode positive electrode materials or are simply not stable at the high potentials of the positive electrode. The Electrolyte Genome is being used to help find an electrolyte that enables the stripping of Mg^{2+} from magnesium metal and the desolvation of Mg^{2+} from solution to plate magnesium metal, all while being stable at the high-voltage-positive electrode [75–77]. The first step in this process is understanding why certain electrolytes are experimentally unstable. In a 2015 publication, JCESR researchers combined first-principles calculations with Molecular Dynamics to simulate the solvation structures of a matrix of magnesium salt and solvent combinations [75]. Magnesium salts showed a strong tendency to form contact ion pairs, close associations of the anion and cation in solution despite abundant solvent present. The $TFSI^-$ anion, useful at positive electrode potentials, exhibited rapid decomposition at negative electrode potentials, triggered by the one electron reduction of $Mg^{2+} \rightarrow Mg^+$. This consequence of contact ion pair formation is a failure mode for this electrolyte in Mg-ion cells. Meanwhile other salts, such as BH_4^- and BF_4^-, exhibited stability at magnesium-negative electrodes despite ion pairing. Such studies corroborate with experimental evidence and provide a path forward toward future discovery of a stable electrolyte for multivalent intercalation devices.

2.17 Combining the Electrolyte Genome with Techno-economic Modeling

JCESR researchers invented a new computation tool, a Dashboard for materials selection of redox flow chemistries, to aid in understanding the performance characteristics needed to meet certain economic targets [47]. The Electrolyte Genome effectively filters redox-active molecules by reduction and oxidation potential and other properties as desired. However, this can still leave a staggering number of candidate molecules to experimentally test for performance. The Dashboard stores experimentally measured parameters, including energy density, cycle life, and rate capability, in one user-friendly database.

The Dashboard combines TE models on aqueous and nonaqueous redox flow chemistries and redox potentials and molecular weights from the Electrolyte Genome. The user selects the target battery price. When the target price is identified, a graph of equilibrium potential vs equivalent weight, populated with known active materials, shows solubility contour lines. As a user selects a negative electrolyte and a positive electrolyte for study, the Dashboard informs the user of the concentration of the active materials required in solution to meet the battery price target. Molecules that can meet these price targets, and at what specific concentration in solution, are then easily identified. The program also includes a Rate-Capability tool, which uses Stokes–Einstein theory to predict the area-specific resistance in the flow cell, a crucial parameter in redox flow batteries. The Dashboard tool is implemented on materialsproject.org and as of 2016, is free for the public to access [70].

2.18 Prototype Development

To ensure robust cell testing and comparable performance data across member institutions, JCESR developed an advanced prototype platform for the validation of nonaqueous redox chemistries. By testing new redox-active molecules, membranes, or electrolytes with common hardware, JCESR researchers can pursue individual research projects in parallel while measuring against a common standard (Figure 2.12). Moreover, decentralized prototyping or "prototyping everywhere" accelerates research development by providing synthetic chemists with a

Figure 2.12 JCESR nonaqueous redox flow cell prototype designs optimized for various cell chemistries. The cells on the left and right of the photograph are variants of the GEN 2 cell design with the cell on the right being the most current version. The cell in the middle is the first JCESR prototype, the GEN 1 cell, which enabled preliminary materials screening albeit at low power due to high cell resistance. The GEN 2 cell design replaced the GEN 1 cell design.

flexible platform for material characterization that enables rapid turnaround (i.e. fail fast) with minimal materials requirements. The initial prototype platform, the GEN 1 flow cell was built upon a pre-existing nonaqueous flow cell design developed at Pacific Northwest National Laboratory. Simple, easy to assemble, and robust, this design was intended for preliminary materials testing where power performance is not a priority. However, as materials continued to advance within JCESR, a need arose for high-performance platforms that enabled materials characterization under conditions closer to those anticipated in practical application. This motivated the development of the GEN 2 flow cell in collaboration with UTRC, a leader in the design of electrochemical cells.

Building on state-of-the-art vanadium RFB designs, the GEN 2 flow cell incorporates zero-gap interdigitated flow fields with thin carbon paper electrodes to achieve low area-specific resistance enabling high-power operation with low cell polarization. As JCESR is interested in nonaqueous redox chemistries, the GEN 2 was designed with components and ancillaries compatible with nonaqueous solvents. To minimize material requirement characterization and enable easy and safe operation in a glove box environment, the GEN 2 flow cell (2.55 cm^2 active area) is scaled down from UTRC's subscale testing platform (25 cm^2 active area). Performance scalability, from the GEN 2 subscale cells, was validated using a number of aqueous and nonaqueous chemistries. In the case of aqueous vanadium chemistries, the performances of the all-vanadium GEN 2 and subscale cells align with state-of-the-art literature. Furthermore, validation tests, using model nonaqueous redox couples, have shown that the GEN 2 flow cell can meet and even exceed the performance requirements outlined for a 100 \$/kWh prototype

[47, 78, 79]. In September 2016, a GEN 2 flow cell workshop was held to distribute cell hardware and to teach practical experimental skills, as well as relevant theory, to JCESR grid researchers.

2.19 Legacy of JCESR

Bringing together scientists, engineers, and manufacturers across the nation, JCESR sought to leave behind three legacies:

1. A library of fundamental science of the materials and phenomena of energy storage at atomic and molecular levels.
2. Two prototypes, one for transportation and one for the grid, that when scaled to manufacturing are capable of meeting JCESR's aggressive performance and cost targets.
3. A new paradigm for battery research and development that integrates discovery science, battery design, research prototyping, and manufacturing collaboration in a single highly interactive organization.

As of October 2017, JCESR produced 327 publications, 58 invention disclosures, 24 314 first-principles calculations, and four techno-economic models, undoubtedly achieving the first legacy of leaving behind a library of fundamental science for the energy storage community. These results have been disseminated through conferences, seminars, and webinars by JCESR scientists.

JCESR's second legacy is to achieve five times the energy density at one fifth the cost, based on the most promising battery chemistries known at JCESR's launch in 2012. JCESR promised two prototypes but delivered four, two for transportation and two for the electricity grid that met the cost goal (100 \$/kWh at the system or "pack" level). JCESR spun out three startups, one dedicated to commercializing JCESR's prototype for long duration, low-cost storage for the electricity grid (Baseload Renewables, Inc.), the others dedicated to commercializing solid-state electrolytes (Blue Current) and polymer membranes (Sepion) using JCESR intellectual property.

JCESR was unable to achieve five times higher energy density (400 $Wh\,l^{-1}$ at the system or "pack" level), achieving instead three times higher energy density for several of its prototypes. For comparison, the Li-ion batteries introduced in 1991 had twice the energy density of the next best batteries (Ni-metal hydride and Ni–Cd), a transformational step for personal electronics, enabling the now-universal laptops, tablets, and smart phones. JCESR found that, in many cases, increasing energy density comes with a price: it compromises cycle life, an unacceptable trade-off symptomatic of next-generation battery research. The barriers to simultaneously achieving several performance metrics such as energy density, cycle life, fast charging, and safety are not engineering but scientific: many of the required materials simply do not yet exist. Discovering or designing new materials for higher performance is an early stage basic science challenge.

JCESR integrated more than 180 students, postdocs, early career researchers, senior scientists, and engineers from over 20 national laboratories, universities, and industry partners into a single organization, thereby achieving the third legacy. As detailed in this chapter, the management and organization of JCESR represents a new paradigm for conducting research that was able to accelerate the process from discovery science to prototype as compared to the individual members working independently.

2.20 Conclusion

Battery research poses no shortage of highly complex phenomena and questions. JCESR sought to bring together a diverse group of individuals in order to shorten the time frame from discovery in the lab to commercialization. Through extensive collaboration in the network, a suite of tools was developed to accelerate progress across numerous areas resulting in collaborative publications and invention disclosures. Although JCESR was unable to achieve five times higher energy density, instead achieving three, JCESR discovered fundamental knowledge relating to the electrochemical systems in question, which described the trade-off between energy density and cycle life. Perhaps, most relevant in the long term, the novel organizational paradigm is portable and can be applied to a wide variety of pressing and complex technical problems facing society.

Acknowledgments

This work was funded by the Joint Center for Energy Storage Research (JCESR) (DE-AC02-06CH11357), an Energy Innovation Hub funded by the United States Department of Energy, Office of Science, Basic Energy Sciences. TJC acknowledges government support under and awarded by DoD, Air Force Office of Scientific Research, and National Defense Science and Engineering Graduate (NDSEG) Fellowship, 32 CFR 168a. We thank Bradley Ullrick, Sue Marconi, Angelique Scarpa, and Dorothy Loudermilk for valuable assistance and critical feedback.

References

1 U.S. Energy Information Administration (2017). *Monthly Energy Review*. Washington, DC: U.S. Energy Information Administration.

2 Committee on Prospering in the Global Economy of the 21st Century: An Agenda for American Science and Technology, National Academy of Sciences, National Academy of Engineering, and Institute of Medicine (U.S.) (2007). *Rising above the Gathering Storm Energizing and Employing America for a Brighter Economic Future*. Washington: National Academies Press.

3 U.S. Energy Information Administration (2015). *Annual Energy Outlook*, 2015. Washington, DC: U.S. Energy Information Administration.

4 Ehleringer, J.R., Cerling, T.E., and Dearing, M.D. (ed.) (2005). *A History of Atmospheric CO_2 and its Effects on Plants, Animals, and Ecosystems*. New York, NY: Springer.

5 International Energy Agency (2012). *Energy Technology Perspectives 2012: Pathways to a Clean Energy System*. Paris, France: International Energy Agency.
6 Denholm, P., Ela, E., Kirby, B., and Miligan, M. (2010). *The Role of Energy Storage with Renewable Electricity Generation*. Golden, CO: National Renewable Energy Laboratory.
7 US Energy Information Administration (2017). *Annual Energy Outlook*. Washington, DC: U.S. Energy Information Administration.
8 US Department of Energy (2013). *Grid Energy Storage*. Washington, DC: U.S. Department of Energy.
9 Whittingham, M.S. (2012). History, evolution, and future status of energy storage. *Proc. IEEE* 100 (Special Centennial Issue): 1518–1534.
10 Goodenough, J.B. and Park, K.-S. (2013). The Li-ion rechargeable battery: a perspective. *J. Am. Chem. Soc.* 135 (4): 1167–1176.
11 Crabtree, G., Kócs, E., and Trahey, L. (2015). The energy-storage frontier: lithium-ion batteries and beyond. *MRS Bull.* 40 (12): 1067–1078.
12 Shao, Y., Liu, T., Li, G. et al. (2013). Coordination chemistry in magnesium battery electrolytes: how ligands affect their performance. *Sci. Rep.* 3 (1).
13 Vardar, G., Sleightholme, A.E.S., Naruse, J. et al. (2014). Electrochemistry of magnesium electrolytes in ionic liquids for secondary batteries. *ACS Appl. Mater. Interfaces* 6 (20): 18033–18039.
14 Barile, C.J., Spatney, R., Zavadil, K.R., and Gewirth, A.A. (2014). Investigating the reversibility of in situ generated magnesium organohaloaluminates for magnesium deposition and dissolution. *J. Phys. Chem. C* 118 (20): 10694–10699.
15 Barile, C.J., Barile, E.C., Zavadil, K.R. et al. (2014). Electrolytic conditioning of a magnesium aluminum chloride complex for reversible magnesium deposition. *J. Phys. Chem. C* 118 (48): 27623–27630.
16 Liao, C., Sa, N., Key, B. et al. (2015). The unexpected discovery of the $Mg(HMDS)_2/MgCl_2$ complex as a magnesium electrolyte for rechargeable magnesium batteries. *J. Mater. Chem. A* 3 (11): 6082–6087.
17 Pan, B., Zhang, J., Huang, J. et al. (2015). A Lewis acid-free and phenolate-based magnesium electrolyte for rechargeable magnesium batteries. *Chem. Commun.* 51 (28): 6214–6217.
18 Kim, C., Phillips, P.J., Key, B. et al. (2015). Direct observation of reversible magnesium ion intercalation into a spinel oxide host. *Adv. Mater.* 27 (22): 3377–3384.
19 Barile, C.J., Nuzzo, R.G., and Gewirth, A.A. (2015). Exploring salt and solvent effects in chloride-based electrolytes for magnesium electrodeposition and dissolution. *J. Phys. Chem. C* 119 (24): 13524–13534.
20 Wetzel, D.J., Malone, M.A., Haasch, R.T. et al. (2015). Passivation dynamics in the anisotropic deposition and stripping of bulk magnesium electrodes during electrochemical cycling. *ACS Appl. Mater. Interfaces* 7 (33): 18406–18414.
21 Xiao, J., Hu, J.Z., Chen, H. et al. (2015). Following the transient reactions in lithium–sulfur batteries using an in situ nuclear magnetic resonance technique. *Nano Lett.* 15 (5): 3309–3316.

22 Yu, X., Pan, H., Zhou, Y. et al. (2015). Direct observation of the redistribution of sulfur and polysufides in Li–S batteries during the first cycle by in situ X-ray fluorescence microscopy. *Adv. Energy Mater.* 5 (16): 1500072.

23 Fan, F.Y., Carter, W.C., and Chiang, Y.-M. (2015). Mechanism and kinetics of Li_2S precipitation in lithium–sulfur batteries. *Adv. Mater.* 27 (35): 5203–5209.

24 Wu, H.-L., Huff, L.A., Esbenshade, J.L., and Gewirth, A.A. (2015). In situ EQCM study examining irreversible changes the sulfur–carbon cathode in lithium–sulfur batteries. *ACS Appl. Mater. Interfaces* 7 (37): 20820–20828.

25 Cao, R., Chen, J., Han, K.S. et al. (2016). Effect of the anion activity on the stability of Li metal anodes in lithium–sulfur batteries. *Adv. Funct. Mater.* 26 (18): 3059–3066.

26 Doris, S.E., Ward, A.L., Frischmann, P.D. et al. (2016). Understanding and controlling the chemical evolution and polysulfide-blocking ability of lithium–sulfur battery membranes cast from polymers of intrinsic microporosity. *J. Mater. Chem. A* 4 (43): 16946–16952.

27 Huff, L.A., Rapp, J.L., Baughman, J.A. et al. (2015). Identification of lithium–sulfur battery discharge products through ^{6}Li and ^{33}S solid-state MAS and ^{7}Li solution NMR spectroscopy. *Surf. Sci.* 631: 295–300.

28 Liao, C., Guo, B., Sun, X.-G., and Dai, S. (2015). Synergistic effects of mixing sulfone and ionic liquid as safe electrolytes for lithium sulfur batteries. *ChemSusChem* 8 (2): 353–360.

29 Wu, H.-L., Huff, L.A., and Gewirth, A.A. (2015). In situ Raman spectroscopy of sulfur speciation in lithium–sulfur batteries. *ACS Appl. Mater. Interfaces* 7 (3): 1709–1719.

30 Park, H., Koh, H.S., and Siegel, D.J. (2015). First-principles study of redox end members in lithium–sulfur batteries. *J. Phys. Chem. C* 119 (9): 4675–4683.

31 Chen, J., Wu, D., Walter, E. et al. (2015). Molecular-confinement of polysulfides within mesoscale electrodes for the practical application of lithium sulfur batteries. *Nano Energy* 13: 267–274.

32 Li, C., Ward, A.L., Doris, S.E. et al. (2015). Polysulfide-blocking microporous polymer membrane tailored for hybrid Li–sulfur flow batteries. *Nano Lett.* 15 (9): 5724–5729.

33 Gerber, L.C.H., Frischmann, P.D., Fan, F.Y. et al. (2016). Three-dimensional growth of Li_2S in lithium–sulfur batteries promoted by a redox mediator. *Nano Lett.* 16 (1): 549–554.

34 Cheng, L., Curtiss, L.A., Zavadil, K.R. et al. (2016). Sparingly solvating electrolytes for high energy density lithium–sulfur batteries. *ACS Energy Lett.* 1 (3): 503–509.

35 Fan, F.Y., Pan, M.S., Lau, K.C. et al. (2016). Solvent effects on polysulfide redox kinetics and ionic conductivity in lithium–sulfur batteries. *J. Electrochem. Soc.* 163 (14): A3111–A3116.

36 Eroglu, D., Zavadil, K.R., and Gallagher, K.G. (2015). Critical link between materials chemistry and cell-level design for high energy density and low cost lithium–sulfur transportation battery. *J. Electrochem. Soc.* 162 (6): A982–A990.

37 Gallagher, K.G., Goebel, S., Greszler, T. et al. (2014). Quantifying the promise of lithium–air batteries for electric vehicles. *Energy Environ. Sci.* 7: 1555–1563.

38 Adams, B.D., Black, R., Williams, Z. et al. (2015). Towards a stable organic electrolyte for the lithium oxygen battery. *Adv. Energy Mater.* 5 (1): 1400867.

39 Bruce, P.G., Freunberger, S.A., Hardwick, L.J., and Tarascon, J.-M. (2011). Li–O_2 and Li–S batteries with high energy storage. *Nat. Mater.* 11 (1): 19–29.

40 Safari, M., Adams, B.D., and Nazar, L.F. (2014). Kinetics of oxygen reduction in aprotic Li–O_2 cells: a model-based study. *J. Phys. Chem. Lett.* 5 (20): 3486–3491.

41 Welland, M.J., Lau, K.C., Redfern, P.C. et al. (2015). An atomistically informed mesoscale model for growth and coarsening during discharge in lithium-oxygen batteries. *J. Chem. Phys.* 143 (22): 224113.

42 Cheng, L., Redfern, P., Lau, K.C. et al. (2017). Computational studies of solubilities of LiO_2 and Li_2O_2 in aprotic solvents. *J. Electrochem. Soc.* 164 (11): E3696–E3701.

43 Wei, X., Cosimbescu, L., Xu, W. et al. (2015). Towards high-performance nonaqueous redox flow electrolyte via ionic modification of active species. *Adv. Energy Mater.* 5 (1): 1400678.

44 Smith, K.C., Brunini, V.E., Dong, Y. et al. (2014). Electroactive-zone extension in flow-battery stacks. *Electrochim. Acta* 147: 460–469.

45 Huang, J., Cheng, L., Assary, R.S. et al. (2015). Liquid catholyte molecules for nonaqueous redox flow batteries. *Adv. Energy Mater.* 5 (6): 1401782.

46 Huang, J., Su, L., Kowalski, J.A. et al. (2015). A subtractive approach to molecular engineering of dimethoxybenzene-based redox materials for non-aqueous flow batteries. *J. Mater. Chem. A* 3 (29): 14971–14976.

47 Dmello, R., Milshtein, J.D., Brushett, F.R., and Smith, K.C. (2016). Cost-driven materials selection criteria for redox flow battery electrolytes. *J. Power Sources* 330: 261–272.

48 Darling, R.M., Gallagher, K.G., Kowalski, J.A. et al. (2014). Pathways to low-cost electrochemical energy storage: a comparison of aqueous and nonaqueous flow batteries. *Energy Environ. Sci.* 7 (11): 3459–3477.

49 Hopkins, B.J., Smith, K.C., Slocum, A.H., and Chiang, Y.-M. (2015). Component-cost and performance based comparison of flow and static batteries. *J. Power Sources* 293: 1032–1038.

50 Ha, S. and Gallagher, K.G. (2015). Estimating the system price of redox flow batteries for grid storage. *J. Power Sources* 296: 122–132.

51 Crawford, A., Thomsen, E., Reed, D. et al. (2016). Development and validation of chemistry agnostic flow battery cost performance model and application to nonaqueous electrolyte systems: chemistry agnostic flow battery cost performance model. *Int. J. Energy Res.* 40 (12): 1611–1623.

52 Darling, R., Gallagher, K., Xie, W. et al. (2016). Transport property requirements for flow battery separators. *J. Electrochem. Soc.* 163 (1): A5029–A5040.

53 Montoto, E.C., Nagarjuna, G., Moore, J.S., and Rodríguez-López, J. (2017). Redox active polymers for non-aqueous redox flow batteries: validation of the size-exclusion approach. *J. Electrochem. Soc.* 164 (7): A1688–A1694.

54 Nemani, V.P. and Smith, K.C. (2017). Uncovering the role of flow rate in redox-active polymer flow batteries: simulation of reaction distributions with simultaneous mixing in tanks. *Electrochim. Acta* 247: 475–485.
55 Iyer, V.A., Schuh, J.K., Montoto, E.C. et al. (2017). Assessing the impact of electrolyte conductivity and viscosity on the reactor cost and pressure drop of redox-active polymer flow batteries. *J. Power Sources* 361: 334–344.
56 Montoto, E.C., Nagarjuna, G., Hui, J. et al. (2016). Redox active colloids as discrete energy storage carriers. *J. Am. Chem. Soc.* 138 (40): 13230–13237.
57 Gossage, Z.T., Schorr, N.B., Hernández-Burgos, K. et al. (2017). Interrogating charge storage on redox active colloids via combined Raman spectroscopy and scanning electrochemical microscopy. *Langmuir* 33 (37): 9455–9463.
58 Nagarjuna, G., Hui, J., Cheng, K.J. et al. (2014). Impact of redox-active polymer molecular weight on the electrochemical properties and transport across porous separators in nonaqueous solvents. *J. Am. Chem. Soc.* 136 (46): 16309–16316.
59 Kucharyson, J.F., Cheng, L., Tung, S.O. et al. (2017). Predicting the potentials, solubilities and stabilities of metal-acetylacetonates for non-aqueous redox flow batteries using density functional theory calculations. *J. Mater. Chem. A* 5 (26): 13700–13709.
60 Ward, A.L., Doris, S.E., Li, L. et al. (2017). Materials genomics screens for adaptive ion transport behavior by redox-switchable microporous polymer membranes in lithium–sulfur batteries. *ACS Cent. Sci.* 3 (5): 399–406.
61 Staszak-Jirkovský, J., Subbaraman, R., Strmcnik, D. et al. (2015). Water as a promoter and catalyst for dioxygen electrochemistry in aqueous and organic media. *ACS Catal.* 5 (11): 6600–6607.
62 Tepavcevic, S., Liu, Y., Zhou, D. et al. (2015). Nanostructured layered cathode for rechargeable Mg-ion batteries. *ACS Nano* 9 (8): 8194–8205.
63 Connell, J.G., Genorio, B., Lopes, P.P. et al. (2016). Tuning the reversibility of Mg anodes via controlled surface passivation by H_2O/Cl^- in organic electrolytes. *Chem. Mater.* 28 (22): 8268–8277.
64 Wei, X., Xu, W., Huang, J. et al. (2015). Radical compatibility with nonaqueous electrolytes and its impact on an all-organic redox flow battery. *Angew. Chem. Int. Ed.* 54: 8684–8687.
65 Carino, E.V., Staszak-Jirkovsky, J., Assary, R.S. et al. (2016). Tuning the stability of organic active materials for nonaqueous redox flow batteries via reversible, electrochemically mediated Li^+ coordination. *Chem. Mater.* 28 (8): 2529–2539.
66 Cheng, L., Assary, R.S., Qu, X. et al. (2015). Accelerating electrolyte discovery for energy storage with high-throughput screening. *J. Phys. Chem. Lett.* 6 (2): 283–291.
67 Jain, A., Ong, S.P., Chen, W. et al. (2015). FireWorks: a dynamic workflow system designed for high-throughput applications: FireWorks: a dynamic workflow system designed for high-throughput applications. *Concurr. Comput. Pract. Exp.* 27 (17): 5037–5059.
68 Qu, X., Jain, A., Rajput, N.N. et al. (2015). The electrolyte genome project: a big data approach in battery materials discovery. *Comput. Mater. Sci.* 103: 56–67.

69 Pelzer, K.M., Cheng, L., and Curtiss, L.A. (2017). Effects of functional groups in redox-active organic molecules: a high-throughput screening approach. *J. Phys. Chem. C* 121 (1): 237–245.

70 Jain, A., Ong, S.P., Hautier, G. et al. (2013). The materials project: a materials genome approach to accelerating materials innovation. *APL Mater.* 1 (1): 011002. doi: 10.1063/1.4812323.

71 Canepa, P., Jayaraman, S., Cheng, L. et al. (2015). Elucidating the structure of the magnesium aluminum chloride complex electrolyte for magnesium-ion batteries. *Energy Environ. Sci.* 8 (12): 3718–3730.

72 Liu, M., Rong, Z., Malik, R. et al. (2015). Spinel compounds as multivalent battery cathodes: a systematic evaluation based on *ab initio* calculations. *Energy Environ. Sci.* 8 (3): 964–974.

73 Hannah, D.C., Sai Gautam, G., Canepa, P. et al. (2017). Magnesium ion mobility in post-spinels accessible at ambient pressure. *Chem. Commun.* 53 (37): 5171–5174.

74 Liu, M., Jain, A., Rong, Z. et al. (2016). Evaluation of sulfur spinel compounds for multivalent battery cathode applications. *Energy Environ. Sci.* 9 (10): 3201–3209.

75 Rajput, N.N., Qu, X., Sa, N. et al. (2015). The coupling between stability and ion pair formation in magnesium electrolytes from first-principles quantum mechanics and classical molecular dynamics. *J. Am. Chem. Soc.* 137 (9): 3411–3420.

76 Sa, N., Rajput, N.N., Wang, H. et al. (2016). Concentration dependent electrochemical properties and structural analysis of a simple magnesium electrolyte: magnesium bis(trifluoromethane sulfonyl)imide in diglyme. *RSC Adv.* 6 (114): 113663–113670.

77 Qu, X., Zhang, Y., Rajput, N.N. et al. (2017). Computational design of new magnesium electrolytes with improved properties. *J. Phys. Chem. C* 121 (30): 16126–16136.

78 Laramie, S.M., Milshtein, J.D., Breault, T.M. et al. (2016). Performance and cost characteristics of multi-electron transfer, common ion exchange non-aqueous redox flow batteries. *J. Power Sources* 327: 681–692.

79 Milshtein, J.D., Barton, J.L., Carney, T.J. et al. (2017). Towards low resistance nonaqueous redox flow batteries. *J. Electrochem. Soc.* 164 (12): A2487–A2499.

3

Determination of Redox Reaction Mechanisms in Lithium–Sulfur Batteries

Kevin H. Wujcik [1,2], *Dunyang R. Wang* [2,3], *Alexander A. Teran* [4], *Eduard Nasybulin* [4], *Tod A. Pascal* [5], *David Prendergast* [5], *and Nitash P. Balsara* [1,2,6]

[1] *University of California, Department of Chemical and Biomolecular Engineering, 201 Gilman Hall, Berkeley, CA 94720-1462, USA*
[2] *Materials Sciences Division, Lawrence Berkeley National Laboratory, 1 Cyclotron Road, Berkeley, CA 94720, USA*
[3] *University of California, Department of Materials Science and Engineering, 210 Hearst Mining Building, Berkeley, CA 94720-1760, USA*
[4] *Blue Current, 600 Bancroft Way, Ste D, Berkeley, CA 94710, USA*
[5] *Lawrence Berkeley National Laboratory, The Molecular Foundry, 1 Cyclotron Road, Building 67, Berkeley, CA 94720, USA*
[6] *Lawrence Berkeley National Laboratory, Environmental Energy Technologies Division, 1 Cyclotron Road, Berkeley, CA 94720, USA*

3.1 Basics of Lithium–Sulfur Chemistry

There is considerable interest in developing reversible lithium–sulfur (Li–S) batteries. It is known that the redox reactions that occur in these cells produce "polysulfides." This term generally refers to both lithium polysulfide dianions (Li_2S_x) and lithium polysulfide radical anions (LiS_x). Polysulfide radical anions are a subject of long-standing interest [1, 2]. These radicals are long-lived in the naturally occurring blue mineral lapis lazuli. When pulverized, the mineral provides a pigment known as ultramarine, which was used by European Renaissance painters in the 1800s and in the preparation of jewelry, ornaments, and sculptures found in Afghanistan, Egypt, Iraq, and Pakistan [3, 4]. The color of ultramarine arises because of the optical absorption spectra of $S_2^{\bullet-}$ and $S_3^{\bullet-}$ radical anions. The radical species are trapped in sodalite cages ($Na_8[Al_6Si_6O_{24}]Cl_2$), which are zeolitic in structure. The color of the mineral varies depending on the ratio of trisulfide to disulfide radical species: a high ratio of $S_3^{\bullet-}$:$S_2^{\bullet-}$ gives a deep blue-violet color, whereas a low ratio leads to a light blue-green color [2, 4, 5]. Recent studies show that these radical anions can also be formed in Li–S cells.

Lithium–sulfur cells were first developed in the 1960s [6–8]. The reactions at the electrodes are as follows:

$$\text{Li} \rightarrow 2\text{Li}^+ + 2\text{e}^- \quad \text{(negative electrode or anode)} \tag{R1}$$

and

$$(1/8)\ \text{S}_8 + 2\text{Li}^+ + 2\text{e}^- \rightarrow \text{Li}_2\text{S} \quad \text{(positive electrode or cathode)} \tag{R2}$$

Electrochemical Engineering: The Path from Discovery to Product,
First Edition. Edited by Richard C. Alkire, Philip N. Bartlett, and Marc T. Koper.

The Li^+ ion generated at the anode travels through the electrolyte to reach the S cathode, whereas the e^- travels through the external circuit. The overall cell reaction is

$$2Li + (1/8)\ S_8 \rightarrow Li_2S \tag{R3}$$

The theoretical energy density based on Reaction (R3) can, in principle, be calculated from the standard free energies of formation, ΔG_f^0, of the species in the reaction. To our knowledge, there are no experimental measurements of ΔG_f^0 of Li_2S. This is stated explicitly in Ref. [9], where they calculate ΔG_f^0 using computer simulations. They report four values, which we average to obtain $-422\ kJ\ mol^{-1}$. The values of ΔG_f^0 for Li and S_8 are 0. The decrease in the standard free energy upon completion of Reaction (R3) ($\Delta G^0 = -422\ kJ$) divided by the mass of the reactants (or products) gives the theoretical specific energy of the cell: $2552\ Wh\ kg^{-1}$. This value is comparable to the theoretical specific energy obtained by reacting octane with oxygen (combustion).

$$C_8H_{18} + \frac{25}{2}\ O_2 \rightarrow 8\ CO_2 + 9H_2O \tag{R4}$$

The standard free energies of formation of the compounds of interest are +17.32 (C_8H_{18}), 0 (O_2), 394.35 (CO_2), and $-237.12\ kJ\ mol^{-1}$ (H_2O). The specific free energy of the combustion reaction, obtained by dividing the decrease in standard free energy by the mass of the reactants, is $2866\ Wh\ kg^{-1}$.

The thermodynamic cell potential, U, is given by

$$U = \frac{-\Delta G_f^0}{nF} \tag{3.1}$$

where n is the number of electrons in Reactions (R1) and (R2) ($n = 2$ in our case) and F is the Faraday's constant. This gives $U = 2.19\ V$.

A charged and discharged lithium–sulfur cell is schematically shown in Figure 3.1. The schematic shows a standard porous cathode, where pores are filled with the electrolyte providing pathways for Li^+ transport and carbon providing pathways for e^- transport. Such electrodes are prepared by casting a slurry containing S_8, carbon, a binder, and a solvent and subsequently drying the solvent. The active particles must be in the form of a fine powder to mitigate the effect of poor Li^+ and e^- transport in S_8 and Li_2S. The electrolyte separator is typically an electrically insulating porous polymer. The cathode pores and electrolyte separator are filled with electrolyte when the cell is assembled. A lithium foil serves as the anode and current collector. One thus needs excess lithium in this electrode. The schematic in Figure 3.1 shows only the amount of lithium that would be consumed if the discharge reaction were to go to completion. All of the reactants and products shown in Reactions (R1), (R2), and (R3) are crystalline at room temperature, with densities of 0.534, 2.07, and $1.66\ g\ cm^{-3}$ for lithium metal, elemental sulfur (α-S_8), and Li_2S, respectively. The capacities of the lithium and sulfur electrodes are obtained by noting the ability to accept or reject electronic charge per gram material: giving $3860\ mAh\ g^{-1}$ for the Li electrode and $1672\ mAh\ g^{-1}$ for the S electrode. For a balanced cell, the mass ratio of Li to S is 0.434 and the volume ratio of Li to S is 1.68. The variable ΔV is defined

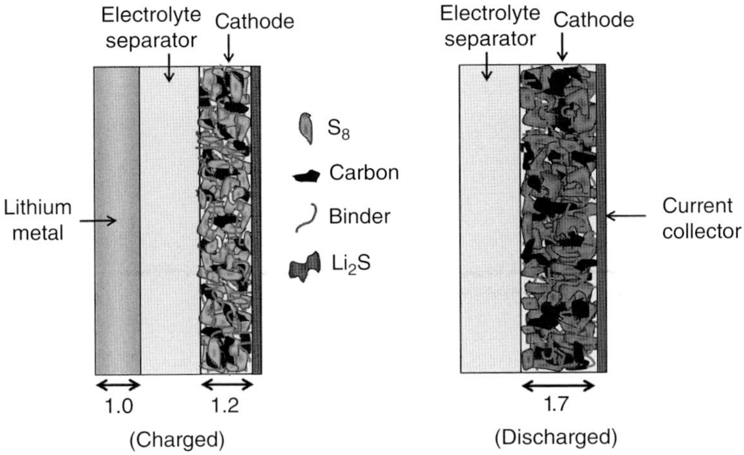

Figure 3.1 Schematic of a charged and discharged balanced Li–S cell. The cathode comprises 50 vol.% S. The relative thicknesses of the electrodes are indicated.

as the net change in the volume of active materials (Li and S) in the discharged state relative to the charged state, and it accounts for the expansion of the sulfur electrode upon lithiation (discharge) and the concomitant disappearance of the lithium electrode. The fractional volume change on discharge ($\Delta V/V_{charge}$) is defined as the ratio of the change in the volume of the active materials after discharge to the volume of the active materials in the original, charged cell. For a balanced Li–S cell, the fractional volume change upon discharge ($\Delta V/V_{charge}$) is – 0.332. The fractional change in the volume of real cells will be smaller due to the presence of inactive components (separator, electrolyte, current collectors, etc.) that are not affected by the charge and discharge processes.

The Li–S cell has several advantages. In addition to its high theoretical specific energy, sulfur is cheap and abundant. Additionally, the high specific energy of the cell is derived from high capacity rather than high cell voltage. This eases the electrolyte stability requirements (current lithium ion cells have open-circuit potentials in the vicinity of 4 V and few electrolytes have a suitably wide electrochemical stability window).

There are, however, significant challenges. Transport of electrons and lithium ions in and out of S_8 and Li_2S is challenging because of low mobilities of both charged species; the compounds are insulating solids. Additionally, the reaction mechanism is much more complex than the overall reaction shown in (R3). The discharge/charge processes involve a series of reactions through which lithium polysulfide (Li_2S_x and LiS_x; $2 \leq x \leq 8$) reaction intermediates are formed. While the end products (Li_2S and S_8) are sparingly soluble in conventional lithium battery electrolytes, some polysulfide reaction intermediates are highly soluble. As a result, polysulfide species formed during charge and discharge can diffuse out of the cathode. This may result in a loss of active material (i.e. cell capacity) if the polysulfides do not diffuse back into the cathode. Diffusion of polysulfide species to the cell anode can lead to parasitic reduction that gives rise to a polysulfide shuttle wherein mid-length polysulfides formed at the anode diffuse back to the cathode and undergo further chemical and electrochemical reactions. If the dissolved polysulfides are fully reduced at the anode, an insulating Li_2S (or, perhaps Li_2S_2) layer is formed on the anode. These problems must be solved in order to make practical lithium–sulfur cells that can be used in applications such as stationary power and electric vehicles. These issues also make fundamental studies of the Li–S reaction mechanisms challenging.

Numerous papers have shown that the cycle life of Li–S cells can be improved by confining the polysulfides in nanoscale carbon-based structures [10–13]. These approaches are important as they give hope for obtaining practical rechargeable Li–S cells. In this review, however, we do not cover them. Instead, we explore the fundamental reaction mechanisms and chemical processes that govern the Li–S charge/discharge processes. It is our belief that a fundamental understanding of what happens in the sulfur cathode is an essential first step for designing long-lasting lithium sulfur batteries.

3.2 End Products of Electrochemical Reactions in the Sulfur Cathode

As all of the reactants and products listed in Reactions (R1)–(R3) are crystalline under standard battery-operating conditions, their formation can be studied using X-ray diffraction (XRD). The XRD characteristics of Li, S_8 (under standard conditions), and Li_2S are given in Figure 3.2, where X-ray scattering intensity, I, is plotted as a function of the scattering angle (2θ).

Given that all three compounds are crystalline solids at room temperature and that they are sparingly soluble in conventional electrolytes, one expects the end

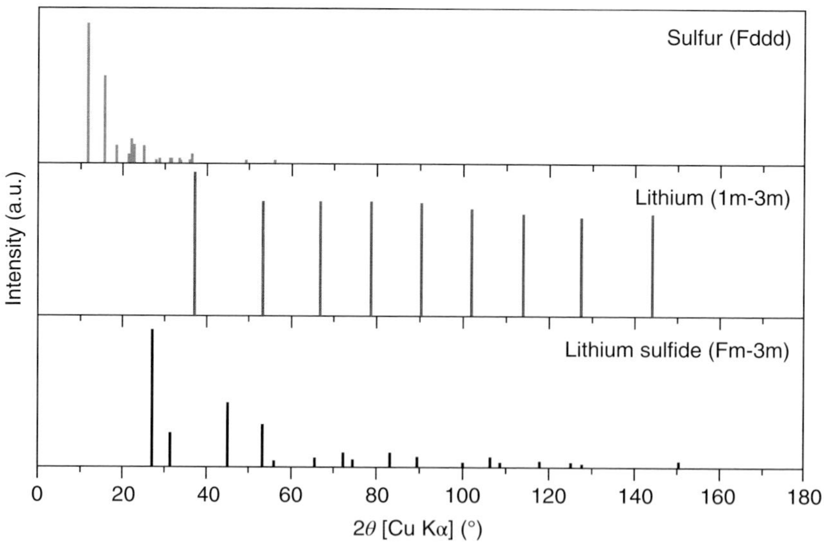

Figure 3.2 X-ray diffraction patterns for elemental sulfur, lithium metal, and lithium sulfide.

points of the charge and discharge in a lithium sulfur cell to be noncontroversial. This is untrue. In three separate studies, Cañas et al., Lowe et al., and Nelson et al. used XRD to probe Li–S electrodes during discharge [14–16]. Cañas et al. observed that Li_2S (the expected end product of discharge) formation started at a depth of discharge of 60% (i.e. the discharge process was only 60% complete) [14]. Lowe et al. observed that Li_2S formation began only at the very end of the discharge process [15]. In contrast, Nelson et al. did not observe any crystalline Li_2S formation during discharge (even when the cathode was completely discharged) [16]. The detection of crystalline sulfur after charging has also been controversial. Nelson et al. and Cañas et al. both observed the formation of crystalline sulfur at the end of charging [14, 16], but other authors who performed ex situ XRD on charged cells found no evidence of crystalline sulfur [17–19]. The disparity in the results produced by these studies can be attributed to an array of factors: difficulties in performing in situ and ex situ experiments without introducing artifacts, dissimilarities between in situ and ex situ results, differences in cell performance (i.e. the obtained capacity), differences in the size of active material particles [20–22], and differences in cell composition. Nonetheless, a collective analysis of literature data suggests that Reactions (R1), (R2), and (R3) are valid. That is, if one were to deliver lithium ions and electrons into a sulfur electrode, the species that forms upon complete conversion is Li_2S. Similarly, if one were to extract lithium ions and electrons from a Li_2S electrode, the species that forms upon complete conversion is S_8.

3.3 Intermediate Products of Electrochemical Reactions in the Sulfur Cathode

It is widely understood that in typical Li–S cells, Reactions (R1) and (R2) are not elementary. The elementary steps generally involve lithium polysulfide dianions (Li_2S_x, $2 \leq x \leq 8$) and radical anions (LiS_x, $3 \leq x \leq 5$). There is, however, no agreement on the elementary steps and the chemical formulae and concentrations of the polysulfides formed as S_8 is reduced. Nonelementary steps are also often proposed. In this review, we discuss a few of the proposed reaction steps taken from Refs. [23–34].

3.3.1 Reactions of S_8

Many researchers assume that the first step in the reduction of sulfur involves two lithium ions as shown in Reaction (R5) [23–28, 30, 33].

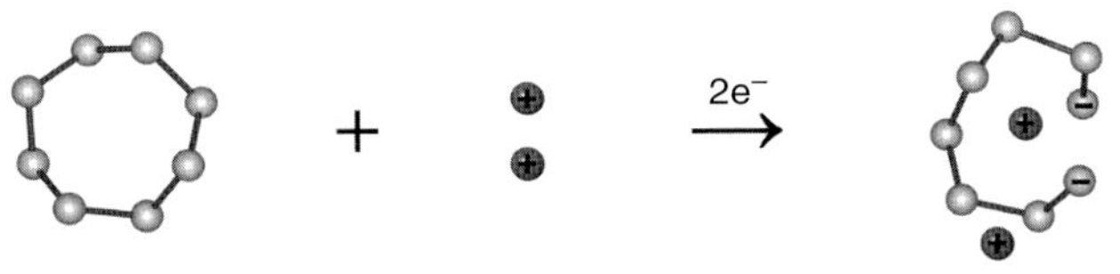

$$S_8 + 2e^- + 2Li^+ \rightarrow Li_2S_8 \qquad \text{(R5)}$$

Here, sulfur is represented by yellow atoms and lithium is represented by purple ions. Although Equation (R5) is generally the most agreed upon first reaction step, other reactions have also been proposed. Wang et al., Mikhaylik et al., and Kawase et al. have proposed a four-electron process as shown in Equation (R6) for the direct formation of Li_2S_4 [25, 32, 34]. This implies that the lithium ions and electrons attack the sulfur ring on diametrically opposite sites.

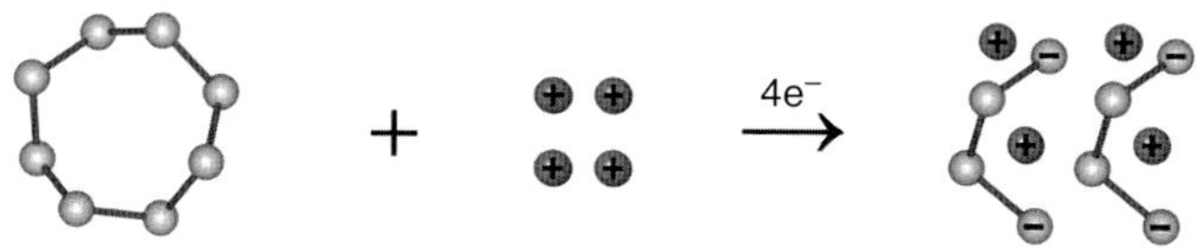

$$S_8 + 4e^- + 4Li^+ \rightarrow 2Li_2S_4 \qquad \text{(R6)}$$

Kawase et al. suggested that the lithium ions and electrons do not necessarily attack on diametrically opposite sites of the sulfur ring [25]. They thus proposed the formation of two different polysulfides Li_2S_x and $Li_2S_{(8-x)}$. For concreteness, one such possibility is shown in (R7).

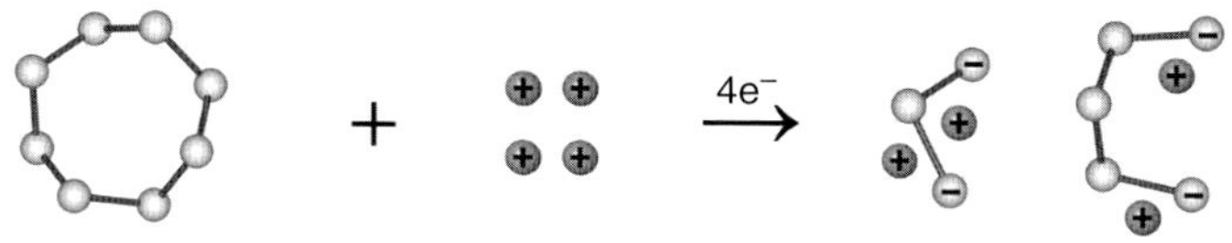

$$S_8 + 4e^- + 4Li^+ \rightarrow Li_2S_{8-x} + Li_2S_x \ (x \leq 7) \qquad \text{(R7)}$$

Thus, the first reduction step may result in the formation of a wide variety of polysulfide dianions, through Reactions (R5), (R6), and (R7). For simplicity, in the discussion below, we assume that the first reduction step results in the formation of Li_2S_8 and Li_2S_4, where Li_2S_4 is a proxy for intermediate chain length polysulfide dianions.

3.3.2 Reactions of Li_2S_8

The reduction of Li_2S_8 is most frequently said to occur through a two-electron reduction to form two Li_2S_4 molecules according to Eq. (R8) [24, 25, 27, 28].

$$Li_2S_8 + 2e^- + 2Li^+ \rightarrow 2Li_2S_4 \qquad \text{(R8)}$$

As suggested by Eq. (R7), there is not necessarily a preference for the formation of two Li_2S_4 molecules; the reaction may form a combination of Li_2S_x and $Li_2S_{(8-x)}$, where $2 \leq x \leq 6$. A surprising number of researchers have suggested an alternative reaction for the consumption of Li_2S_8, shown in Reactions (R9a) and (R9b) [23, 24, 26–29]. Here, the lithium ions and electrons must attack four Li_2S_8 molecules at precisely the same location as shown in Reaction (R9a). The four resulting S_2 fragments must then recombine to form S_8 (Reaction (R9b)).

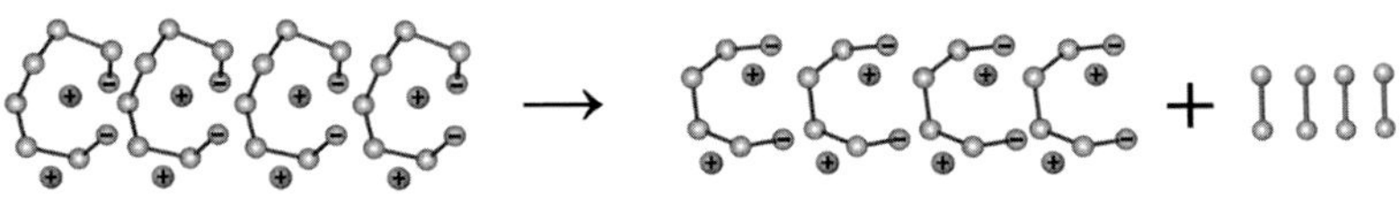

$$4Li_2S_8 \rightarrow 4Li_2S_6 + 4S_2 \tag{R9a}$$

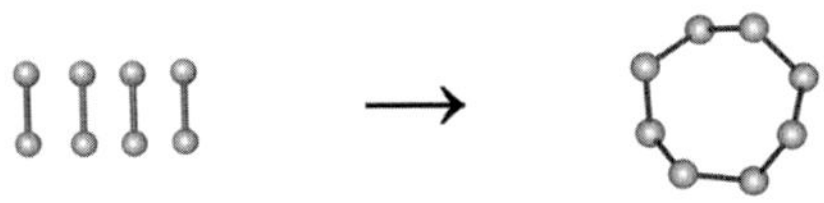

$$4S_2 \rightarrow S_8 \tag{R9b}$$

Although Reaction (R9) is popular in the literature, the number of species involved and the complexity of the reactions suggest that this pathway is unlikely in the liquid state wherein entropy dominates. In addition, there is no reason to believe that the second sulfur–sulfur bond from the end is the weakest link in the chain. Reaction (R9b) must also involve several highly improbable transition states. The presence of high concentrations of other reactive polysulfides (that could react with S_2, for instance) would also make Reaction (9b) highly improbable. We therefore conclude that the consumption of Li_2S_8 occurs by Reaction (R8), and it leads to the formation of Li_2S_4, which again is a proxy for intermediate chain length polysulfide dianions.

3.3.3 Reactions of Li_2S_4

Many researchers agree that the reduction of Li_2S_4 occurs by Reaction (R10) [25, 30, 34].

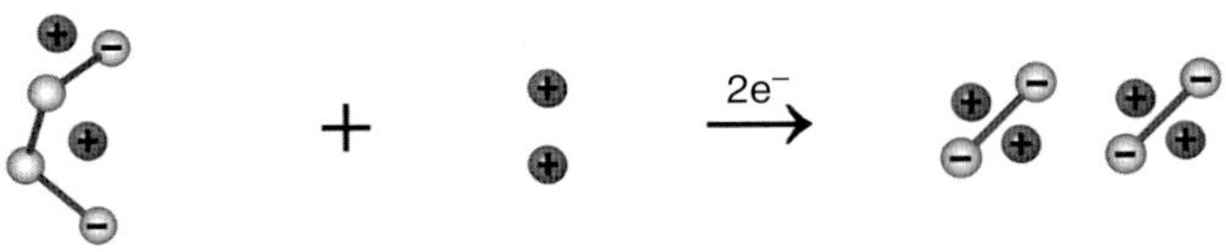

$$Li_2S_4 + 2e^- + 2Li^+ \rightarrow 2Li_2S_2 \tag{R10}$$

However, in Ref. [23], an alternative reaction has been proposed [23].

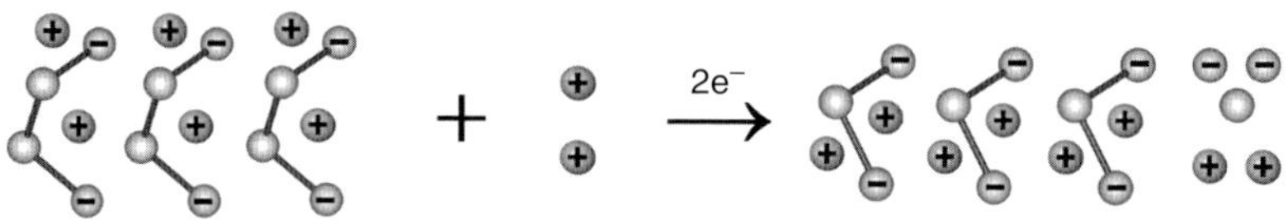

$$3Li_2S_4 + 2e^- + 2Li^+ \rightarrow 3Li_2S_3 + 2S^- + S + 2Li^+ \quad \text{(R11a)}$$

$$2S^- + S + 2Li^+ \rightarrow Li_2S_3 \quad \text{(R11b)}$$

This is at least a two-step reaction wherein the lithium ions and electrons attack the ends of three Li_2S_4 molecules, resulting in the formation of Li_2S_3 dianions (R11a) and two sulfur ions and one sulfur atom. The three separate sulfur species must then recombine to give Li_2S_3 (R11b). (One could construct alternate scenarios but they would be more complicated.) In contrast, Reaction (R10) involves one step and one reactive polysulfide molecule. We therefore conclude that the reduction of Li_2S_4 occurs through R10 and not R11.

3.3.4 Reactions of Li_2S_2

Most researchers agree on the reaction step for conversion of Li_2S_2 to Li_2S, given by Reaction (R12) [23, 25, 27, 30, 33, 34]. After formation via Reaction (R10), Li_2S_2 is reduced to form lithium sulfide (Li_2S).

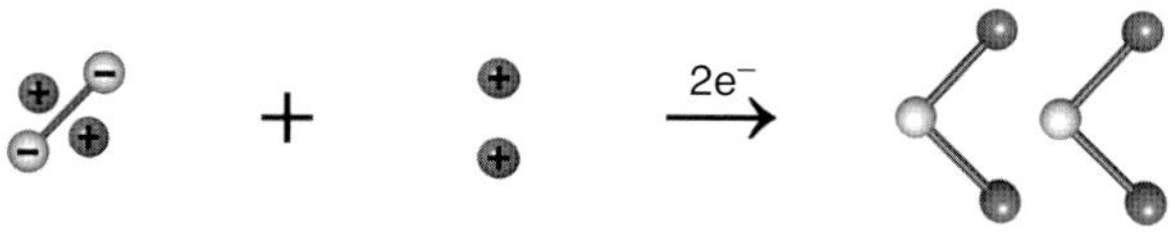

$$Li_2S_2 + 2e^- + 2Li^+ \rightarrow 2Li_2S \quad \text{(R12)}$$

It is likely that the reduction pathways of other intermediate chain length polysulfide dianions such as Li_2S_3 and Li_2S_5 are more complex than the reaction pathway for Li_2S_4. It is also possible that polysulfide dianion species of short chain length can be reduced by the general reaction [33]:

$$Li_2S_x + 2\,e^- + 2Li^+ \rightarrow Li_2S + Li_2S_{x-1} \quad (2 \leq x \leq 5) \quad \text{(R13)}$$

In addition to electrochemical reactions, polysulfide dianions can spontaneously undergo chemical reactions such as recombination and disproportionation:

$$Li_2S_3 + Li_2S_5 \rightarrow 2Li_2S_4 \quad \text{(R14)}$$

$$2Li_2S_6 \rightarrow Li_2S_5 + Li_2S_7 \quad \text{(R15)}$$

There is no consensus on the relative importance of chemical vs electrochemical reactions in sulfur cathodes.

3.3.5 Production of Radical Anions

Polysulfide intermediates may also take the form of radical anions. Yamin et al. suggested that the initial reduction of elemental sulfur may proceed through Reaction (R16) that forms the LiS_8 radical anion [29].

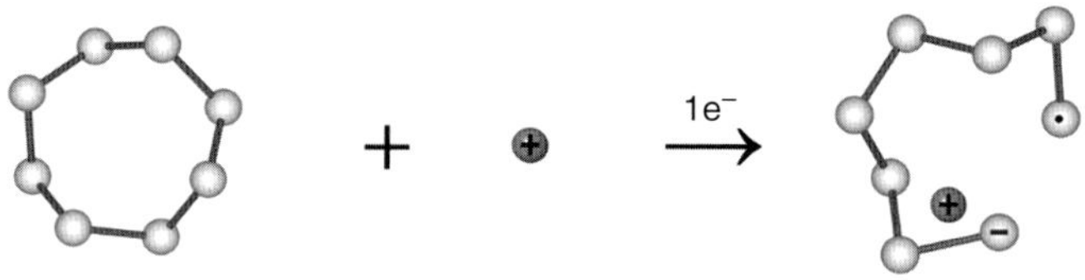

$$S_8 + 1e^- + 1Li^+ \rightarrow LiS_8 \quad \text{(R16)}$$

Reaction (R16) involves only one electron and one lithium ion as opposed to Reaction (R5), which involves two electrons and two lithium ions. If two electrons and two lithium ions do not arrive simultaneously at the active site, then one may consider Reaction (R16) as a necessary first step in the sulfur reduction process.

The formation and dissolution of polysulfide radicals depend on the solvent's ability to stabilize the molecules. Polysulfide radicals have been shown to be stable in high concentrations in solvents that have high electron pair donor numbers and high dielectric constants (dimethylformamide (DMF) and dimethylacetamide (DMA)) and at lower concentrations in ether-based solvents [24, 35–41]. Polysulfide radical anions can participate in both electrochemical and chemical reactions such as R17, R18, and R19:

$$LiS_x + 1e^- + 1Li^+ \rightarrow Li_2S_x \quad \text{(R17)}$$

$$Li_2S_8 \rightarrow LiS_x + LiS_{(8-x)} \quad (x = 3, 4) \quad \text{(R18)}$$

$$Li_2S_6 \rightarrow 2LiS_3 \quad \text{(R19)}$$

The formation of the trisulfur radical anion (LiS_3) via Reaction (R19) is frequently cited in studies of sulfur redox chemistry [23, 24, 26–28]. Single-electron reactions such as R17 (with $x = 8$) may proceed more rapidly than two electron/lithium ion reactions such as R8 if the difference between the free energies of the reactants and products is comparable.

3.4 Fingerprinting Lithium Polysulfide Intermediates

Reaction mechanisms such as those shown in Reactions (R5)–(R19) have been a focus of Li–S battery research for over four decades [26, 29, 31, 42–46]. The reactants and products in these reactions, dianions and radical anions, are not crystalline and thus not detected by in situ XRD. This motivates the use of spectroscopic techniques such as X-ray absorption spectroscopy (XAS), nuclear magnetic resonance (NMR) spectroscopy, UV–Visible (UV–Vis) absorption spectroscopy (UV–Vis), and Raman spectroscopy.

The challenge of performing spectroscopic experiments has been that there are no widely agreed upon spectral standards for the lithium polysulfide reaction intermediates. Lithium polysulfides can only exist as dissolved species and cannot be isolated from solution. Additional complexity arises because polysulfides can transform into other compounds spontaneously by chemical reactions depending on the nature of the solvent, temperature, and concentration. It is, perhaps, not surprising that studies of the Li–S redox reaction mechanisms have produced widely varying results. Spectroelectrochemical techniques have nonetheless taught researchers a great deal about Li–S redox reactions, and such studies are still ongoing. Below we examine the spectroscopy techniques that have been employed to study Li–S reaction chemistry.

3.4.1 X-ray Absorption Spectroscopy

XAS is a technique that has gained much attention over the past few years as a probe of Li–S reaction mechanisms. This approach is capable of detecting chemicals that are crystalline or amorphous, concentrated or dilute [47–49]. It also offers the benefit of being element-specific and fairly penetrative in the presence of materials typically found in Li–S cells. However, attempts to obtain unambiguous spectral standards for polysulfide intermediates have been problematic. We illustrate these difficulties in Figure 3.3 using the reported XAS

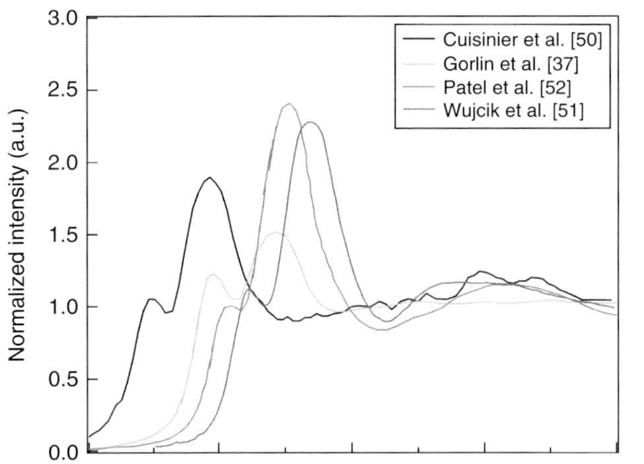

Figure 3.3 Proposed sulfur K-edge XAS spectral standards for Li_2S_4 in the literature [37, 50–52]. There is no agreement on either the locations or the heights of the main spectral features.

signatures of a particular polysulfide dianion, Li_2S_4, in four different publications [37, 50–52].

The spectra shown in Figure 3.3 are vastly different. As XAS is mainly sensitive to the excitation energies of core electrons, factors such as solvent type and aggregation state are not expected to affect the spectra. The differences in fingerprints shown in Figure 3.3 apply to all polysulfide species and all techniques (UV–Vis, Raman, FTIR, and EPR (electron paramagnetic resonance)); fingerprints proposed by different groups differ radically from each other. It is obvious that correct interpretation of data obtained from in situ spectroscopy in the absence of well-established fingerprints is perilous. Reactions such as R9 and R11, which seem unjustifiably complex, have been proposed due to the detection of fingerprints of certain polysulfide species and the presumed absence of other species.

To overcome experimental difficulties, Pascal et al. obtained X-ray spectra of individual polysulfide dianion species through first-principles calculations [53]. The resulting spectra from these calculations are shown in Figure 3.4. The calculated polysulfide spectra exhibit two peaks, similar to those shown in Figure 3.3. The longer polysulfides exhibit a peak around 2472.6 eV, referred to as the main edge peak due to its occurrence near the elemental sulfur absorption edge, and a peak at 2470.5 eV, referred to as the pre-edge peak due to its occurrence below the main edge [53, 54]. The peaks in Li_2S_3 move to higher energies, and Li_2S_2 only exhibits one peak. Pascal et al. showed that these features arise from the $1s \rightarrow 3p$ transitions; the "main edge" feature at 2472.6 eV arises from internal, neutral sulfur atoms within the polysulfide dianion chain, whereas the terminal,

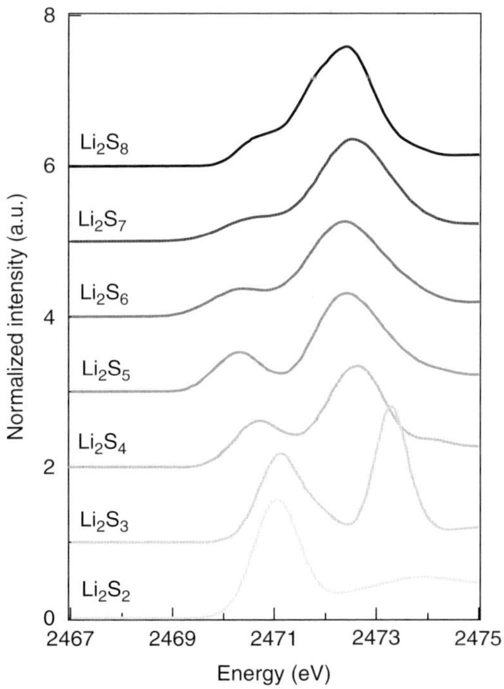

Figure 3.4 First-principles simulation of sulfur K-edge XAS spectra of lithium polysulfide dianion species dissolved in tetraethylene glycol dimethyl ether, $x = 2$–8. Source: Pascal et al. 2014 [53]. Reproduced with permission of American Chemical Society.

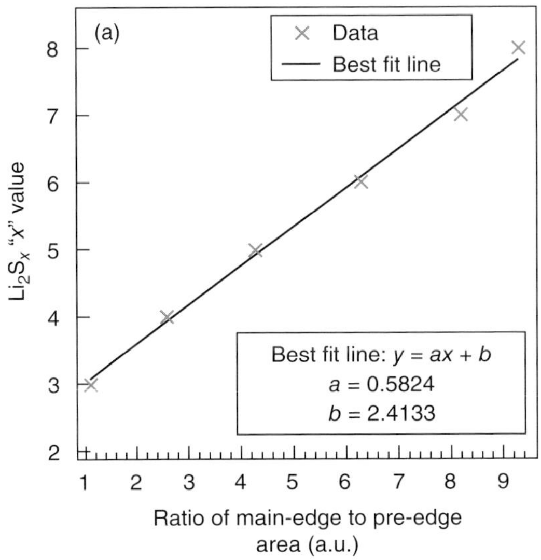

Figure 3.5 Ratio of main edge peak area to pre-edge peak area vs x, chain length of polysulfide dianion (Li_2S_x). Areas determined from Gaussian fits through the theoretical X-ray absorption spectra obtained from first-principles calculations (see Figure 3.4).

negatively charged sulfur atoms give rise to the pre-edge peak at 2470.5 eV [53]. Additionally, they found that the ratio of the main edge peak area to pre-edge peak area scales linearly with the polysulfide chain length "x" value, as shown in Figure 3.5. The molecular origin of the two peaks suggests such a relationship.

Figure 3.5 can be used as a calibration that relates the main edge to pre-edge peak ratio obtained in sulfur K-edge XAS experiment to a single polysulfide chain length "x." However, as a result of disproportionation reactions, polysulfide dianions are typically present in the form of an equilibrium mixture of polysulfide species with different chain lengths. Thus, the "x" value obtained from the calibration curve in Figure 3.5 may more accurately be described as the probed distribution's average polysulfide dianion chain length (x_{av}). If the mixture of interest contains a few polysulfide species, then it may be possible to use statistical approaches such as principal component analysis to determine the chemical structures and compositions of the polysulfides in solution [51]. It appears, however, that redox reactions in Li–S cells produce complex mixtures comprising a large number of dianions and radicals, and in such cases, determination of the species in solution based on statistical approaches may not be possible [24, 37, 52, 55, 56]. Nonetheless, the average polysulfide dianion chain length is a direct indicator of the amount of lithium that has reacted with sulfur due to redox reactions. Such an indicator may be of great use for in situ studies of Li–S battery cathodes.

In addition to polysulfide dianion species, Pascal et al. also calculated spectra of polysulfide radical anions [54]. The spectra of LiS_3, LiS_4, and LiS_5 radicals are shown in Figure 3.6. The most distinguishable difference between spectra of polysulfide radical anion species and those of the dianions is the presence of spectral features at energies below 2470 eV. These features arise due to $1s \rightarrow \pi^*$ transitions of terminal sulfur atoms of the polysulfide chain. However, the contribution of this feature to the overall XAS signal is small due to the presence

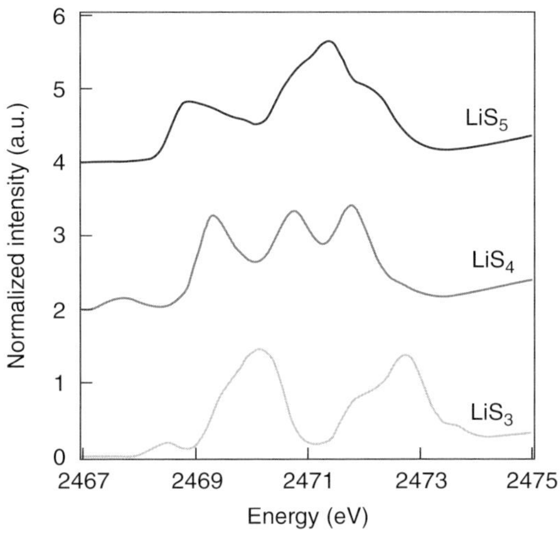

Figure 3.6 First-principles simulation of sulfur K-edge XAS spectra of lithium polysulfide radical anion species dissolved in tetraethylene glycol dimethyl ether, $x = 3$–5, obtained by Pascal and coworkers. Source: Wujcik et al. 2015 [33]. Reproduced with permission of John Wiley & Sons.

of $1s \rightarrow \sigma^*$ transitions of both terminal and internal S atoms and $1s \rightarrow \pi^*$ transitions of internal S atoms, which occur at higher energies (2470–2474 eV). The unique spectral fingerprint of radical anions can thus only be observed in concentrated systems. Based on the spectra in Figure 3.6 and typical XAS resolution limits, we anticipate that XAS is only effective for detecting radicals in systems wherein at least 20 mol% of S atoms in the system are present in the form of radicals.

Unlike the spectra of polysulfide dianion species, the spectra of lithium polysulfide radical species do not share common pre-edge and main edge peaks. This is perhaps valuable, as it means that different polysulfide radical species can be distinguished from one another. However, when polysulfide radicals are present in Li–S cells, they always coexist with dianions. This is due to the reversible nature of dissociation and recombination reactions (e.g. R19). The formation of radical polysulfide species in Li–S battery cathodes adds another element of complexity to spectral analysis, as the calibration curve shown in Figure 3.5 only applies to systems that are dominated by dianions.

The sulfur K-edge XAS spectra of lithium sulfide and elemental sulfur are well established and are shown in Figure 3.7. The spectrum of elemental sulfur contains one single main edge peak at 2472.6 eV. Thus, when probing Li–S cathodes with XAS, the presence of a peak below the main edge peak of elemental sulfur indicates that lithium has reacted with sulfur to form polysulfide species. Examining the spectrum of Li_2S, we see that all of the peaks occur at energies greater than 2472.6 eV. It is obvious that when Li_2S or S_8 are in the cathode, the signal from these molecules must be subtracted before the calibration curve shown in Figure 3.5 is used.

3.4.2 Electron Paramagnetic Resonance Spectroscopy

Another technique used to study Li–S battery reaction mechanisms is electron paramagnetic resonance spectroscopy ((EPR). Generally speaking, EPR is used

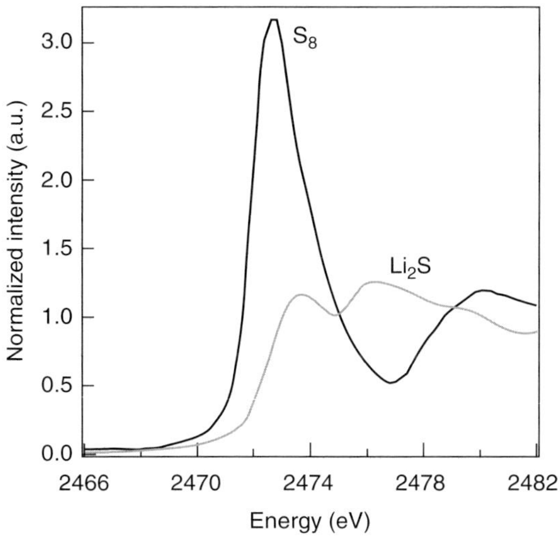

Figure 3.7 Sulfur K-edge XAS standards for elemental sulfur (S_8) and lithium sulfide (Li_2S), obtained experimentally.

to detect radical-containing chemicals, and thus its purpose in Li–S battery research has been to examine lithium polysulfide radical anions. EPR is particularly advantageous as a technique to probe lithium polysulfide radical species because the measured signal will not contain any contribution from polysulfide dianions. However, the complexity of EPR equipment makes in situ spectroelectrochemical experiments challenging: samples for EPR must fit in small capillary tubes that have diameters typically <5 mm. For that reason, EPR has mostly been used to study chemically synthesized mixtures of polysulfide species, or mixtures of polysulfide species extracted from Li–S cathodes [5, 36, 38–41, 57].

The lithium trisulfur radical anion (LiS_3) is the most frequently occurring radical species in Li–S spectroscopic studies. The EPR signature of LiS_3 is well known and occurs at a g-factor of approximately 2.029 [27, 38, 57, 58]. A study performed by Levillain et al. showed that the LiS_4 radical may also be present (in very low concentrations) in chemically synthesized mixtures of polysulfide species in DMF, with a g-factor of 2.033 [39].

3.4.3 UV–Vis Spectroscopy

UV–Vis spectroscopy is, in principle, a simple approach for detecting polysulfides. The proposed spectral fingerprints of dianions, however, vary widely from study to study. Differences in peak assignments may stem from differences in how ex situ mixtures of polysulfide species were synthesized, what solvent the polysulfide species were dissolved in, and how in situ and ex situ spectra were interpreted. Table 3.1 summarizes the spectral peak assignments presented by several groups. In these studies, three methods were used to prepare the polysulfide species: (i) reaction of elemental S with Li metal, $2Li + \frac{x}{8}S_8 \rightarrow Li_2S_x$; (ii) reaction of elemental S with lithium sulfide (Li_2S), $Li_2S + \frac{x-1}{8}S_8 \rightarrow Li_2S_x$; and (iii) electrochemical reduction of S_8.

We see in Table 3.1 that fingerprints of particular polysulfide species depend on solvent type: compare Li_2S_8 signatures in DMA [43] and DMF [59]. This is

Table 3.1 UV–Vis absorption peak assignments in nonaqueous solvents in the literature; method A, $2Li + \frac{x}{8}S_8 \rightarrow Li_2S_x$; B, $Li_2S + \frac{x-1}{8}S_8 \rightarrow Li_2S_x$; C, electrochemical reduction of S_8.

Solvent (dielectric constant)	Method	Absorbance peak wavelength (nm)					
		S_8	Li_2S_8	Li_2S_6	Li_2S_4	Li_2S_2	LiS_3
DMSO (46.7)	A [31]		492	475	420		618
	C [26]		490, 380	450, 340	310–325		605–625
DMA (37.8)	C [43]		518, 365	450–460			618
DMF (36.7)	C [59]		490, 355	450	420	280	600
	C [60]		500				617
THF (7.58)	A [44]	265			435, 285		620–625
TEGDME (7)	B [23]	270–280	560	470, 350, 300, 260	420		617
	B [45]				503, 415, 300	338, 236	
	C [61]	265		332	425		615
Ionic liquid	C [62]			460, 350	440		620

not surprising. What is, however, surprising is that different studies of particular polysulfide species in the same solvent report different fingerprints: compare Li_2S_4 signatures in tetraethylene glycol dimethyl ether (TEGDME), reported in Refs. [23, 45]. It is generally agreed that light absorption in the 400–500 nm range is related to the polysulfide dianion species: Li_2S_4, Li_2S_6, and Li_2S_8 [63, 64]. Many authors have proposed that the absorption wavelength increases with increasing polysulfide chain length [61, 63]. This is based on the fact that as chain length increases, the average negative charge density along the chain decreases [53], and this decreases the gap between the highest occupied molecular orbital and the lowest unoccupied molecular orbital (HOMO–LUMO gap). Despite claims to the contrary, we believe that assignment of particular UV–Vis peaks to specific polysulfide dianions is impossible at the present time.

The UV–Vis fingerprints of polysulfide radical anions are much clearer. An absorption peak at 610–620 nm, which is well separated from the dianion peaks, has been attributed to the trisulfur radical anions, regardless of the solvent (see LiS_3 column in Table 3.1) [23, 28, 38, 41, 61]. In early work, this peak was assigned to neutral molecules S_x ($x = 2–4$) [65, 66]. In pioneering work, Gruen et al. [67] and Chivers and Drummond [35, 68] used a combination of EPR, UV–Vis, infrared, and resonance Raman spectroscopy to show that the blue color of alkali polysulfide solutions is due to the presence of the trisulfur radical anion, $S_3^{\bullet -}$. These results highlight the need for employing several independent techniques to establish polysulfide fingerprints.

Rauh et al. used UV–Vis spectroscopy to study the solubility of polysulfides in different solvents and found that their solubility increased with the basicity of the solvent, which can be quantified by the electron pair donor number [63]. They also concluded that the solvent dielectric constant (given in Table 3.1) plays a major role in determining the stability of polysulfide radicals. Polysulfide radicals are usually produced and stabilized in high dielectric constant solvents such as DMF. This raises an interesting question: are polysulfide radicals stable in ether-based solvents (often used in Li–S cells) in spite of their low dielectric constants and low donor numbers? Some researchers have detected radical species in ether-based electrolytes [23, 33, 61, 69], whereas others have not [15, 50–52, 70, 71]. Recently, Wujcik et al. used a combination of UV–Vis and EPR spectroscopy to show that radical species (namely the trisulfur radical anion) are present in low concentrations in chemically synthesized polysulfide mixtures in ether-based electrolytes, TEGDME and PEO. To ascertain the assignment of the UV–Vis peak at 617 nm to polysulfide radicals, the peak area of the 617 nm absorption peak was compared to the integrated peak area of the measured EPR signal for each solution [41]. The results of this comparison are shown in Figure 3.8.

A few studies have examined the relationship between the total concentration of polysulfide species in a given solution and radical concentration [38, 63]. These studies have shown that the dissociation of dianions to form radicals (e.g. Reaction (R19)) is favored at low concentrations.

One way to solve the problem of obtaining conclusive spectral signatures of polysulfide dianion species is through computational methods. To our knowledge, the only computational calculation of UV–Vis polysulfide spectra is that presented by Kawase et al. [25]. This work used time-dependent density functional theory to simulate UV–Vis spectra for lithium polysulfide molecules. The simulations were performed on polysulfides suspended in a vacuum environment, an environment that stabilizes linear chain conformations. In contrast, Pascal et al., who performed molecular dynamics (MD) simulations of lithium polysulfides in an explicit solvation environment with TEGDME, found that the sulfur chains bend into a claw-shaped structure, wrapping around one of the lithium ions [53]. Molecular conformations and interactions between polysulfides and solvent are expected to affect excitation states (and therefore

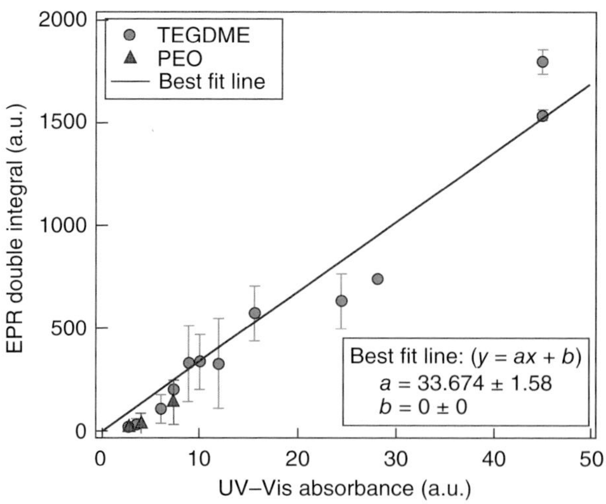

Figure 3.8 UV–Vis peak area (617 nm) vs the double integral of the EPR spectra for TEGDME and PEO solutions. Peak areas obtained from UV–Vis and the EPR double integrals are proportional, indicating a direct relationship between the 617 nm UV–Vis feature and the presence of radical anions. Source: Wujcik et al. 2016 [41]. Reproduced with permission of American Chemical Society.

UV–Vis spectra). Theoretical calculations of UV–Vis spectra based on MD simulations seem warranted. Nevertheless, the work of Kawase et al. provides molecular insight into the UV–Vis spectral features of polysulfides [25].

3.4.4 Raman Spectroscopy

Raman spectroscopy of Li–S cells has been used by a number of researchers in an attempt to differentiate the various polysulfides (both dianions and radical anions), sulfur, and lithium sulfide [58, 72–77]. In an attempt to circumvent the issues associated with obtaining spectral standards of polysulfides in solution, some authors have used crystalline solid analogs of lithium polysulfides such as sodium and potassium polysulfides, as well as other related solids such as ultramarine. It has, however, been shown that the Raman spectra of crystalline polysulfides differ substantially from those of polysulfides in solution [77–80]. In addition, the Raman peak positions obtained for a given polysulfide change by several cm^{-1} as temperature and solvent are changed [58, 75]. Hagen used DFT to calculate the spectra for polysulfide dianions and radical species [75]. They found that spectra from different polysulfides contained broad, overlapping features, making fingerprinting difficult. Hagen et al. pointed out that experimental Raman spectra of polysulfides contained broad spectral features, consistent with the theoretical calculations [75]. The experimentally obtained Raman spectra thus cannot be used to identify particular polysulfide dianions.

In contrast to the dianions, the LiS_3 radical anion has been experimentally detected by a Raman peak at 535 cm^{-1}, consistent with the calculations of Hagen et al. [5, 58, 73–75]. Similarly, the Raman peaks at 155, 220, and 475 cm^{-1} are signatures of S_8 [72–74, 81–88].

3.4.5 Nuclear Magnetic Resonance Spectroscopy

In the context of Li–S batteries, ^6Li, ^7Li, ^{17}O, and ^{33}S nuclei have all been used to study Li–S battery chemistry by NMR spectroscopy [40, 50, 89, 90]. Employment of NMR as a means to distinguish various polysulfide molecules rests on the idea that the probed element (Li or S) has a chemical shift that is different for each polysulfide. Lithium NMR (^6Li and ^7Li) has been the most useful type of NMR spectroscopy used [50, 89]. Results of such studies suggest that while lithium in the form of Li_2S has a chemical shift that is clearly distinguishable from lithium polysulfides, the chemical shifts of various polysulfides are unfortunately not different enough for the polysulfide species to be distinguished from one another. The chemical shift for ^6Li and ^7Li isotopes were found to be similar for Li_2S, occurring at roughly 2.30 ppm [50, 89]. Huff et al. showed that the ^6Li chemical shift for various polysulfide dianion species occurs in the range of −0.25 to −0.75 ppm, but could not detect separate peaks corresponding to particular polysulfide species [89]. It is evident that NMR can be used to determine the state of lithium in the cathode (Li_2S vs lithium polysulfides) [89]. NMR is thus a powerful probe for identifying the time at which Li_2S is formed during cell discharge and when it is consumed during charge.

3.5 In Situ Spectroscopic Studies of Li–S Batteries

3.5.1 X-ray Absorption Spectroscopy

The first use of XAS for studying Li–S chemistry was the work of Gao et al. [91]. Here, the researchers performed XAS on a number of cathodes taken from Li–S cells that contained different electrolytes. The results notably showed that carbonate-based electrolytes react irreversibly with sulfur-containing species in the battery. More recently, XAS has been used to probe Li–S reaction mechanisms in a number of in situ spectroelectrochemical studies [15, 24, 33, 37, 50, 52, 55, 92]. We have shown in Section 3.4 that XAS, after correcting the signal for the presence of radicals, S_8 and Li_2S, provides a robust measure of the average length of polysulfide dianion chains. XAS is thus an unambiguous way to detect the overall amount of lithium that has reacted with sulfur. It can also be used to detect the presence of radical polysulfides and to determine whether or not lithium sulfide has formed.

The detection of radical species in Li–S batteries has gained attention in Li–S spectroscopic studies. Cuisinier et al. performed in situ XAS on two separate Li–S cells: one that contained an ether-based solvent (DOL:DME) and the other that contained DMA [24]. The spectra for the cell that contained DMA clearly showed the presence of radical polysulfide species as an intermediate formed during discharge and charge (a peak at 2468–2469 eV). The spectra of the cell that contained DOL:DME displayed no presence of radical species. Additionally, comparing the discharge and charge voltage profiles for the DMA-containing cell to the one that contained TEGDME, it was evident that the TEGDME cell voltage profile exhibited significant overpotentials compared to the DMA cell. The DMA cell reached a first discharge capacity of 1575 $mAh\,g^{-1}$, whereas the TEGDME cell reached 1300 $mAh\,g^{-1}$. These results suggest that the radical species may aid in the discharge and charge processes.

The lack of radical species in the DOL:DME cell probed by Cuisinier et al. contrasts the results of Wujcik et al., who probed the electrolyte separator of Li–S cells that had been discharged to various depths [33]. This work showed that radical species were generated via disproportionation reactions involving polysulfide species that were formed through the discharge reaction pathway. The detection of radical species in ether-based solvents has been a recent point of contention. Although some spectroscopic studies have provided evidence for radical species, other studies have not. Recent work of Wujcik et al., as described in the above UV–Vis section, likely suggests that radical species are present in equilibrium with polysulfide dianion species in ether-based polysulfide mixtures, but at concentrations much lower than what is observed in solvents like DMA, DMF, and DMSO (that have high electron pair donor numbers). It is possible that radical species were present in the DOL:DME cell probed by Cuisinier et al., but in considerably lower concentrations relative to the other sulfur-containing species in the cathode [50]. It is also possible that the approach Wujcik et al. used to probe Li–S cell enabled the detection of the radical species. Because the X-rays in this study probed the electrolyte separator in addition to the cathode, obtained spectra may represent polysulfide species that diffused out of the cathode and underwent disproportionation reactions. Had the X-rays directly probed the Li–S cathode, it is possible that the radical anions may not have been detected. Note that the unique XAS fingerprint of radical anions at 2470 eV is weak and can thus only be detected if the concentration of the species exceeds a threshold (about 20 mol%) [54].

Gorlin et al. have also performed XAS on a discharging Li–S cell and notably detected the formation of Li_2S at early stages of the second voltage plateau [37]. More recently, Gorlin et al. have performed in situ XAS of an Li–Li_2S cell [55]. Interestingly, the first charging of the cell showed little to no evidence of polysulfide reaction intermediates, and the cell charge voltage profile was characterized by significant overpotentials. Upon second charging of the cell, however, XAS detected a considerable level of polysulfide dianion species, and the cell charge voltage profile did not display the same overpotentials as seen during first charging. This suggests that polysulfide intermediates may aid in the oxidation of Li_2S during charging.

3.5.2 Electron Paramagnetic Resonance Spectroscopy

While EPR of polysulfide solutions dates back to the 1960s, it was not until recently that researchers were able to probe an operating Li–S cell in situ [27]. Wang et al. built an Li–S electrochemical cell that fit tightly into a cuvette and performed time-resolved EPR on the cell during cyclic voltammetry. Their results showed that the LiS_3 radical anion was present during both the first discharge and charge processes; these experiments were conducted at room temperature. The cell contained an ether-based electrolyte of DOL:DME (1 : 1, v/v) and 1M LiTFSI. The results obtained by Wang et al. are shown in Figure 3.9. The EPR signal reached a maximum shortly after the midpoint of the discharge processes. The concentration of radical species during the two adjacent cycles was substantially different.

Additionally, Wang et al. performed EPR on a chemically synthesized polysulfide solution having a stoichiometric ratio of lithium to sulfur given by an Li_2S_x with "x" value equal to 8, dissolved in DOL:DME. Initial measurement of the solution detected the presence of the LiS_3 radical anion in solution. Wang et al. then added small amounts of propylene carbonate (PC) solvent with 1 M $LiPF_6$ salt to the polysulfide solution. With the addition of the PC ($LiPF_6$) solvent solution, the measured EPR signal decreased and eventually disappeared. It had previously been reported by Gao et al. that PC may react with lithium polysulfide intermediates [91]. The work of Wang et al., in conjunction with those of Gao et al., suggests that the degradation of PC in Li–S cells involves reactions with polysulfide radical anions.

EPR measurements are often carried out at low temperatures to enhance the detectability of radical species. Leghie et al. showed that decreasing temperature results in a decrease in the polysulfide radical anion concentration [38]. This is not surprising as equilibrium constants are known to be temperature dependent, and we expect an equilibrium between radical anions and dianions. The results of Leghie et al. indicate that the dissociation of Li_2S_6 to form LiS_3 is favored at higher temperatures. Thus, it is important that EPR be performed at temperatures that are relevant to Li–S-cell-operating conditions, as done by Wang et al. [27]

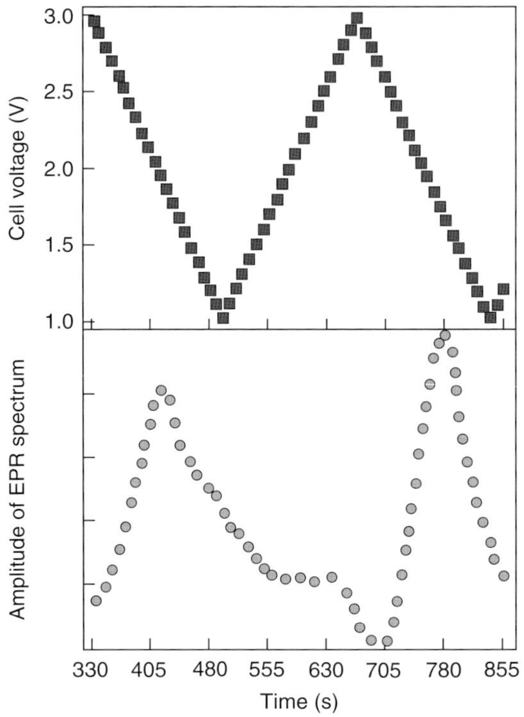

Figure 3.9 In situ EPR of a Li–S cell during cyclic voltammetry. (Top) Cell voltage as a function of time. (Bottom) Amplitude of obtained EPR spectrum. Based on the data presented in Ref. [27].

3.5.3 UV–Vis Spectroscopy

In situ UV–Vis of Li–S reactions is challenging due to strong absorption by the polysulfide species, reducing the number of photons that can pass through the cathode. In addition, typical materials used in Li–S cathodes (aluminum current collector, sulfur, carbon, etc.) are not transparent. It is thus not surprising that most of the in situ UV–Vis experiments probe species in the electrolyte. To probe the reactions that occur in the cathode, Patel et al. obtained UV–Vis spectra in reflection mode using a light beam incident on the cathode [52, 70]. They observed changes in peak positions in the 400–650 nm range. At the beginning of discharge, they observed a shift of an absorbance peak toward higher wavelengths, indicating the presence of long-chain polysulfide dianions (see Table 3.1). The peak then shifted to shorter wavelengths as discharge proceeded, indicating the formation of short-chain polysulfide dianions. During charging, the peak shifted back to longer wavelengths, as expected. However, the spectrum obtained at the end of the discharge–charge cycle was shifted toward lower wavelength, compared to the initial spectrum, indicating that some short-chain polysulfides were not converted back to sulfur.

3.5.4 Raman Spectroscopy

Hagen et al. performed in situ and ex situ Raman spectroscopy experiments on an Li–S cell [75]. The measured spectra showed that a variety of polysulfide dianions and radical anions coexisted throughout the course of discharge and charge, but it was difficult to clearly distinguish between different species. They were able to unambiguously show that crystalline sulfur formed during the charge process. Interestingly, the Raman measurements made a few minutes after charging ended indicated a decrease in S_8 concentration. This suggested that S_8 had reacted chemically (not electrochemically) with leftover polysulfide species. The importance of chemical reactions (as opposed to electrochemical reactions) is not widely appreciated. Such reactions may affect cell potential in a partially discharged (or partially charged) cell. It may also complicate interpretation of ex situ experiments.

3.5.5 Nuclear Magnetic Resonance Spectroscopy

Huff et al. used ^{6}Li NMR to probe cathodes extracted from Li–S cells after stopping at different depths of discharge [89]. Their results showed that cathodes stopped at 2.15, 2.03, and 2.00 V had single NMR peaks at −0.53, −0.29, and −0.76 ppm, respectively. These peaks were considerably different than the shifts observed for LiTFSI powder (a single peak at −1.22 ppm), LiTFSI in DOL:DME (a single peak at −1.27 ppm), and lithium sulfide powder (a single peak at 2.40 ppm). Although this indicates that NMR may be used to detect the presence of polysulfides, the non-monotonic changes in peak position suggest a nonintuitive relationship between ^{6}Li NMR peak position and polysulfide chain length. The spectrum obtained from a cathode that was fully discharged to 1.50 V displayed two peaks: one signifying the presence of Li_2S and another at 0.12 ppm. The authors suggested that the second peak was due to the presence of Li_2CO_3, which has a chemical shift of 0.11 ppm, formed due to degradation reactions that involve the electronically conducting carbon.

Huff et al. also attempted ^{33}S NMR on lithium sulfide and the cathode after discharge. Their results showed that single peaks were obtained for both samples at chemical shifts of −342.7 and −344.1 ppm for lithium sulfide powder and the discharged cathode, respectively. While the peaks were nearly overlapping, Huff et al. noted that the peak of the cathode was broader, suggesting that a distribution of species may have been present in the cathode. Huff et al. concluded that the chemical shifts in both ^{6}Li and ^{33}S NMR were not sensitive to polysulfide chain lengths.

In another study, Xiao et al. built an electrochemical NMR cell to probe the Li–S discharge and charge reactions in situ [93]. Their conclusions based on ^{7}Li NMR were similar to those obtained by Huff et al.; they were not able to clearly distinguish between different polysulfide species. They obtained broad NMR peaks indicating the presence of a distribution of polysulfide species at all times. The design of the cell that Xiao et al. used to perform in situ NMR is notable because it allowed them to probe reactions in both the cathode and the anode. This allowed them to observe the detrimental reactions between polysulfide species and the lithium anode.

In a recent in situ NMR study, See et al. studied the first discharge of a Li–S cell and found signatures of Li_2S formation in the early stages of discharge [90]. They proposed a Li–S–electrolyte ternary phase diagram that was consistent with their results. They did not provide any independent support for the proposed phase diagram. Thermodynamic measurements of the phase behavior of the S_8/polysulfide/Li_2S mixtures in electrolytes would be useful at this juncture.

3.6 Practical Considerations

The important parameters that affect the performance of sulfur cathodes (see Figure 3.1) are (i) the sulfur loading, SL, (ii) the carbon loading, CL, and (iii) the electrolyte to sulfur ratio ES [94–97]. SL must be maximized while CL and ES must be minimized for high specific energy. However, Hagen et al. showed that high first discharge capacities are only obtained in cells with high ES. An Li–S cell with ES equal to 8 obtained a first discharge-specific capacity of 1200 mAh g^{-1} S, whereas a cell with ES equal to 3 only obtained 50 mAh g^{-1} S [94]. This is attributed to the presence of polysulfides in the electrolyte; high concentrations of polysulfide species in the electrolyte in low ES systems slow lithium ion transport. Hagen and coworkers also noted that none of the cells with high capacity and long cycle life reported in the literature had SL, CL, and ES values appropriate for high specific energy cells. There are many papers in the Li–S literature that claim to have designed Li–S cells of practical value where these parameters are not even mentioned.

Addressing the cycle life issue is much more challenging than specific energy; note that the previous paragraph only addressed first cycle capacity. There are severe problems delivering electrons to active particles that are insulators such as S_8, polysulfide intermediates, and Li_2S. It is unlikely that electrons can penetrate more than 1 nm in these compounds. Using a "conventional" composite lithium battery cathode, one in which electrons and lithium ions are transported to more-or-less static active particles, is out of the question. The (assumed) liquid nature of polysulfides formed by the reduction of S_8 and oxidation of Li_2S is essential for ensuring contact between Li^+, e^-, and the active molecules. While ES is beneficial to high first-cycle capacities, the problem is that if the active material dissolves, then there is nothing that holds the electronically conducting network in place. The importance of the collapse of these networks in Li–S cells was clearly shown in Ref. [98]. Thus, while the heightened mobility of dissolved polysulfides enables high first capacity Li–S cells, the dissolution of the active material and collapse of conductive networks over time will lead to capacity fade and failure in these cells.

The dissolution of polysulfides and collapse of the electronically conductive pathways are also detrimental to fundamental studies of the Li–S reaction mechanism. Dissolution of reaction intermediates from reaction sites will undoubtedly affect the reaction pathway. Additionally, dissolution leads to chemical disproportionation reactions, which, in turn, makes it difficult to determine whether a species was formed electrochemically or chemically.

The ideal design of a cathode material for properly studying the Li–S reaction pathways is shown in Figure 3.10. Here, the active materials are housed in a conductive shell that prevents the dissolution of polysulfide molecules while allowing the transport of lithium ions and electrons. Numerous researchers claim

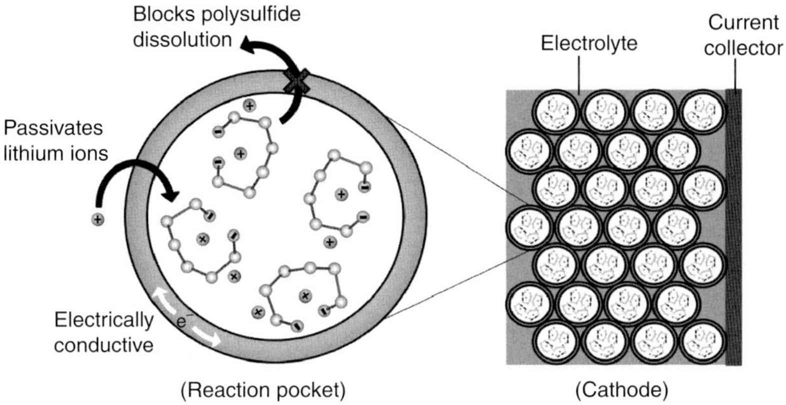

Figure 3.10 Hypothetical reaction "pocket" and ideal cathode for Li–S cells that enable long cycle life and studies of Li–S redox reactions that are void of the issues invoked by polysulfide dissolution.

to have engineered systems that are similar to the schematic shown in Figure 3.10 [11, 13, 87]. The change in volume during reaction has been ignored in this discussion; we assume that the shells are elastic and that they accommodate the necessary volume changes, or that there is no problem with partially filled pockets.

In one attempt to create such pockets, Fu et al. studied an Li–S cell wherein sulfur was infused into a microporous carbon structure with an average pore size of only 1 nm [99]. The sub-nanometer confinement may, in principle, address two main problems with Li–S chemistry: the electronically insulating nature of the reactants and products and the solubility of the polysulfides. It is likely that the electronic conductivity of the active materials is enough to enable redox reactions across such length scales. It is also possible that the products of redox reactions under such severe confinement are fundamentally different from those discussed in Section 3.3. In Figure 3.11, the discharge profile of this cell with sub-nano-confined sulfur is contrasted with that obtained from a typical cell with a liquid electrolyte, taken from Ref. [100], where polysulfides are allowed to diffuse. The sharp features seen in the conventional cell, which are taken to be signatures of the formation of certain polysulfides, are not evident in the cell with the sub-nano-confined sulfur. Also shown in Figure 3.11 is an all-solid Li–S cell with $80Li_2S\text{-}20P_2S_5$ glass used as both the electrolyte and the ionic conductor in the cathode [101]. The pockets of sulfur in this cell are likely to be in the micron range (or larger) as conventional blending was used to make the cathode. The glass electrolyte appears to be impermeable to polysulfides. The discharge profile of this solid-state cell is also smooth as is the case with the sub-nano-confined sulfur cell. The potential of the all-solid cell is intermediate between that of the conventional cell and the sub-nano-confined sulfur cell. A noteworthy feature of the cell is that the second cycle capacity is larger than the first cycle capacity, indicating that access to active materials is promoted by cycling. In contrast, the sub-nano-confined sulfur cell shows more conventional

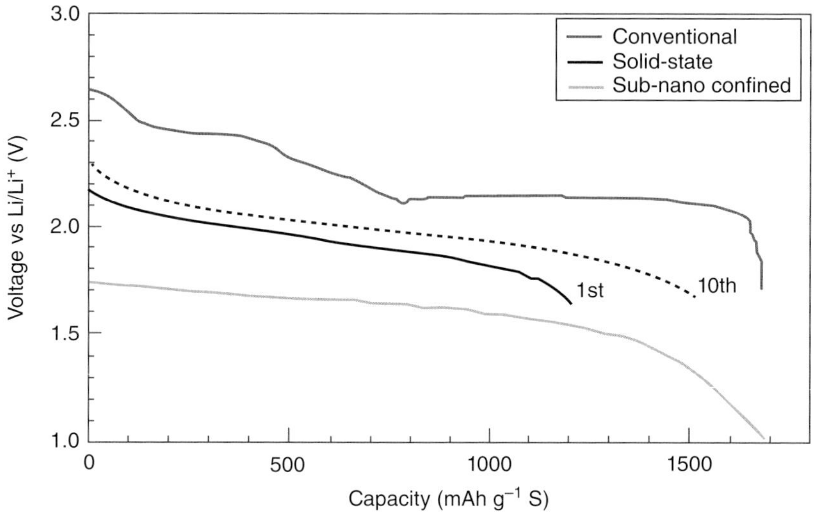

Figure 3.11 Discharge profile of three Li-S cells: the red trace ("Conventional"), adopted from Jeong et al. [100], is the first discharge profile of a conventional Li–S cell with a liquid-like electrolyte. The black traces ("solid-state"), adopted from Nagao et al. [101], show the 1st and 10th discharges of a fully solid Li–S cell, in the solid and dotted lines, respectively. The blue trace ("sub-nano confined"), adopted from Fu et al. [99], shows the first discharge of a Li–S cell with a liquid electrolyte and a "sub-nano-confined" cathode.

capacity fade, typical of traditional cells; the fade of the traditional cell is not shown in Figure 3.11 as it is documented across the Li–S literature. The cells in Figure 3.11 had different values of SL, CL, and ES, and thus, some of the differences seen in Figure 3.11 may be because of this fact. The large variation in open-circuit voltage is particularly noteworthy in light of the absence of an experimental measure of the Gibbs free energy of formation of Li_2S.

Returning to the ideal electrode shown in Figure 3.10, there is a necessary balance between the size of the pocket and cathode capacity (mAh g^{-1} of cathode); large pockets will enable high capacity. Charging and discharging will stop as soon as a film of an insulating solid coats the inner surface of the shell [102]. The size of the "pockets" that can be cycled will depend on many factors, such as the morphology of the deposited products of electrochemical reactions, mobility of polysulfide intermediates, etc. It may be possible to cycle large pockets if the mobility of the polysulfides is sufficiently high and if the insoluble solids stay suspended in the pockets. Electrodes like those shown in Figure 3.10 are necessary for both the proper examination of Li–S reaction mechanisms and for practical, high-performance Li–S batteries.

3.7 Concluding Remarks

The objective of this review is to describe what we know about the redox pathways that occur in the cathode of a Li–S cell as it is charged and discharged. After a brief introduction, we began our discussion by describing some of the redox reactions that have been proposed to occur in the sulfur cathode of a Li–S cell. While the end products of the redox reactions, S_8 and Li_2S, are simple crystalline solids at room temperature and atmospheric pressure, the reaction intermediates (Li_2S_x and LiS_x) are amorphous compounds that have not yet been successfully isolated (Reactions (R5)–(R19)). Simple entropy-based arguments suggest that some of the popular reactions (e.g. Reaction (R9)), proposed on the basis of in situ measurements, are not likely to occur.

This was followed by a summary of efforts to assemble spectral fingerprints that may be used to identify the reaction intermediates. Based on both experiments and first-principles calculations, it is clear that XAS is sensitive to the average length of the polysulfide dianions due to the presence of separate peaks corresponding to the charged dianion chain ends (centered at 2470.5 eV) and the uncharged internal sulfur atoms in the chain (centered at 2472.6 eV). The clearest signature of lithium polysulfide radicals is obtained by EPR measurements (g-factor of LiS_x are in the vicinity of 2.03), and UV–Vis spectroscopy (peak centered around 617 nm). While an XAS peak in the 2467.5–2468.5 eV region is also a definitive signature of radical anions, most of the electronic transitions responsible for the measured XAS spectra have higher characteristic energies that overlap with those obtained from dianions. XAS can thus only be used to infer the presence of radicals when they are present in high concentrations. It is our view that NMR and the Raman fingerprints of polysulfide dianions and radical anions have not yet been established, as reported in Refs. [40, 50, 72–75, 89].

We then discussed application of the fingerprints to study the proposed reaction mechanisms and side reactions. Many techniques such as XRD, XAS, Raman, NMR, and UV–Vis spectroscopy can be used to distinguish between the final products of charge and discharge and polysulfide intermediates. However, there is no evidence that the spectroscopic techniques used thus far to study Li–S chemistry have unique signatures of individual polysulfide species. EPR is best suited for detecting polysulfide radical anions while sulfur K-edge XAS is best for detecting polysulfide dianions. In spite of the widespread use of XAS, our proposed relationship between spectral features and polysulfide chain length is not widely accepted, see Figure 3.3. While UV–Vis peaks corresponding to polysulfide radical anions and polysulfide dianions are separated, there is no basis for identifying particular species or their chain length. The same can be said for Raman and NMR spectra.

Although the generally assumed reactions in Li–S cells do not directly involve solvent molecules (all the reactions we have proposed, R5–R19, fall into this category), it is not clear if this is really true. If this assumption is correct, as is the case in traditional lithium-ion cells, then the amount of electrolyte can be systematically reduced to improve specific energy with no regard for the effect on the intrinsic cell chemistry. However, all of the indirect evidence suggests that this assumption is incorrect as fundamentally different behavior is obtained when the electrolyte to active material ratio is changed. Given this complexity, and the limitations of our approaches for determining the nature of the intermediate products, it seems prudent to examine Li–S chemistry with the same parameters (i.e. keeping SL, CL, and ES fixed) using complimentary spectroscopic methods. For example, one could use XAS to quantify reaction pathways and average polysulfide chain length, but only after subtracting contributions from S_8, Li_2S, and radical anions to the measured signal. These contributions could be determined by independent in situ XRD, EPR, NMR, and UV–Vis experiments. NMR is an invaluable tool for identifying the products of side reactions, especially those that may be catalyzed by polysulfides and resulting in products that do not contain sulfur (e.g. Li_2CO_3). Identifying all of the reactions that occur in any given cell is a worthwhile and challenging goal. Finding systems wherein the Li–S chemistry is truly independent of solvent loading will make achieving the goal more manageable.

We hypothesize that polysulfides are liquids in the pure form and that the mobility of polysulfide molecules is crucial for the cyclability of practical Li–S cells. It is difficult to either prove or disprove our hypothesis as pure lithium polysulfides cannot be prepared in the laboratory. With few exceptions [103], we have virtually no knowledge about the mobility of polysulfides. Perhaps, computer simulations may be used to quantify polysulfide mobility.

It is known that lithium polysulfides are soluble to varying degrees in typical solvents used as electrolytes in traditional lithium-ion batteries. Studies of reaction pathways are necessarily complicated if the reactants and products diffuse to areas of the cell where electrons and lithium ions might not be available. Such studies may also be irrelevant for high specific energy Li–S cells if low sulfur loadings are used to enable high capacity and reasonable cycle life. To our knowledge, all in situ studies on Li–S cells [15, 23–25, 27, 33, 50, 52, 70, 71, 74, 75, 92, 104, 105], including work done in the laboratory of the principle investigator of this study, suffer from these limitations. The extent to which diffusion is a problem depends on practical considerations such as the electrolyte to sulfur ratio, sulfur loading, and cathode thickness. As different groups use different cathode designs, the extent to which diffusion has affected the results in Refs. [15, 23–25, 27, 33, 50, 52, 70, 71, 74, 75, 92, 104, 105] remains to be established.

In Figure 3.10, we show a schematic of an ideal sulfur cathode for studying the fundamentals of redox chemistry. It comprises "pockets" that contain the redox species (S_8, Li_2S, LiS_x, and Li_2S_x) in a matrix that conducts lithium ions and electrons, but is impermeable to all of the sulfur species. Ideally, the lithium ion and electronic conductivities must be matched as all of the redox reactions involve equal numbers of ions and electrons. In such a cell, redox reactions would be limited to the perimeter of the pocket due to the electronically insulating nature of all species, perhaps a thickness of 1–2 nm. High conversion would only occur because of the mobility of polysulfide species, i.e. their liquid-like nature either as pure species or as mixtures. Given that the final products are crystalline solids comprising immobile molecules, it is unlikely that either charge or discharge reactions would go to completion. The accessible capacity would be limited by the points when redox reactions cease due to the complex interplay between molecular mobility, electronic transport, and perhaps volume changes. An ideal matrix would be elastic enough to accommodate volume changes upon lithiation and delithiation; if this is not the case, then voids will form during charge as the volume of active material shrinks. This complication will not have a major effect on either reaction pathways or the ability to conduct in situ spectroscopic analysis but will limit practical cell-specific energy.

The overall objective of current research in the lithium–sulfur chemistry is to enable the design of long-lived batteries with high specific energy by exploiting the high capacity of sulfur for lithium and electrons. To achieve this, it is necessary to create cells with low electrolyte to sulfur ratios much $<5\,\mu l\,mg^{-1}$ [97] and preferably $<3\,\mu l\,mg^{-1}$ [94, 97], cathodes with high sulfur loading ($>10\,mg\,cm^{-2}$) [95], and sulfur volume fractions in the cathode that are at least 60–70 vol.% sulfur [96, 97]. In addition, large format applications such as electric vehicles and grid-scale energy storage require cycle life in the 1000–10 000 cycle range. Batteries with sparingly soluble redox-active species can be cycled, as is the case of the lead–acid battery, but it is unlikely to lead to the cycle-life needed for large format applications. It is our contention that designing ideal electrodes like those depicted in Figure 3.10 is important for both fundamental studies of lithium–sulfur redox chemistry and practical lithium–sulfur batteries. There remains a discrepancy between the lithium–sulfur redox processes determined by the in situ spectroelectrochemical studies shown in Section 3.5 and the likely reaction mechanisms described in Section 3.3. In situ experiments on ideal sulfur electrodes may overcome this discrepancy.

Although we have focused on Li–S cells, there are many other electrochemical systems wherein the intermediate steps for completing the desired redox reactions have not yet been established. In alkaline fuel cells, electrolyte degradation often interferes with the desired electrode reactions [106]. Similarly, normal carbonate- and ether-based electrolytes interfere with the desired reactions in lithium–oxygen cells [107]. In such cases, it is important to establish fingerprints of the expected compounds and search for them in in situ experiments. Fingerprints that are independent of the environment are particularly helpful for detecting the species of interest in complex milieus. The electrodes in the in situ experiments must be designed so that all of the active molecules (or elements) in the electrode have unimpeded access to the species of interest (electrons, lithium ions, etc.). This ensures that the measurements reflect reaction mechanisms, not transport bottlenecks. Time-dependent measurements of voltage and current place constraints on the compounds that are formed. In this context, the work of McCloskey and coworkers on the lithium–air chemistry is exemplary [107]. They took advantage of the gaseous nature of the compounds of interest and used mass spectrometry to follow reactions. However, by keeping track of the number of electrons delivered to the electrodes, they could report on differences between the expected rate of the redox reaction of interest and the actual rate of reaction. By judicious use of isotopic labeling, they were able to determine the origin of certain elements in their products. For example, they noted the unexpected formation of CO_2 (not a primary redox product) and were able to show that much of the oxygen in the CO_2 came from the electrolyte, not from the oxygen supplied to the cathode [108]. Advanced electrochemical devices that have the potential to deliver clean energy provide challenging test beds for modern analytical tools.

Acknowledgment

This work was supported by the Assistant Secretary for Energy Efficiency and Renewable Energy, Office of Vehicle Technologies of the US Department of Energy under Contract DE-AC02-05CH11231 under the Battery Materials Research program. The authors would also like to acknowledge Philip Ross for helpful discussions and insight.

References

1 Chivers, T. and Elder, P.J.W. (2013). Ubiquitous trisulfur radical anion: fundamentals and applications in materials science, electrochemistry, analytical chemistry and geochemistry. *Chem. Soc. Rev.* 42: 5996–6005.
2 Steudel, R. (2003). *Elemental Sulfur and Sulfur-Rich Compounds II*, vol. 231. Berlin, Heidelberg: Springer.
3 Fleet, M.E. and Liu, X. (2010). X-ray absorption spectroscopy of ultramarine pigments: a new analytical method for the polysulfide radical anion $S3^-$ chromophore. *Spectrochim. Acta B At. Spectrosc.* 65: 75–79.
4 Reinen, D. and Lindner, G.-G. (1999). The nature of the Chalcogen Colour Centres in ultramarine-type solids. *Chem. Soc. Rev.* 28: 75–84.
5 Clark, R.J.H. and Cobbold, D.G. (1978). Characterization of sulfur radical anions in solutions of alkali polysulfides in dimethylformamide and hexamethylphosphoramide and in the solid state in ultramarine blue, green, and red. *Inorg. Chem.* 17: 3169–3174.
6 Bhaskara, R.M.L. (1968) Organic Electrolyte Cells. US3413154 A.
7 Birk, J.R. and Steunenberg, R.K. (1975). Chemical investigations of lithium–sulfur cells. In: *New Uses of Sulfur*, vol. 140, 186–202. American Chemical Society.
8 Danuta, H. and Juliusz, U. (1962). Electric dry cells and storage batteries. US3043896 A.
9 Park, H., Koh, H.S., and Siegel, D.J. (2015). First-principles study of redox end members in lithium–sulfur batteries. *J. Phys. Chem. C* 119: 4675–4683.
10 Wang, H., Yang, Y., Liang, Y. et al. (2011). Graphene-wrapped sulfur particles as a rechargeable lithium–sulfur battery cathode material with high capacity and cycling stability. *Nano Lett.* 11: 2644–2647.
11 Jayaprakash, N., Shen, J., Moganty, S.S. et al. (2011). Porous hollow carbon@sulfur composites for high-power lithium–sulfur batteries. *Angew. Chem.* 123: 6026–6030.
12 Ji, L., Rao, M., Zheng, H. et al. (2011). Graphene oxide as a sulfur immobilizer in high performance lithium/sulfur cells. *J. Am. Chem. Soc.* 133: 18522–18525.
13 Ji, X., Lee, K.T., and Nazar, L.F. (2009). A highly ordered nanostructured carbon-sulphur cathode for lithium–sulphur batteries. *Nat. Mater.* 8 (6): 500.

14 Cañas, N.A., Wolf, S., Wagner, N., and Friedrich, K.A. (2013). In-situ X-ray diffraction studies of lithium–sulfur batteries. *J. Power Sources* 226: 313–319.
15 Lowe, M.A., Gao, J., and Abruna, H.D. (2014). Mechanistic insights into operational lithium–sulfur batteries by in situ X-ray diffraction and absorption spectroscopy. *RSC Adv.* 4: 18347–18353.
16 Nelson, J., Misra, S., Yang, Y. et al. (2012). In operando X-ray diffraction and transmission X-ray microscopy of lithium sulfur batteries. *J. Am. Chem. Soc.* 134: 6337–6343.
17 Cheon, S.-E., Ko, K.-S., Cho, J.-H. et al. (2003). Rechargeable lithium sulfur battery: I. Structural change of sulfur cathode during discharge and charge. *J. Electrochem. Soc.* 150: A796–A799.
18 Ryu, H.-S., Ahn, H.-J., Kim, K.-W. et al. (2006). Discharge process of Li/Pvdf/S cells at room temperature. *J. Power Sources* 153: 360–364.
19 Yuan, L., Qiu, X., Chen, L., and Zhu, W. (2009). New insight into the discharge process of sulfur cathode by electrochemical impedance spectroscopy. *J. Power Sources* 189: 127–132.
20 Liu, Y., Zhang, J., Liu, X. et al. (2014). Nanosulfur/polyaniline/graphene composites for high-performance lithium–sulfur batteries: one pot in-situ synthesis. *Mater. Lett.* 133: 193–196.
21 Kulisch, J., Sommer, H., Brezesinski, T., and Janek, J. (2014). Simple cathode design for Li–S batteries: cell performance and mechanistic insights by in operando X-ray diffraction. *Phys. Chem. Chem. Phys.* 16: 18765–18771.
22 Zhao, H., Peng, Z., Wang, W. et al. (2014). Reduced graphene oxide with ultrahigh conductivity as carbon coating layer for high performance sulfur@reduced graphene oxide cathode. *J. Power Sources* 245: 529–536.
23 Barchasz, C., Molton, F., Duboc, C. et al. (2012). Lithium/sulfur cell discharge mechanism: an original approach for intermediate species identification. *Anal. Chem.* 84: 3973–3980.
24 Cuisinier, M., Hart, C., Balasubramanian, M. et al. (2015). Radical or not radical: revisiting lithium–sulfur electrochemistry in nonaqueous electrolytes. *Adv. Energy Mater.* doi: 10.1002/aenm.201401801.
25 Kawase, A., Shirai, S., Yamoto, Y. et al. (2014). Electrochemical reactions of lithium–sulfur batteries: an analytical study using the organic conversion technique. *Phys. Chem. Chem. Phys.* 16: 9344–9350.
26 Kim, B.S. and Park, S.M. (1993). In situ spectroelectrochemical studies on the reduction of sulfur in dimethyl sulfoxide solutions. *J. Electrochem. Soc.* 140: 115–122.
27 Wang, Q., Zheng, J., Walter, E. et al. (2015). Direct observation of sulfur radicals as reaction media in lithium sulfur batteries. *J. Electrochem. Soc.* 162: A474–A478.
28 Zou, Q. and Lu, Y.-C. (2016). Solvent-dictated lithium sulfur redox reactions: an operando UV–Vis spectroscopic study. *J. Phys. Chem. Lett.* 7: 1518–1525.
29 Yamin, H., Gorenshtein, A., Penciner, J. et al. (1988). Lithium sulfur battery: oxidation/reduction mechanisms of polysulfides in THF solutions. *J. Electrochem. Soc.* 135: 1045–1048.

30 Kumaresan, K., Mikhaylik, Y., and White, R.E. (2008). A mathematical model for a lithium–sulfur cell. *J. Electrochem. Soc.* 155: A576–A582.

31 Martin, R.P., Doub, W.H., Roberts, J.L., and Sawyer, D.T. (1973). Electrochemical reduction of sulfur in aprotic solvents. *Inorg. Chem.* 12: 1921–1925.

32 Mikhaylik, Y.V. and Akridge, J.R. (2004). Polysulfide shuttle study in the Li/S battery system. *J. Electrochem. Soc.* 151: A1969–A1976.

33 Wujcik, K.H., Pascal, T.A., Pemmaraju, C.D. et al. (2015). Characterization of polysulfide radicals present in an ether-based electrolyte of a lithium–sulfur battery during initial discharge using in situ X-ray absorption spectroscopy experiments and first-principles calculations. *Adv. Energy Mater.* 5 (16): 1500285.

34 Wang, L., Zhang, T., Yang, S. et al. (2013). A quantum-chemical study on the discharge reaction mechanism of lithium–sulfur batteries. *J. Energy Chem.* 22: 72–77.

35 Chivers, T. and Drummond, I. (1972). Characterization of the trisulfur radical anion S3- in blue solutions of alkali polysulfides in hexamethylphosphoramide. *Inorg. Chem.* 11: 2525–2527.

36 Giggenbach, W. (1968). On the nature of the blue solutions of sulfur. *J. Inorg. Nucl. Chem.* 30: 3189–3201.

37 Gorlin, Y., Siebel, A., Piana, M. et al. (2015). Operando characterization of intermediates produced in a lithium–sulfur battery. *J. Electrochem. Soc.* 162: A1146–A1155.

38 Leghie, P., Levillain E., Lelieur J. P., Lorriaux A, Spectroscopic study of lithium hexasulfide solutions in DMF. *N. J. Chem.* 1996, 20, 1121-1130.

39 Levillain, E., Leghie, P., Gobeltz, N., and Lelieur, J.P. (1997). Identification of the S4-radical anion in solution. *N. J. Chem.* 21: 335–341.

40 Vijayakumar, M., Govind, N., Walter, E. et al. (2014). Molecular structure and stability of dissolved lithium polysulfide species. *Phys. Chem. Chem. Phys.* 16: 10923–10932.

41 Wujcik, K.H., Wang, D.R., Raghunathan, A. et al. (2016). Lithium polysulfide radical anions in ether-based solvents. *J. Phys. Chem. C* 120 (33): 18403–18410.

42 Merritt, M.V. and Sawyer, D.T. (1970). Electrochemical reduction of elemental sulfur in aprotic solvents. formation of a stable S8- species. *Inorg. Chem.* 9: 211–215.

43 Paris, J. and Plichon, V. (1981). Electrochemical reduction of sulphur in dimethylacetamide. *Electrochim. Acta* 26: 1823–1829.

44 Tobishima, S.-I., Yamamoto, H., and Matsuda, M. (1997). Study on the reduction species of sulfur by alkali metals in nonaqueous solvents. *Electrochim. Acta* 42: 1019–1029.

45 Marmorstein, D. (2002). *Solid State Lithium/Sulfur Batteries for Electric Vehicles: Electrochemical and Spectroelectrochemical Investigations.* Berkeley: University of California.

46 Xu, R., Lu, J., and Amine, K. (2015). Progress in mechanistic understanding and characterization techniques of Li–S batteries. *Adv. Energy Mater.* 5 (16): 1500408.

47 Jalilehvand, F. (2006). Sulfur: not a "silent" element any more. *Chem. Soc. Rev.* 35: 1256–1268.

48 Newville, M. (2004). *Fundamentals of Xafs Consortium for Advanced Radiation Sources*. Chicago, IL: University of Chicago.

49 Yano, J. and Yachandra, V.K. (2009). X-ray absorption spectroscopy. *Photosynth. Res.* 102: 241–254.

50 Cuisinier, M., Cabelguen, P.-E., Evers, S. et al. (2013). Sulfur speciation in Li–S batteries determined by operando X-ray absorption spectroscopy. *J. Phys. Chem. Lett.* 4: 3227–3232.

51 Wujcik, K.H., Velasco-Velez, J., Wu, C.H. et al. (2014). Fingerprinting lithium–sulfur battery reaction products by X-ray absorption spectroscopy. *J. Electrochem. Soc.* 161: A1100–A1106.

52 Patel, M.U.M., Arčon, I., Aquilanti, G. et al. (2014). X-ray absorption near-edge structure and nuclear magnetic resonance study of the lithium–sulfur battery and its components. *ChemPhysChem* 15 (5): 894–904.

53 Pascal, T.A., Wujcik, K.H., Velasco-Velez, J. et al. (2014). X-ray absorption spectra of dissolved polysulfides in lithium–sulfur batteries from first-principles. *J. Phys. Chem. Lett.* 5: 1547–1551.

54 Pascal, T.A., Pemmaraju, C.D., and Prendergast, D. (2015). X-ray spectroscopy as a probe for lithium polysulfide radicals. *Phys. Chem. Chem. Phys.* 17: 7743–7753.

55 Gorlin, Y., Patel, M.U., Freiberg, A. et al. (2016). Understanding the charging mechanism of lithium–sulfur batteries using spatially resolved operando X-ray absorption spectroscopy. *J. Electrochem. Soc.* 163: A930–A939.

56 Wujcik, K.H., Wang, D.R., Pascal, T.A. et al. (2017). In situ X-ray absorption spectroscopy studies of discharge reactions in a thick cathode of a lithium sulfur battery. *J. Electrochem. Soc.* 164: A18–A27.

57 Seel, F., Güttler, H.J., Simon, G., and Wieckowski, A. (1977). Colored sulfur species in Epd-solvents A2. In: *Non-Aqueous Solutions–5* (ed. J.B. Gill), 45–54. Pergamon.

58 Dubois, P., Lelieur, J.P., and Lepoutre, G. (1988). Identification and characterization of lithium polysulfides in solution in liquid-ammonia. *Inorg. Chem.* 27: 73–80.

59 Han, D.-H., Kim, B.-S., Choi, S.-J. et al. (2004). Time-resolved in situ spectroelectrochemical study on reduction of sulfur in *N,N'*-dimethylformamide. *J. Electrochem. Soc.* 151: E283–E290.

60 Levillain, E., Gaillard, F., Leghie, P. et al. (1997). On the understanding of the reduction of sulfur (S8) in dimethylformamide (DMF). *J. Electroanal. Chem.* 420: 167–177.

61 Cañas, N.A., Fronczek, D.N., Wagner, N. et al. (2014). Experimental and theoretical analysis of products and reaction intermediates of lithium–sulfur batteries. *J. Phys. Chem. C* 118: 12106–12114.

62 Manan, N.S.A., Aldous, L., Alias, Y. et al. (2011). Electrochemistry of sulfur and polysulfides in ionic liquids. *J. Phys. Chem. B* 115: 13873–13879.

63 Rauh, R.D., Shuker, F.S., Marston, J.M., and Brummer, S.B. (1977). Formation of lithium polysulfides in aprotic media. *J. Inorg. Nucl. Chem.* 39: 1761–1766.

64 Teran, A.A. and Balsara, N.P. (2011). Effect of lithium polysulfides on the morphology of block copolymer electrolytes. *Macromolecules* 44: 9267–9275.

65 Seel, F., Guttler, H.-J., Simon, G., and Wieckowski, A. (1977). Colored sulfur species in EPD-solvents. *Pure Appl. Chem.* 49 (1): 45–54.

66 Lux, H. and Anslinger, H. (1961). Uber Die Blauen Losungen Des Schwefels. *Chemische Berichte-Recueil* 94: 1161–1172.

67 Gruen, D.M., McBeth, R.L., and Zielen, A.J. (1971). Nature of sulfur species in fused salt solutions. *J. Am. Chem. Soc.* 93 (24): 6691–6693.

68 Chivers, T. and Drummond, I. (1971). Blue solutions of sulphur in hexamethylphosphoramide. *J. Chem. Soc. D Chem. Commun.* (24): 1623a.

69 Dokko, K., Tachikawa, N., Yamauchi, K. et al. (2013). Solvate ionic liquid electrolyte for Li–S batteries. *J. Electrochem. Soc.* 160: A1304–A1310.

70 Patel, M.U.M., Demir-Cakan, R., Morcrette, M. et al. (2013). Li–S battery analyzed by UV/Vis in operando mode. *ChemSusChem* 6: 1177–1181.

71 Patel, M.U.M. and Dominko, R. (2014). Application of in operando UV/Vis spectroscopy in lithium–sulfur batteries. *ChemSusChem* 7: 2167–2175.

72 Yao, H., Yan, K., Li, W. et al. (2014). Improved lithium–sulfur batteries with a conductive coating on the separator to prevent the accumulation of inactive S-related species at the cathode-separator interface. *Energy Environ. Sci.* 7: 3381–3390.

73 Wu, H.-L., Huff, L.A., and Gewirth, A.A. (2015). In situ Raman spectroscopy of sulfur speciation in lithium–sulfur batteries. *ACS Appl. Mater. Interfaces* 7: 1709–1719.

74 Yeon, J.-T., Jang, J.-Y., Han, J.-G. et al. (2012). Raman spectroscopic and X-ray diffraction studies of sulfur composite electrodes during discharge and charge. *J. Electrochem. Soc.* 159: A1308–A1314.

75 Hagen, M., Schiffels, P., Hammer, M. et al. (2013). In-situ Raman investigation of polysulfide formation in Li–S cells. *J. Electrochem. Soc.* 160: A1205–A1214.

76 Chivers, T. and Lau, C. (1982). Raman spectroscopic identification of the $S4N^-$ and $S3^-$ ions in blue solutions of sulfur in liquid ammonia. *Inorg. Chem.* 21: 453–455.

77 Janz, G., Downey, J. Jr.,, Roduner, E. et al. (1976). Raman studies of sulfur-containing anions in inorganic polysulfides. Sodium polysulfides. *Inorg. Chem.* 15: 1759–1763.

78 Kelly, B. and Woodward, P. (1976). Crystal structure of dipotassium pentasulphide. *J. Chem. Soc. Dalton Trans.* 1314–1316.

79 Diao, Y., Xie, K., Xiong, S., and Hong, X. (2012). Insights into Li–S battery cathode capacity fading mechanisms: irreversible oxidation of active mass during cycling. *J. Electrochem. Soc.* 159: A1816–A1821.

80 El Jaroudi, O., Picquenard, E., Demortier, A. et al. (2000). Polysulfide anions II: structure and vibrational spectra of the $S4^{2-}$ and $S5^{2-}$ anions.

Influence of the cations on bond length, valence, and torsion angle. *Inorg. Chem.* 39: 2593–2603.
81 Meyer, B. (1976). Elemental sulfur. *Chem. Rev.* 76: 367–388.
82 Qiu, L., Zhang, S., Zhang, L. et al. (2010). Preparation and enhanced electrochemical properties of nano-sulfur/poly (pyrrole-co-aniline) cathode material for lithium/sulfur batteries. *Electrochim. Acta* 55: 4632–4636.
83 Wang, J., Chen, J., Konstantinov, K. et al. (2006). Sulphur-polypyrrole composite positive electrode materials for rechargeable lithium batteries. *Electrochim. Acta* 51: 4634–4638.
84 Wang, J., Chew, S., Zhao, Z. et al. (2008). Sulfur–mesoporous carbon composites in conjunction with a novel ionic liquid electrolyte for lithium rechargeable batteries. *Carbon* 46: 229–235.
85 Yu, X., Xie, J., Li, Y. et al. (2005). Stable-cycle and high-capacity conductive sulfur-containing cathode materials for rechargeable lithium batteries. *J. Power Sources* 146: 335–339.
86 Yu, X.-G., Xie, J.-Y., Yang, J. et al. (2004). Lithium storage in conductive sulfur-containing polymers. *J. Electroanal. Chem.* 573: 121–128.
87 Wang, J.-Z., Lu, L., Choucair, M. et al. (2011). Sulfur–graphene composite for rechargeable lithium batteries. *J. Power Sources* 196: 7030–7034.
88 Zhang, S.-C., Zhang, L., Wang, W.-K., and Xue, W.-J. (2010). A novel cathode material based on polyaniline used for lithium/sulfur secondary battery. *Synth. Met.* 160: 2041–2044.
89 Huff, L.A., Rapp, J.L., Baughman, J.A. et al. (2015). Identification of lithium–sulfur battery discharge products through ^{6}Li and ^{33}S solid-state MAS and ^{7}Li solution NMR spectroscopy. *Surf. Sci.* 631: 295–300.
90 See, K.A., Leskes, M., Griffin, J.M. et al. (2014). Ab initio structure search and in situ ^{7}Li NMR studies of discharge products in the Li–S battery system. *J. Am. Chem. Soc.* 136: 16368–16377.
91 Gao, J., Lowe, M.A., Kiya, Y., and Abruña, H.D. (2011). Effects of liquid electrolytes on the charge–discharge performance of rechargeable lithium/sulfur batteries: electrochemical and in-situ X-ray absorption spectroscopic studies. *J. Phys. Chem. C* 115: 25132–25137.
92 Cuisinier, M., Cabelguen, P.E., Adams, B.D. et al. (2014). Unique behaviour of nonsolvents for polysulphides in lithium–sulphur batteries. *Energy Environ. Sci.* 7: 2697–2705.
93 Xiao, J., Hu, J.Z., Chen, H. et al. (2015). Following the transient reactions in lithium–sulfur batteries using an in situ nuclear magnetic resonance technique. *Nano Lett.* 15: 3309–3316.
94 Hagen, M., Fanz, P., and Tübke, J. (2014). Cell energy density and electrolyte/sulfur ratio in Li–S cells. *J. Power Sources* 264: 30–34.
95 Hagen, M., Dörfler, S., Fanz, P. et al. (2013). Development and costs calculation of lithium–sulfur cells with high sulfur load and binder free electrodes. *J. Power Sources* 224: 260–268.
96 Eroglu, D., Zavadil, K.R., and Gallagher, K.G. (2015). Critical link between materials chemistry and cell-level design for high energy density and low cost lithium–sulfur transportation battery. *J. Electrochem. Soc.* 162: A982–A990.

97 McCloskey, B.D. (2015). Attainable gravimetric and volumetric energy density of Li–S and Li ion battery cells with solid separator-protected Li metal anodes. *J. Phys. Chem. Lett.* 6: 4581–4588.

98 Cheon, S.-E., Ko, K.-S., Cho, J.-H. et al. (2003). Rechargeable lithium sulfur battery: II. Rate capability and cycle characteristics. *J. Electrochem. Soc.* 150: A800–A805.

99 Fu, C., Wong, B.M., Bozhilov, K.N., and Guo, J. (2016). Solid state lithiation–delithiation of sulphur in sub-nano confinement: a new concept for designing lithium–sulphur batteries. *Chem. Sci.* 7: 1224–1232.

100 Jeong, S.S., Lim, Y.T., Choi, Y.J. et al. (2007). Electrochemical properties of lithium–sulfur cells using PEO polymer electrolytes prepared under three different mixing conditions. *J. Power Sources* 174: 745–750.

101 Nagao, M., Hayashi, A., and Tatsumisago, M. (2011). Sulfur–carbon composite electrode for all-solid-state Li/S battery with Li2s–P2s5 solid electrolyte. *Electrochim. Acta* 56: 6055–6059.

102 Barchasz, C., Leprêtre, J.-C., Alloin, F., and Patoux, S. (2012). New insights into the limiting parameters of the Li/S rechargeable cell. *J. Power Sources* 199: 322–330.

103 Wang, D.R., Wujcik, K.H., Teran, A.A., and Balsara, N.P. (2015). Conductivity of block copolymer electrolytes containing lithium polysulfides. *Macromolecules* 48: 4863–4873.

104 Gaillard, F. and Levillain, E. (1995). Visible time-resolved spectroelectrochemistry: application to study of the reduction of sulfur (S8) in dimethylformamide. *J. Electroanal. Chem.* 398: 77–87.

105 Gaillard, F., Levillain, E., Dhamelincourt, M.C. et al. (1997). Polysulphides in dimethylformamide: a micro-Raman spectroelectrochemical study. *J. Raman Spectrosc.* 28: 511–517.

106 Varcoe, J.R. and Slade, R.C.T. (2005). Prospects for alkaline anion-exchange membranes in low temperature fuel cells. *Fuel Cells* 5: 187–200.

107 McCloskey, B.D., Scheffler, R., Speidel, A. et al. (2011). On the efficacy of electrocatalysis in nonaqueous Li–O_2 batteries. *J. Am. Chem. Soc.* 133: 18038–18041.

108 McCloskey, B.D., Bethune, D.S., Shelby, R.M. et al. (2011). Solvents' critical role in nonaqueous lithium–oxygen battery electrochemistry. *J. Phys. Chem. Lett.* 2: 1161–1166.

4

From the Lab to Scaling-up Thin Film Solar Absorbers

Hariklia Deligianni[1], Lubomyr T. Romankiw[1], Daniel Lincot[2], and Pierre-Philippe Grand[3]

[1] *Thomas J. Watson Research Center, IBM Research, 1101 Kitchawan Road, Yorktown Heights, NY 10598, USA*
[2] *CNRS, Institut Photovoltaïque Ile de France (IPVF), 30 route départementale 128, 91120 Palaiseau, France*
[3] *EDF Recherche et Développement, Institut Photovoltaïque Ile de France, 30 route 128, 91120 Palaiseau, France*

4.1 Introduction

Photovoltaics (PVs) is an emerging technology that represents 227 GW of production capacity installed worldwide, of which more than 50 GW was installed in 2016. The most important PV markets are China (44 GW), Germany (39.7 GW), Japan (34.4 GW), and the United States (25.6 GW), while India, South America, and Africa are large emerging markets [1]. Presently, only about 1.3% of the electricity consumed in the world is using PV. In some leading countries such as Germany, Italy, and Greece, 10% of the total electricity production is from solar energy conversion. It is now anticipated that the growth of PV will not only continue but even accelerate reaching up to about 5 TW by 2050 as projected by the International Energy Association (IEA).

The main reason for this development is due to the decrease in the cost of PV electricity from about \$1 kWh^{-1}, 10 years ago, down to less than 10 cents in many countries. This trend makes PV electricity competitive with classical fossil or nuclear electricity. In recent years, the cost has reached only about 3–4 cents kWh^{-1}. The decrease in cost has resulted from scaling, an increase in volume production. Typically, the production cost of a module, expressed in dollar per Watt, decreases by a factor of 20% when the cumulative production doubles. In addition, due to an improvement in materials and methods, an increase in solar energy conversion efficiency leads to lower production costs.

Research and development plays a major role in this process. Innovation and scientific breakthroughs in existing PV technologies are continuously needed to introduce new technologies, which have the potential to replace existing technologies. This is remarkably exemplified in the case of PV by considering the evolution of record efficiencies at the cell level for all technologies, from established

Electrochemical Engineering: The Path from Discovery to Product,
First Edition. Edited by Richard C. Alkire, Philip N. Bartlett, and Marc T. Koper.

to emerging, recorded in the general table made by the US National Renewable Energy Lab (NREL) and displayed in Figure 4.1. Let us consider the case of silicon wafer technologies, which currently represent more than 90% of the PV market. The PV efficiency progress is related to a succession of high efficiency (vertical steps) followed by periods of plateaus, corresponding to the introduction of innovations and market consolidation. Recently, efficiencies up to 26.6% have been achieved for single-crystalline solar cells [2]. For polycrystalline silicon, the efficiency curve follows a similar trend but it is lower, with a best efficiency at 21.3%, because of intrinsic cell limitations with lower carrier lifetime. The rest of the market (about 9%) is represented by thin film solar cells, first cadmium telluride (CdTe) and copper indium gallium diselenide and/or disulfide (CIGS) followed by thin film silicon (amorphous and microcrystalline).

The recent efficiency improvement of the chalcogenide technologies (CdTe and CIGS) has been impressive, exceeding the 20% efficiency level, up to 22.6% for CIGS, which represents the highest efficiency for polycrystalline thin film solar cells [3]. These two technologies are now surpassing the record of polycrystalline Si wafer solar cells, a breakthrough that occurred for the first time ever. It is, however, important to compare cell efficiency of different technologies at equivalent or similar scale. As an example, the best certified Solar Frontier 30 cm × 30 cm sub-module was measured at 19.2%, a value that can be compared with the already mentioned 26.6% for mono-crystalline silicon, with a designated area of 180 cm^2.

Figures 4.2 and 4.3 represent the structures of silicon and of CIGS devices with record efficiency. In the case of silicon, wafers of about 250 μm are used and are processed to create the p–n junction and contacts in various configurations based on technologies derived from microelectronics. In the case of CIGS cells, the structure consists of polycrystalline layers, with micron scale thicknesses, deposited directly on glass, plastic, or metal foil substrates.

Although the race for higher efficiencies at the module level, pushed by record cell efficiencies, is very important to further increase the competitiveness of PV, the other very important trends are low production costs and high reliability. Scaling effects have played an enormous role in decreasing the production costs, but further progress along with innovative technologies will be needed to lower costs.

The comparison between thin film technologies and wafer-based technologies is typical of this paradigm and the subject of lively debates for many years. Thin film technologies are basically coating technologies that are inherently better adapted for large area processing as required for large-scale PV (thousands of square kilometers) compared to slicing individual crystalline wafers and assembling them as a mosaic. However, the cost reduction of the wafer technology with scaling and the industrial maturity of silicon technology are so rapid that this technology is cost leading. This is the main reason why the introduction of thin film technologies in the market is limited. Demonstrating and increasing the competitiveness of thin film technologies is the objective of many dedicated working groups globally.

An alternative to vacuum-based deposition methods (co-evaporation and sputtering), which are presently used for producing CIGS layers, is to use

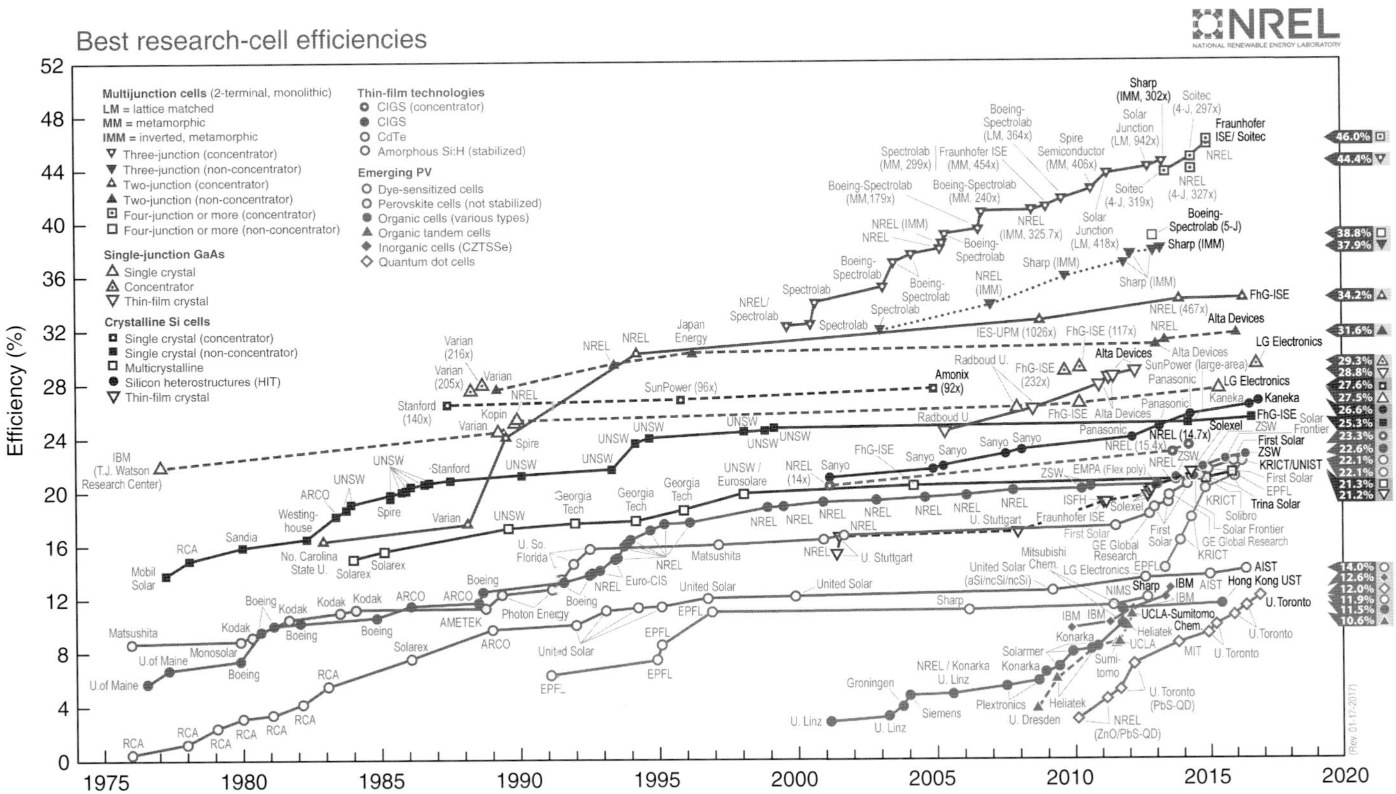

Figure 4.1 Chart of record efficiencies of laboratory solar cells for all PV technologies. Source: National Renewable Energy Lab (NREL), March 2017.

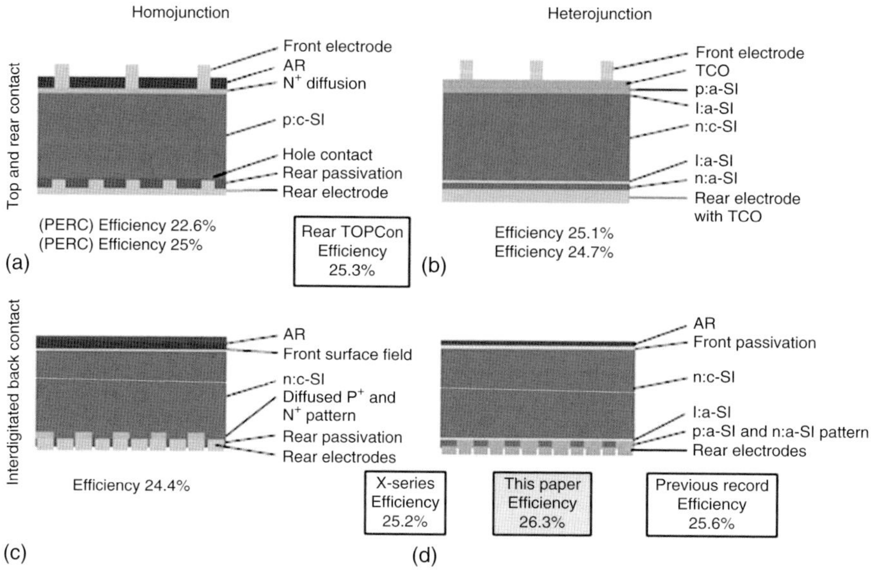

Figure 4.2 Structure of silicon solar cells with record efficiencies. Source: Yoshikawa et al. 2017 [2]. Reproduced with permission of Springer Nature.

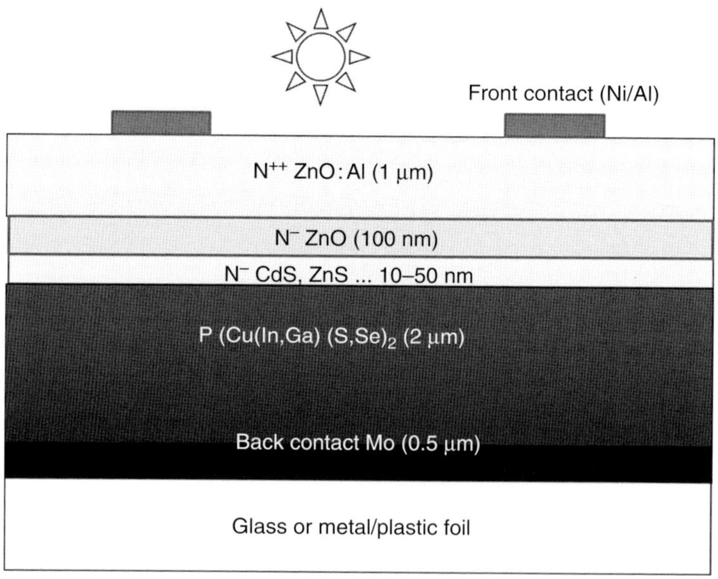

Figure 4.3 Typical structure of a CIGS thin film solar cells.

room temperature methods such as electrodeposition or printing. Such methods have been shown to be of low cost and scalable, in other industrial applications [4].

The aim of this chapter is to focus on recent development of electrodeposition for CIGS solar cells. It provides a remarkable scientific and technological path from the laboratory research to early industrial development. Electrodeposition for CIGS is an innovative technology that can substantially lower production costs. More specifically what follows in the first part of the chapter is a discussion on the different fabrication approaches for the electrodeposited precursor CIGS materials, fundamental of electrodeposition, and development of solution chemistries for copper, indium, and gallium and emerging earth-abundant electrodeposited kesterite precursor materials such as $Cu_2ZnSn(Se,S)$. Fabrication of solar cells and annealing of the precursor materials is described. In the latter part of the chapter, a comprehensive description of the scale-up of the electrodeposited CIGS precursor layers is described from the laboratory scale of a rotating disk electrode to 15 cm × 15 cm glass plates to 30 cm × 60 cm modules and finally to a full size 60 cm × 120 cm module and a 1 m^2 min^{-1} industrial scale production of solar panels.

4.2 State-of-the-art Electrodeposition for Photovoltaics

Electrodeposition for solar cells is an active area of research, and it has been shown to provide improved benefits in several solar technologies. For the silicon solar cell contacts, electrodeposited metal contacts result in solar cell efficiency improvement due to a lower processing temperature with better yield than screen printing, lower resistance, and the ability to fabricate small contacts that reflect less sunlight. Copper electrodeposition is now emerging as an industrial alternative to silver or aluminum paste [4]. This follows the successful introduction of copper plating in microelectronics, known as the damascene process.

For the chalcopyrite ($CuInGa(Se,S)_2$ or CIGS) and the kesterite ($Cu_2ZnSn(Se,S)_4$ or CZTS) thin films, electrodeposition offers very low cost of manufacturing with a continuous reel-to-reel or roll-to-roll process. Rigid glass or flexible metallic substrates can be used with these thin films, potentially opening up new applications for solar energy. Applications span from energy generation for energy utilities to building integrated materials, solar-powered electric cars, and internet of things (IoT) solar-powered electronic devices.

The CIGS electrodeposition-based method was developed independently by separate groups. One such group was at the French IRDEP (Institut de la Recherche et Developement sur l'Energie Photovoltaique), where a record cell efficiency for the $CuIn(S,Se)_2$ solar cell of 11.4% was demonstrated. Then, NEXCIS, a spin off from IRDEP, developed in collaboration with IBM on electrodeposition, a module-manufacturing process compatible with high throughput and industrial size. Another group was with the American company SOLOPOWER, which demonstrated record efficiency for CIGS cell of 13.8%. The electrodeposition method is considered low cost, capable of producing high-efficiency thin film solar cells. In the following sections, 30 cm × 60 cm and 60 cm × 120 cm wafers, modules, and associated parameters were obtained in NEXCIS facility based in Rousset, France.

Challenges at the large scale with lateral compositional uniformity due to electrodeposition and thermal treatment are still present. The main goal is to enable fabrication of high-efficiency modules at low cost. This requires a very tight compositional control for the mass deployment of this technology. All efforts to reduce processing time, absorber thickness and costs, and activities to enhance the efficiency have one common goal: to leverage the CIGS production by electrodeposition at large scale (Figure 4.4).

4.2.1 Electrodeposited $CuInGaSe_2$ (CIGS)

4.2.1.1 Metal Layers

The stack electroplating approach, where copper, indium, and gallium are electroplated as individual layers, offers an easier alternative to alloy co-deposition. In this approach, the ratio of the three elements is tightly related to the thickness of

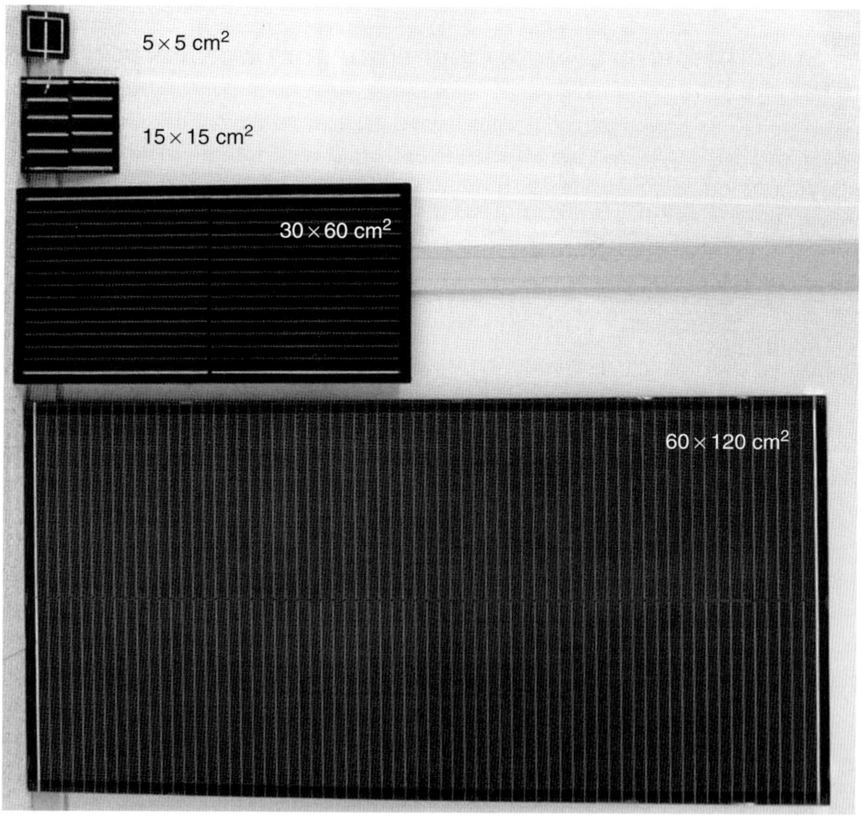

Figure 4.4 Picture of electrodeposited CIGS modules from NEXCIS.

each layer. Reproducible and minimal thickness variations and minimal surface roughness of all layers are therefore critical for a well-controlled absorber.

4.2.1.2 Electrodeposition of Copper

The copper electroplating chemistry has three types of additive classes [5–10]. One of the additives is called accelerator and it also acts as a brightener. Other additives contained in the copper chemistry act as suppressors and levelers. An example suppressor is polyethylene glycol (PEG). The suppressing action to copper electrodeposition reaction is provided by PEG when combined with the chloride ion in solution [11–14]. Most commercially available baths have other polyether types of macro-molecules as suppressors. Suppressors are not readily consumed during electrodeposition. Most suppressors are co-polymers of ethylene and propylene oxides or block co-polymers of other similar polymerized oxides. As the copper bath ages, these polymers break down in smaller fragments. It is not clear although what the impact on this fragmentation is on the final product.

Another class of additives are organic molecules that contain nitrogen [5, 15, 16]. These typically act as inhibitors and levelers. A leveler would get consumed during electrodeposition and would result in ppm amounts of nitrogen and other impurities in the deposit. The rate of arrival of the leveler compound to the cathode surface is typically controlled by mass transport. The leveler is a large polymer as well, typically containing aromatic amines (—NH_2). The leveler polymeric chains break down during bath aging and the impact on the deposit will need to be assessed with a detailed aging study.

The brightener is bis-(sodium-sulfopropyl) disulfide (SPS) [17–26]. Its chemical formula is shown in Figure 4.5. During electroplating of copper or even during the time when the bath is not exercised (idle time), bath aging can occur primarily at the copper phosphorous (CuP) anode. Especially when the CuP anode does not have a well-developed protecting film, SPS will decompose on the anode and oxidize yielding a first and a second oxidation by-product shown in Figure 4.6. The first oxidation by-product of SPS contains only one oxygen with a double bond. The anode film is black, thick, and quite well adherent on the CuP surface. If the film has holes exposing the pink copper surface, then SPS will oxidize on the anode. Oxidation effects of the SPS-type additives are more pronounced in the case of an inert anode such as IrO_2 or platinized titanium. There is no consensus in the scientific literature about whether an inert anode or a CuP anode is better for copper electrodeposition. A Nafion membrane is used to prevent the additives from being consumed at the anode and to prolong the bath life.

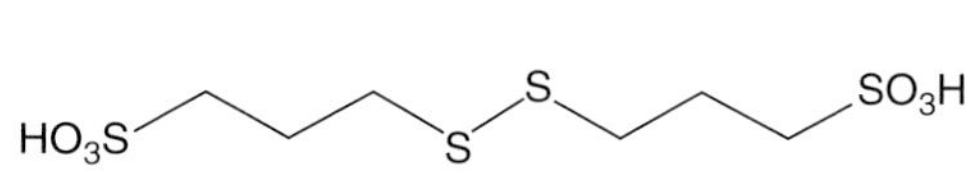

Figure 4.5 Chemical formula of bis-(sodium-sulfopropyl) disulfide (SPS).

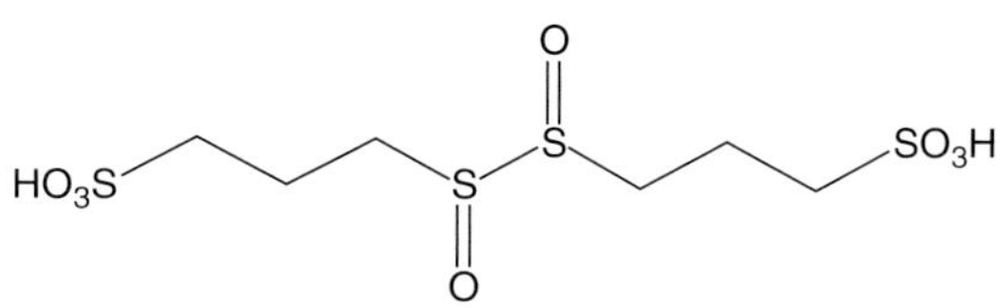

Figure 4.6 Chemical formula of second oxidized or breakdown by-product during bath aging.

The bath temperature for acid copper chemistries is to be controlled within 1 °C. The reason is that the function of the additives degrades or the polymer chains of the suppressor and leveler additives break into smaller fragments at temperatures above 27 °C. In the case of thin PV films, one would expect that the deposited thin films would become rougher and the deposits would have increasingly large microstructure and impurity levels as the additives decompose.

4.2.1.3 Electrodeposition of Indium

The indium electrochemical kinetics was studied in chloride and sulfate solutions [27–29]. The electrodeposition of indium thin film on copper substrates for CIGS applications was studied by Valderrama et al. [30]. These authors studied the indium deposition on Mo/Cu substrates in $InCl_3$ solutions and observed instantaneous nucleation followed by three-dimensional (3D) growth. CuIn alloy phase was observed for some of the as-plated indium films. However, the mechanism of the indium deposition on copper and the CuIn alloy formation has not been well understood. An investigation of indium plating from a sulfate solution on different copper substrates was carried out by Huang et al. [31]. The nucleation and growth of thin indium films on top of copper thin layers was studied using a rotating disk electrode.

Nucleation of indium on copper differs from the nucleation of indium on indium. A two-step nucleation and growth phenomenon was discovered for the indium thin film plating on copper substrates. Firstly, the indium deposits on copper in a layer-by-layer manner due to a under potential deposition (UPD) like behavior. Secondly, the indium starts to form nuclei on the indium surface. This nuclei formation may be related to the film stress, indium surface oxidation, or an enhanced side reaction, which lowers the deposition potential. The nuclei that form grow into islands. Top-down scanning electron microscope (SEM) images and the energy dispersive X-ray (EDX) analysis show that a much higher indium composition is observed for the bumps. The area in between the bumps shows a much lower indium composition, indicating a much thinner continuous layer.

Indium and copper forms $CuIn_2$ alloy even at temperatures well below room temperature. This low temperature alloy formation was found to facilitate a layer-by-layer deposition. In addition, an anodic stripping study showed that a significant amount of alloy forms during plating (Figure 4.7). In addition, Figure 4.8 shows a time-dependent X-ray diffraction (XRD) evolution at room temperature for a plated Cu/In stack structure. All the indium was consumed to form $CuIn_2$ alloy within 24 hours after plating. The observation for the plated films by Huang et al. [31] is consistent with the observations reported for sputtered and evaporated films [32].

Fast inter-diffusion happens in situ while the indium is plated on copper. Therefore, the layer-by-layer plating is extended well beyond a few monolayers, as in most cases of UPD. As a result, a continuous "copper–indium" layer up to ~80 nm forms before "indium" nucleation bumps form. The formation of bumps depends on a threshold indium composition at the top surface during the plating. Bump formation depends on not only the overall amount of the Cu and In that is in the stack but also on how fast indium is plated. In addition, bump formation also

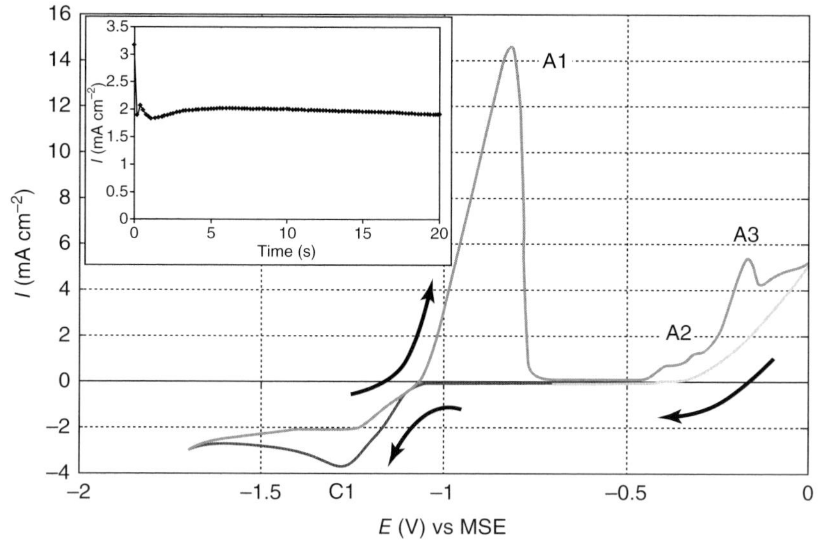

Figure 4.7 Cyclic voltammogram of indium sulfate solution on 30 nm Cu seed and insert is chronoamperometry of indium deposition on Cu seed at −1.2 V vs SMSE [31].

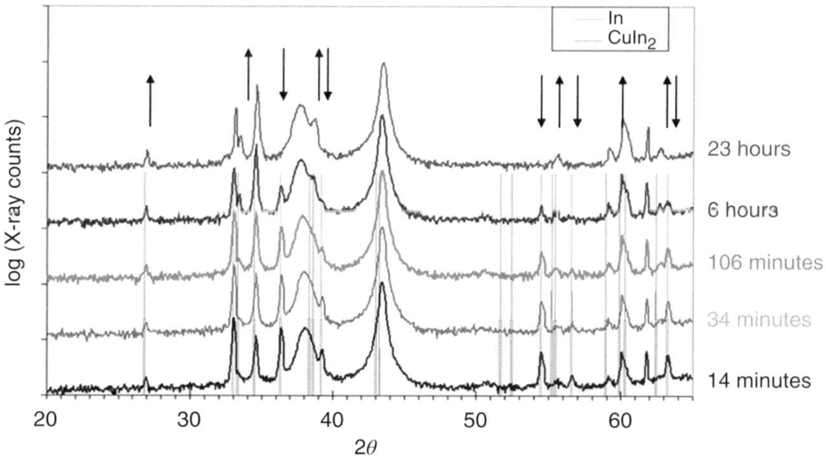

Figure 4.8 X-ray diffraction spectra of 150 nm Cu/200 nm In thin film structure [31].

depends on the Cu—In inter-diffusion behavior, or how fast the indium deposited on the surface diffuses away from the surface.

Figure 4.9 shows a comparison of two indium films plated at two different current densities, i.e. different deposition rates. A more coalescent film or a higher island density is obtained at a higher current density. More importantly, the cross-sectional SEM images of the two samples showed that a thinner continuous layer was observed at a higher current density. This confirms an earlier threshold surface composition and island formation.

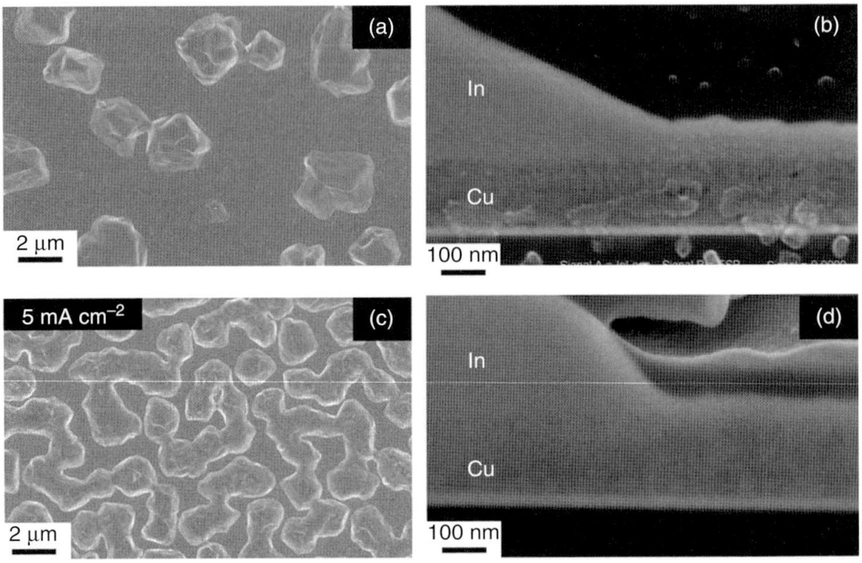

Figure 4.9 Top-down and cross-sectional SEM images of 200 nm indium plated from the indium sulfate solution on 150 nm Cu (a, b) at 0.5 mA cm^{-2} and (c, d) at 5 mA cm^{-2}. Source: Reproduced with the permission of Electrochemical Society.

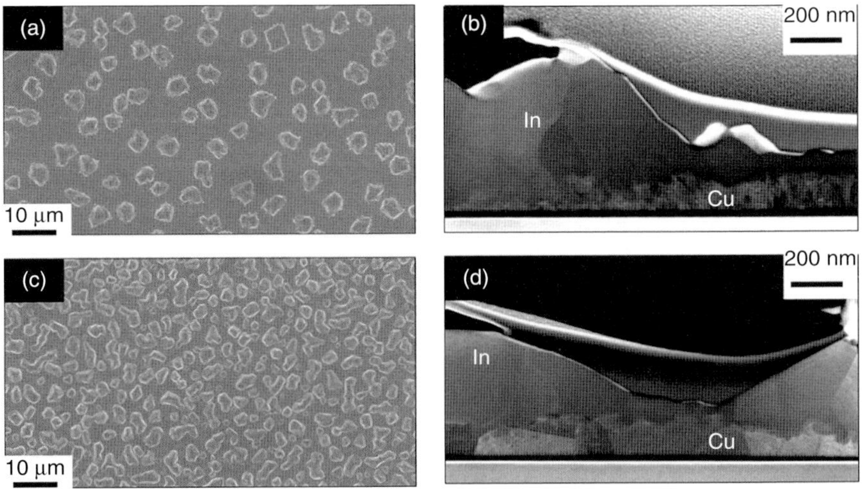

Figure 4.10 Top-down SEM and cross-sectional STEM images of indium plated from the In sulfate solution on 300 nm Cu plated from a damascene copper chemistry. The copper was (a, b) as plated and (c, d) 100 °C one hour annealed in N_2 before the indium plating. Source: Reproduced with the permission of Electrochemical Society.

Another approach to change nucleation and growth is to change the diffusion behavior, shown in Figure 4.10. A damascene copper plating chemistry was used to plate 300 nm copper. The Cu plated from the damascene copper chemistry is known to crystallize with a 100 °C anneal. The grain size of copper increases from 50 nm as plated to as large as 1 μm after annealing. Because solid-state diffusion occurs preferentially along grain boundaries, a comparison between the indium plating on the as-plated and annealed Cu layer is expected to show the effect of Cu—In inter-diffusion.

Figure 4.10a,c clearly show a much higher island density for the indium plating on the annealed copper layer than on the as-plated copper. In the former case, the inter-diffusion rate is believed to be slower between indium and the recrystallized copper and an earlier island formation is therefore expected. In addition, as the inter-diffusion rate is slower, the net nucleation current density, the total plating current density subtracted by the consumption of indium due to the alloy formation, is higher. Therefore, a higher nucleation density would also be expected due to a higher nucleation potential.

Cross-sectional STEM analysis was carried out to analyze the grain structure of the two samples, shown in Figure 4.10b,d as well. The annealed Cu has a large grain size, about 1 μm, while the as-plated copper has very fine grains. An intermixed region is observed for both cases, where the grains in the top indium layer extend into the copper under layer. In other words, the copper dissolves into the indium and a grain structure similar to pure indium is formed. The thickness of this intermixed region depends on the thickness of indium and the grain size of the copper. The intermixed region under the thin continuous indium layer is thinner than that under the indium islands because the amount of indium available for Cu to dissolves into is limited. The most pronounced intermixing is observed under the indium island for the indium plated on the as-plated copper, consistent with the fast diffusion rate expected for fine grain metals.

Three methods can improve the coalescence of indium films: (i) slowing down the indium deposition rate or speeding up the indium–copper inter-diffusion to facilitate the formation of a thicker continuous alloy layer; (ii) increasing the indium deposition rate or slowing down the inter-diffusion to limit the continuous alloy layer and increasing the nucleation density; and (iii) pulse plating nanometric multilayers to avoid the nucleation step of indium plating. The pulse plating method is described in detail in a PCT patent application [33].

4.2.1.4 Electrodeposition of Gallium

Gallium is an element from the group IIIA. It has been used in the semiconductor and the electronics industries due to its unique electrical and thermal properties [34–36]. Gallium forms compounds with aluminum, indium, phosphorus, arsenic, and antimony (e.g. gallium arsenide and indium gallium arsenide phosphide) with valuable semiconductor and optoelectronic properties [37, 38]. Some of these compounds form the basis of electronic devices, for example are used as light-emitting diodes and semiconductor lasers. Gallium plays an important role in achieving high-efficiency solar cells if it is incorporated in $CuInSe_2$ as a partial replacement of indium to form $CuInGaSe_2$ (CIGS) [39–44]. The absorber layer band gap increases from 1.02 eV for $CuInSe_2$ to 1.1–1.2 eV by a partial substitution of In with Ga, leading to substantial increase in efficiency [39]. However, gallium has a low melting point of 29.78 °C and it readily diffuses to form an alloy with other metals, e.g. Cu, In, Sn, Zn, etc., at room temperature [45–49]. In CIGS fabrication, Ga can be incorporated in the film using both vacuum (PVD) [39–41] and electrochemical methods [50–53].

Gallium is generally considered a difficult metal to electrodeposit without excessive hydrogen generation at the cathode. The electrode potential required for gallium electrodeposition overlaps with the hydrogen evolution overpotential. As a result, hydrogen evolution causes low plating efficiency, porosity, surface roughness, and microdefects in the deposited thin film. The efficiency of CIGS solar cell, in addition to the other factors, is also dependent on the solid-state reaction and stability of Ga in the CIGS layer [54].

As an example, Figure 4.11 [55] shows a typical cyclic voltammogram for the electrodeposition of gallium on gold in 0.25 M Ga^{3+}. The potential was scanned from 0.5 to −2.0 V (SCE) at 0.05 V s^{-1}. As the potential is scanned in the negative direction, a cathodic current corresponding to the electrodeposition of gallium was observed at −0.8 V (SCE). Initially, the current is controlled by reaction kinetics but as the potential is increased further, a local maximum (~11 mA cm^{-2}) is reached at −1.6 V. This is due to mass transport control, i.e. diffusion of the Ga-ions to the electrode surface becomes rate limiting. As the scan continues to a potential more negative than −1.65 V, the magnitude of the current decreases. This is due to the depletion of the Ga-ions adjacent to the electrode. The current eventually reaches a plateau at 10.5 mA cm^{-2} when the thickness of the depletion layer (i.e. the region adjacent to the electrode that is depleted of Ga ions) becomes constant. As the potential is more negative, an increase in current is again observed due to the hydrogen gas evolution $2H^+ + 2e^- \rightarrow H_2$. The potential scan is reversed at −2.0 V. At −1.5 V, the rest potential is reached and no further electrodeposition of gallium occurs. As the potential is scanned to values

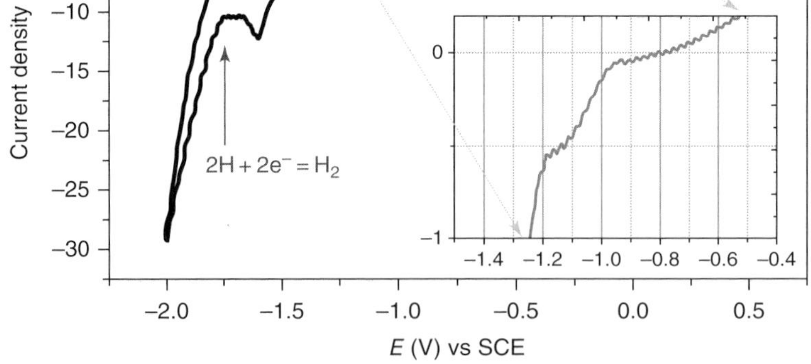

Figure 4.11 Cyclic voltammogram of a polished gold substrate in a Ga bath consisting of 0.25 M Ga^{3+} and 0.5 M sodium citrate with pH ≥ 10 at a scan rate 0.05 V s^{-1} from+0.5 to −2.0 V (SCE). The scaled-up *I–V* in the inset clearly shows the onset of cathodic reduction at −0.8 V vs SCE [55].

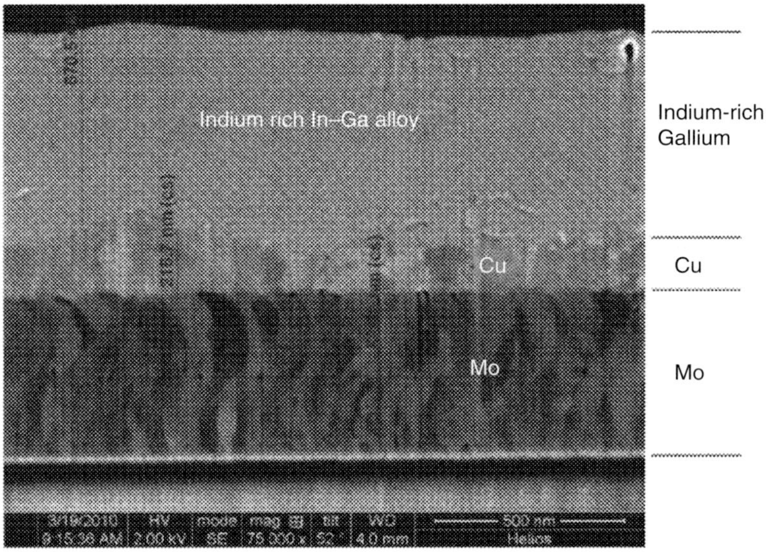

Substrate: Glass / 600 nm Mo / 240 nm Cu / 360 nm In / 150 nm Ga

Figure 4.12 A scanning electron micrograph of a cross-sectional view of a metallic precursor stack. Gallium was electrodeposited onto an indium layer and subsequently annealed to form an indium-rich gallium eutectic layer [57].

more positive than the rest potential, the current becomes anodic and the gallium begins to oxidize $Ga \rightarrow Ga^{3+} + 3e^-$.

A rotating disk electrode was used for the development of the gallium plating chemistry. Gallium can be plated at room temperature from both acidic (pH 0.0–2.6) and alkaline solutions (pH 12.6–14). The plating at pH ~ 1.0 for the acidic solution and at pH > 13.0 for the alkaline solution gives a plating efficiency greater than 90%. In both alkaline and in acidic solutions, a high current density can be used for electroplating and the resultant smooth film has a fine microstructure and is very bright.

IBM developed an acidic and an alkaline chemistry for gallium electroplating. The gallium solution contains 0.2 M Ga^{+3} dissolved in ≥0.5 M methanesulfonic acid (MSA). $Ga_2(SO_4)_3$ dissolves and the solution is neutralized by adding sodium hydroxide (NaOH). In the alkaline bath, NaOH was added until the pH became greater than 12 [56] (Figure 4.12).

Gallium is a low melting (29.7 °C) metal. Due to the sudden exposure of the films to a high temperature (90 °C), gallium melts spontaneously and thus starts to ball-up. This causes discontinuity of the film and thus creates voids on the surface and the film underneath becomes exposed. To homogenize the alloy, balling-up needs to be prevented by allowing considerable time for the gallium diffusion to take place through the layers underneath. To further homogenize the Cu—In—Ga precursor, one can choose one of the two options: either to increase the annealing time or to use a higher temperature to homogenize the alloy. In Figure 4.13, the SEM image shows the surface topography of a Glass/Mo/Cu/In/Ga sample after annealing using a temperature of 150 °C.

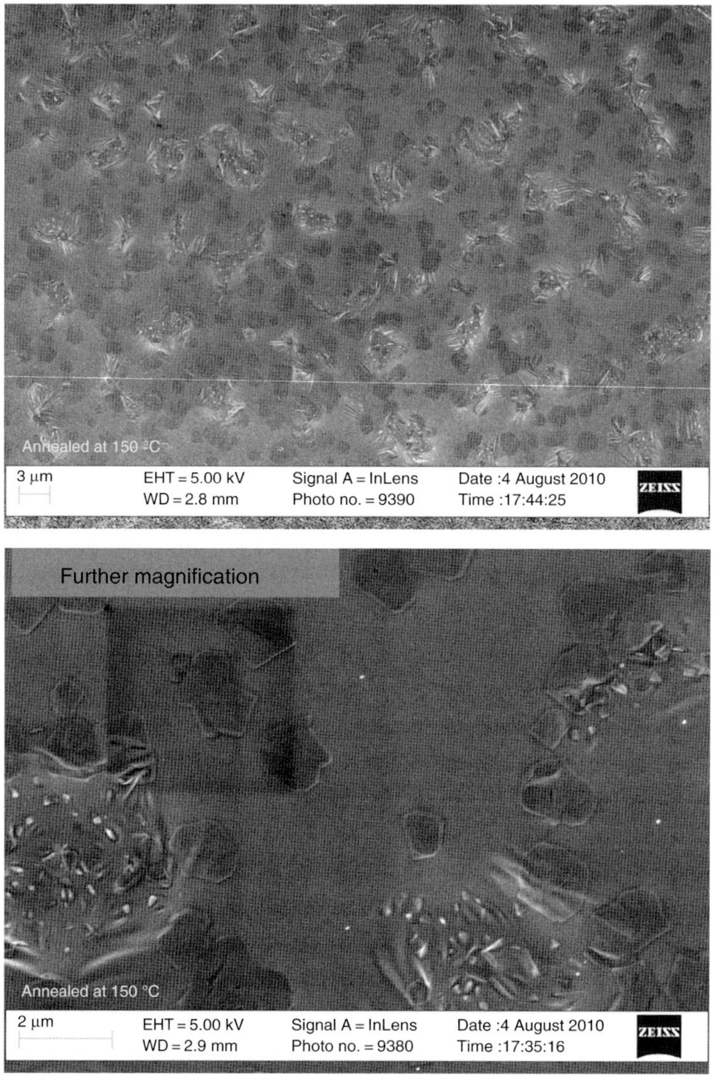

(a) (b)

Figure 4.13 SEM image shows a surface topography of glass/Mo/Cu/In/Ga sample after annealing up to 150 °C.

TEM analysis samples that were annealed using a higher temperature (150 °C) of annealing shows large grains of 0.5–1.5 µm present in the CuInGa layer with composition of 40 at.% Ga and 60 at.% Cu. The composition is consistently Cu_2Ga_3 at a distance 75 nm from the top surface. The compositional profile next to the Mo/alloy interface shows that Ga and Cu diffused into the Mo to about 10 nm. The localized compositional variation in the large Cu_3Ga_2 crystals was measured. A uniform and consistent Cu (60 at.%)–Ga (40 at.%) composition was observed.

4.2.2 Single Cu—In—Ga—Se—O Multicomponent Chemistries

Electrodeposition of compound semiconductors is expected to be very complex owing to the large difference in the deposition potentials of the individual elements and the possibility of the formation of intermediate phases during electrodeposition. The reader is reminded that the standard potentials for deposition of Cu, In, Ga, and Se are

$$Cu^{2+} + 2e^- \rightarrow Cu \quad E\,(V/NHE) + 0.34\,V \tag{4.1}$$

$$In^{3+} + 3e^- \rightarrow In \quad E\,(V/NHE) - 0.34\,V \tag{4.2}$$

$$Ga^{3+} + 3e^- \rightarrow Ga \quad E\,(V/NHE) - 0.53\,V \tag{4.3}$$

$$H_2SeO_3 + 4e^- + 4H^+ \rightarrow Se + 3H_2O \;\; E\,(V/NHE) + 0.74\,V \tag{4.4}$$

4.2.2.1 Cu—In—Se Co-deposition

One-step electrodeposition of $CuInSe_2$ can be obtained from acidic aqueous solutions, using sodium or potassium sulfate as supporting electrolyte and containing Cu(II), In(III), and Se(IV) electroactive species. By the control of the appropriate ratio of $[Se]_{bath}/[Cu]_{bath}$ and $[Cu]_{bath}/[In]_{bath}$ together with the application of a reduction potential at the cathode, $CuInSe_2$ can be deposited along with elemental Se and binary phases as Cu_2Se or In_2Se_3 [58, 59].

The main reaction for the $CuInSe_2$ deposition could be defined as follows:

$$2H_2SeO_3 + Cu^{2+} + In^{3+} + 13e^- + 8H^+ \rightarrow CuInSe_2 + 6H_2O \tag{4.5}$$

As shown in Figure 4.14, appropriate deposition conditions lead to relatively smooth precursor film morphology together with relatively high film density. Usually, these films exhibit the presence of some fine voids (widths of tens of nanometers) through their upper half and an evolution of the grain size is from 50 to 100 nm near the back contact to only 10–20 nm toward the top surface of the films. In Figure 4.15, the Raman spectra for a slightly Cu-rich precursor are shown. Raman spectroscopy shows that this precursor is mostly composed of $CuInSe_2$, elemental Se, and a secondary phase of Cu_xSe [60, 61]. The presence of Cu-deficient domains with chalcopyrite crystallographic order (known

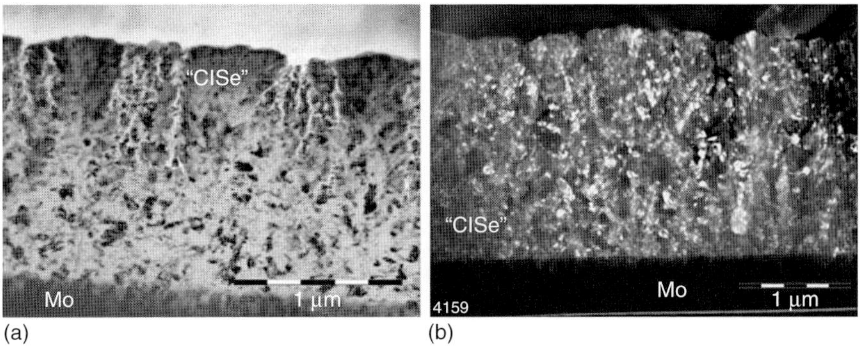

Figure 4.14 Bright and dark-field TEM micrograph of a typical CIS electro-deposited precursor. Crevices are observed and average grain sizes can be estimated.

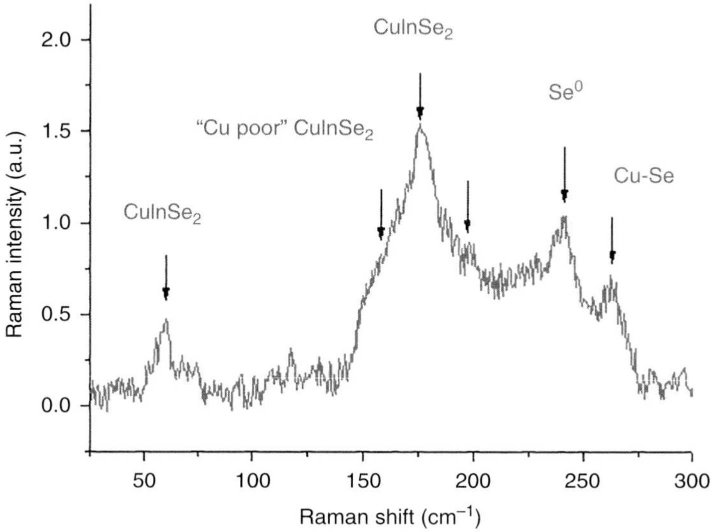

Figure 4.15 Raman spectra for a slightly Cu-rich CIS electrodeposited precursor.

in the literature as ordered vacancy compounds (OVCs) is also highlighted by Raman spectroscopy [62, 63]. Cross-sectional EDX measurements of TEM samples reveal that the overall composition is close to uniform in depth, with a slight Cu enrichment toward the back contact and some Cu/In fluctuations from grain to grain. The characteristic scale of the Cu/In fluctuations is the same as that of the average grain size, i.e. 50–100 nm, close to the back contact and less near the top surface. A detailed examination of the rear contact interface shows an intermixed zone on the order of 10 nm where Mo, Cu, In, and Se seem to be present.

$CuInSe_2$ uniformity was demonstrated on $30 \times 30\,cm^2$ substrates (see Figure 4.16) where the relative lateral compositional variations to be <5%, although thickness variations may be greater (10–20%). The improvement of the thickness uniformity is directly linked with the bath agitation and/or the Mo sheet resistance [64, 65].

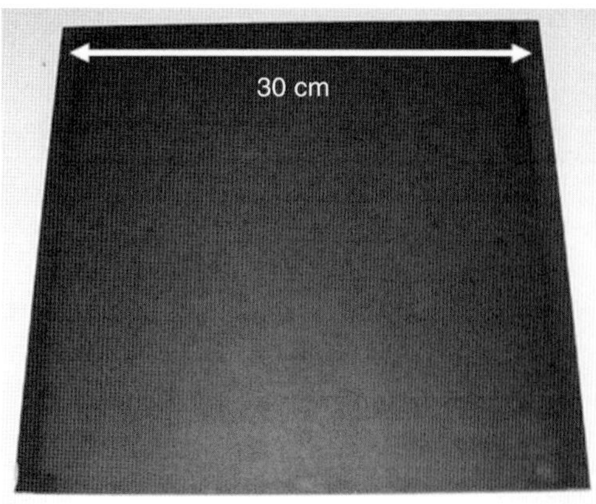

Figure 4.16 Picture of $30 \times 30\,cm^2$ $CuInSe_2$ co-electrodeposited film on glass/molybdenum substrate.

4.2.2.2 Cu—In—Ga—Se Co-deposition

The electrodeposition of compound semiconductors containing gallium is expected to be more complex owing to a large difference in the deposition potentials of the Cu, In, Ga, and Se elements and the possibility of the formation of intermediate phases during electrodeposition. During CIGS film formation, the general reaction occurring at the working electrode is as follows:

$$Cu^{2+} + In^{3+} + Ga^{3+} + 2H_2SeO_3 + 8H^+ + 13e^- \rightarrow Cu(In,Ga)Se_2 + 6H_2O \quad (4.6)$$

This reaction is a combination of electrochemical and chemical reactions that occur at certain pH values. The exact mechanism of CIGS thin film formation is not well established; however, it depends on the redox potential of each chemical species, on the solution composition (as $[Cu]_{bath}/([In]_{bath} + [Ga]_{bath})$ and $[In]_{bath}/[Ga]_{bath}$) and on applied potential (Figure 4.17).

Local pH variations can occur involving oxide and/or hydroxide precipitation.

$$2Ga^{3+} + 6OH^- \rightarrow Ga_2O_3 + 3H_2O \quad (4.7)$$

$$2In^{3+} + 6OH^- \rightarrow In_2O_3 + 3H_2O \quad (4.8)$$

Control of pH in the solution can help in the incorporation of Cu, In, and Ga into the film and, consequently, in the ratios of Cu/(In + Ga) and Ga/(In + Ga) in the film. The use of pH buffers also reduces the occurrence of oxide or hydroxide precipitates that interfere with the formation of Cu–Ga–In–Se precursor thin films and can induce cracks after annealing [67] (Figure 4.18).

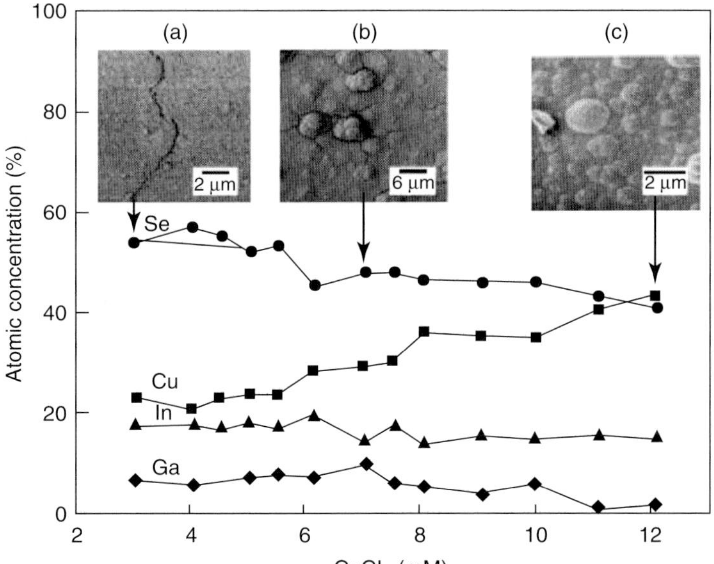

Figure 4.17 Composition of the precursor film as a function of $CuCl_2$. Source: Fernandez and Bhattacharya 2005 [66]. Reproduced with the permission of Elsevier.

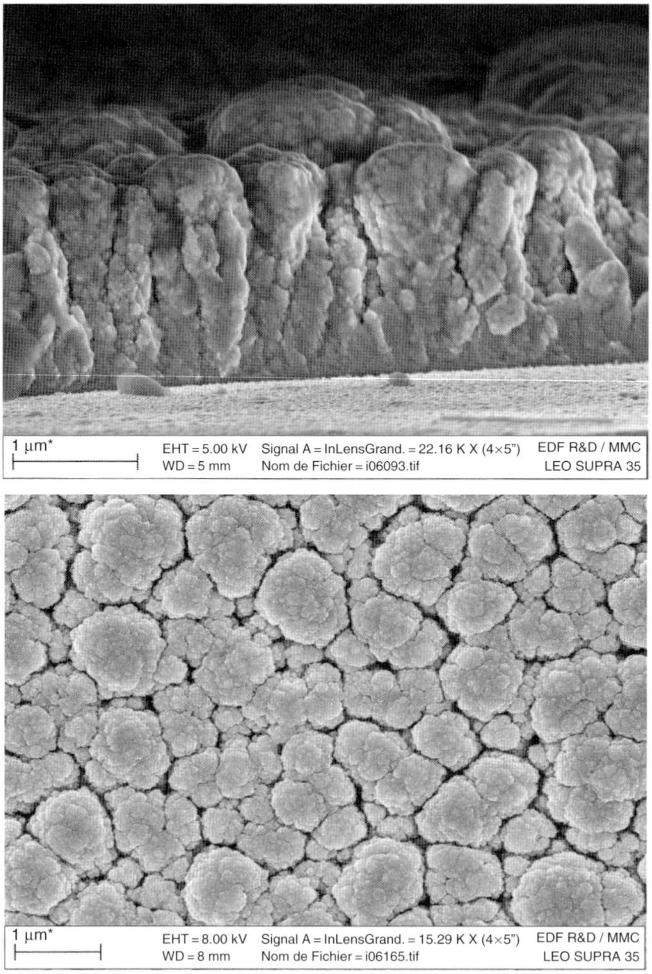

Figure 4.18 SEM cross-sectional and top view of one step electrodeposited Cu(In,Ga)Se_2 film.

4.2.2.3 Cu—In—Ga—O Co-deposition

$$NO_3^{-} + 2H^{+} + 2e^{-} \rightarrow NO_2^{-} + H_2O \tag{4.9}$$

$$Cu^{2+} + NO_3^{-} + H_2O + 2e^{-} \rightarrow Cu(OH)_2 + NO_2^{-} \tag{4.10}$$

$$In^{3+} + \frac{3}{2}NO_3^{-} + \frac{3}{2}H_2O + 3e^{-} \rightarrow In(OH)_3 + \frac{3}{2}NO_2^{-} \tag{4.11}$$

$$Ga^{3+} + \frac{3}{2}NO_3^{-} + \frac{1}{2}H_2O + 3e^{-} \rightarrow GaOOH + \frac{3}{2}NO_2^{-} \tag{4.12}$$

The deposition method for this approach is to produce a pH increase near the electrode surface from the reduction of nitrate ions from reaction (4.9) and the precipitation of a mixed Cu–In–Ga hydroxide film resulting from reactions (4.10), (4.11) and (4.12). Then this precursor film is reduced to a metallic Cu–In–Ga layer by thermal treatment in hydrogen gas, followed by thermal selenization to convert it to CIGS. Efficiencies up to 12% have been achieved with this method [68, 69].

4.2.2.4 Cu—In—Ga Co-deposition

As already mentioned in the previous sections, the main complexity of the deposition of all metallic Cu—In—Ga alloys in one step is the large difference in reduction potentials of the reacting species. The difference of reduction potentials leads to a much higher rate of reduction of copper than with indium or gallium. The imbalance in deposition rates could be reduced by decreasing the reduction potential of Cu. One approach consists on using of selective complex agents that form high-stability complexes with Cu, as for instance thiocyanate, which is a strong complexing agent for Cu^+ [70–72].

The complexing approach allowed to shift the deposition potential of elemental copper considerably (up to −0.51 V vs NHE) and to use the Cu^+ precursor instead of Cu^{2+} in solution. At the same time, thiocyanate complexes of In^{3+} and Ga^{3+} prevent the formation of Ga and In hydroxides. It was found by Kois et al. [70] that the gallium content in the electrodeposited alloys grows at relatively negative potentials (−950 mV). The indium ratio in the deposited alloy films was proportional to $[Cu^+In^{3+}]$ in solution. However, the gallium content decreased with increasing the agitation rate of the solution. This can be explained by the increase in the growth rate of more noble species (H^+ or Cu^+ and In^{3+}) vs Ga^{3+} because of solution agitation. The alloy precursor film morphology has often voids with high porosity, which is highly undesirable.

4.2.3 Annealing Methods

Apart from co-evaporation, where all the elements are evaporated at sufficiently high temperature to form the required chalcopyrite compound, none of the precursor films (deposited by vacuum or wet method) result in an electronic grade chalcopyrite semiconductor. Thermal annealing is always needed to form device quality semiconductor films.

In the *two-step approach* (i.e. precursor formation followed by an annealing step shown in Figure 4.19), uniformity in thermal treatment is very important. The annealing conditions will depend on the nature of the precursor layer. Stacked elemental layers, multilayers, and single multiphasic layer will obviously not recrystallize and homogenize by the same mechanisms during annealing. Annealing under neutral atmosphere or vacuum (400–550 °C) has been proved to markedly improve the structural and optical properties of electrodeposited films; however, the electronic properties are generally not sufficient.

For this, a sufficient high activity of selenium in the vapor phase during annealing is needed during annealing. A dramatic improvement of quality, as shown by electrical transport and luminescence measurements, can be achieved when the partial pressure of Se is increased. Not only the required p-type character of the layer was achieved but also the structural quality of the layer was improved dramatically as compared to annealing treatments under an inert atmosphere. The crystallization is also particularly dependent on the composition of the Cu/(In + Ga) and Ga/(In + Ga) ratios of the precursor and the uniformity of these elements over the entire surface.

Many groups working with annealing treatments of the electrodeposited precursor report the use of a selenium overpressure. Annealing is generally carried out in a closed chamber. In the case of stacked metallic layers, annealing

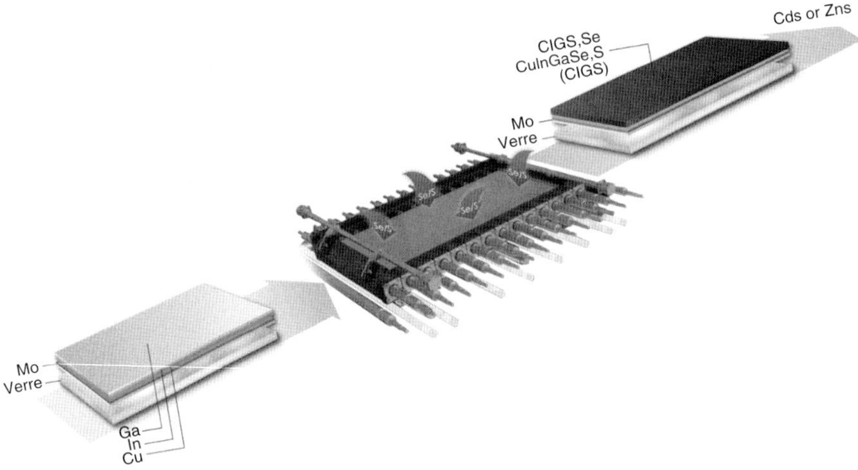

Figure 4.19 Schematic of the two-step approach – metallic precursor and annealing under Se and/or S atmosphere – for the synthesis of the CIGS absorber.

under selenium (or H_2Se) or sulfur (or H_2S) pressure leads to the formation of the desired chalcopyrite film, which can be $Cu(In,Ga)Se_2$, $Cu(In,Ga)S_2$, or $Cu(In,Ga](Se,S)_2$. Annealing treatments are generally carried out at temperatures ranging from 450 to 650 °C from a few minutes to one hour or more. The differences can be explained by the differences between the precursor layers. It is anticipated that forming the single-phase compound from stacked layers is more demanding in terms of temperature or time than a single "electrochemically" premixed layer. A rapid thermal annealing process (RTP) for electrodeposited precursor layers was used for the formation of CIGS from the precursor layers.

The selenium uniformity is insured by controlling the flow of the gas in the furnace to insure a homogeneous partial pressure of selenium over the surface and standard deviation below 3% is achieved as with Cu, In, and Ga electrodeposition. The equipment is used to process $60 \times 120\ cm^2$ plates and resulting CIGSSe absorber as shown in Figure 4.20.

(a) (b)

Figure 4.20 Photographs of NEXCIS $60 \times 120\ cm^2$ RTP equipment (a) and CIGSSe absorber layer (b).

4.2.4 Fabrication of Solar Cells

The CIGS manufacturing process starts with 3-mm-thick soda-lime glass substrates coated with a conductive molybdenum-based back-contact. Molybdenum is covered by a few hundred Angstrom of copper deposited in the same vacuum system in order to provide a platable surface for the electrodeposition process. In the case of NEXCIS, once deposited, the Cu—In—Ga metallic stack is processed in an atmospheric pressure thermal treatment equipment to produce $Cu(In,Ga)(S,Se)_2$ absorbers by reacting with elemental selenium and sulfur. Hydride gases are not used in this process, which would provide the advantage of being less hazardous and less expensive. After formation, the CIGS absorber is covered with a nanometer-thick CdS buffer layer. Finally, a window layer consisting of 80 nm intrinsic ZnO and a few hundreds of nanometer of aluminum-doped zinc oxide is sputter-deposited on top of the buffer. All process steps up to the transparent conductive oxide (TCO) deposition included are performed without any patterning steps, which would interrupt the front-end process (Figure 4.21).

At the end of the front-end process, the full stack glass/Mo/CIGS/CdS/*i*ZnO/ZnO:Al is laser-patterned to discretize the cells. The separated cells are connected together with printed silver-based contacts. The silver paste metallic material used to form the grid fingers that are deposited on top of the TCO to improve the window layer electrical conductivity (Figure 4.22).

In order to test champion cells at NEXCIS, a small piece ($5 \times 5\,cm^2$) was cut out of the full $60 \times 120\,cm^2$ processed substrate. After TCO deposition, a $5 \times 5\,cm^2$ piece was cut and processed at IRDEP for cell-patterning and Ni/Al deposition. From this, $0.5\,cm^2$ cells were mechanically scribed and a Ni/Al grid was evaporated on top of the TCO. As it is standard with champion cells, an anti-reflective coating was deposited on top of the cell to reduce the losses at the air–TCO interface. Current–voltage curves were measured using a continuous illumination AAA simulator for cells and a flash illumination solar simulator for modules. The NEWPORT PV Lab Corporation certified a $0.48\,cm^2$ cell cut from a $60 \times 120\,cm^2$ NEXCIS plate at 17.3% with an open-circuit voltage of 621 mV, short-circuit current density of $37.3\,mA\,cm^{-2}$, and 74.7% fill factor (Figure 4.23).

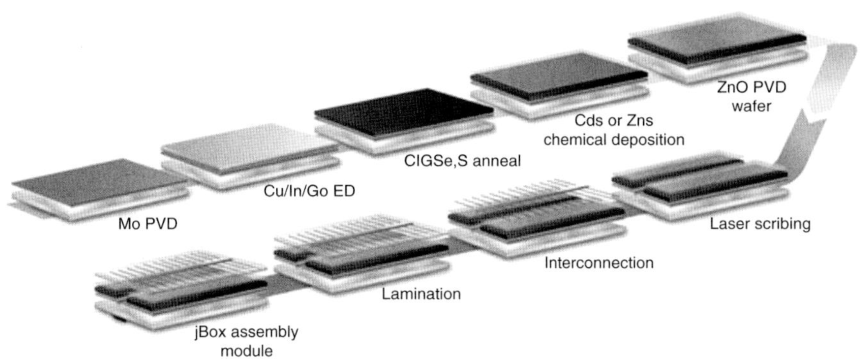

Figure 4.21 General process flow for the synthesis of $60 \times 120\,cm^2$ CIGS modules.

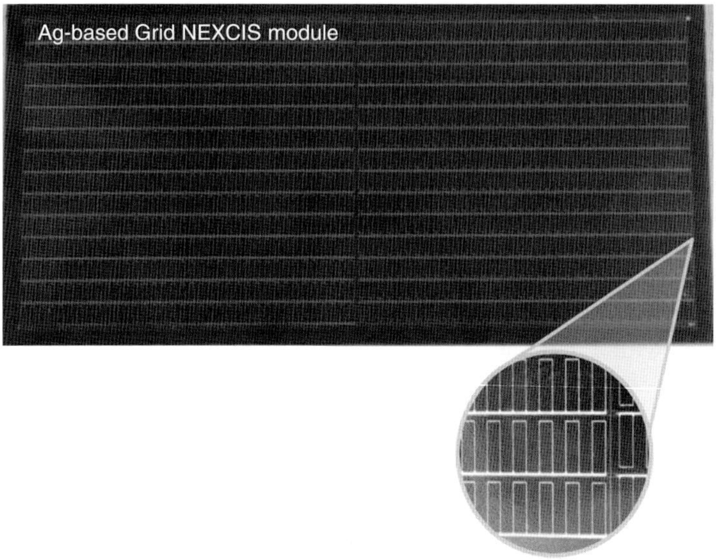

Figure 4.22 60 × 120 cm² CIGS module developed at NEXCIS.

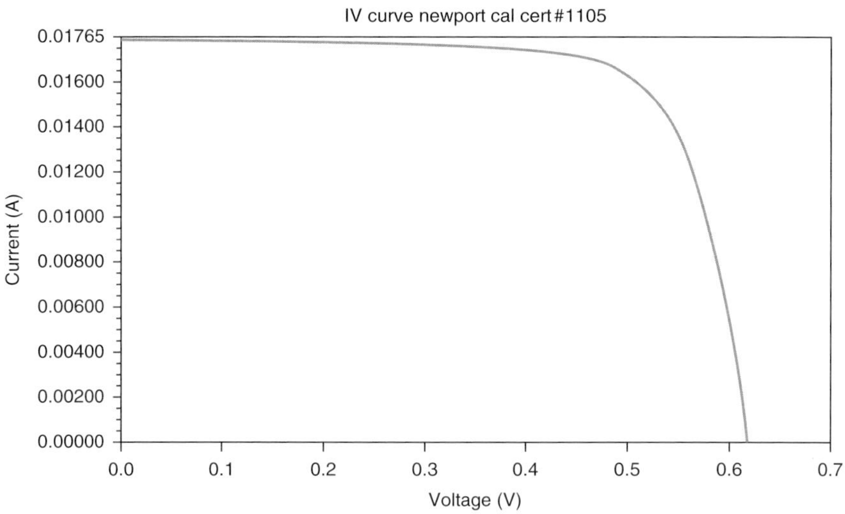

Figure 4.23 Certified IV curve of 60 × 120 cm² NEXCIS plate with 17.3% efficiency.

To the best of our knowledge, the champion cell efficiency obtained from the electrodeposition-based process exceeds all other previously published efficiencies from cells made using non-vacuum processes and demonstrates that electrodeposition-based processes are relevant to achieve high-efficiency cells [73].

Furthermore, the same process also yielded 60 × 120 cm² modules up to 92.8 W as certified by the company CERTISOLIS, which corresponds to a 14.0%

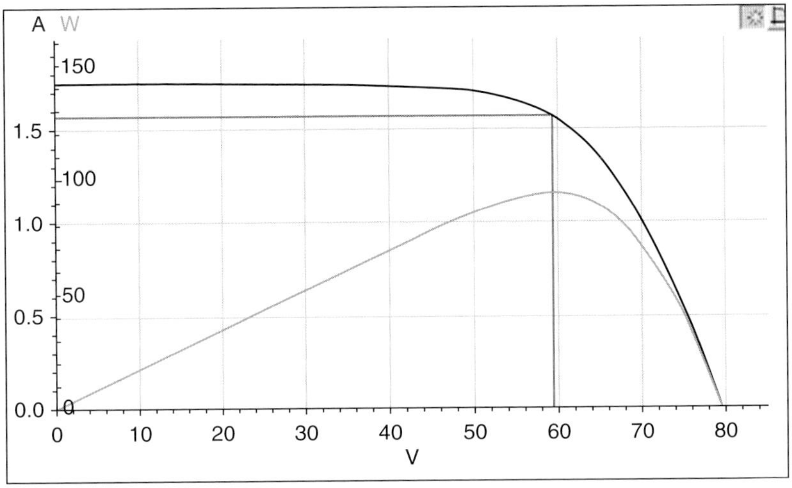

Figure 4.24 Certified IV curve of $60 \times 120\,cm^2$ NEXIS module with 14.0% aperture area efficiency.

aperture area efficiency ($AA = 6610\,cm^2$) as shown in Figure 4.24. This shows that the difference between cell efficiency and module efficiency is low, it is only 3%, while it is usually closer to 5% for other CIGS processes such as co-evaporation. We conclude that the small difference in efficiency is because the champion cell presented here was produced on the same equipment as the $60 \times 120\,cm^2$ modules [73].

4.3 Electrodeposited $Cu_2ZnSn(Se,S)_4$ (CZTS) and Emerging Materials

4.3.1 $Cu_2ZnSn(Se,S)_4$ (CZTS)

The metal stack of copper, tin, and zinc was created by sequential electrodeposition of the layers on $15 \times 15\,cm^2$ molybdenum films sputtered on soda-lime glass substrates. The metals were deposited from commercially available plating solutions at high-speed rates, 180–300 nm min^{-1}, showing great compatibility with industry high-throughput requirements. Copper was electrodeposited from a copper "damascene type" bath containing H_2SO_4, Cl^-, and organic additives. Tin was electrodeposited from a tin electrodeposition chemical bath that contained $SnSO_4$, H_2SO_4, and organic additives. Zinc was electrodeposited from an alkaline zinc electrodeposition solution that contained $Zn(OH)_2$, NaOH, and organic additives.

The copper, tin, and zinc layers were electrodeposited using direct current and current densities of 5, 10, and 20 mA cm^{-2}, respectively. Typical deposition times were very short, 30–90 seconds and were timed to obtain the desired thickness. By varying the thickness, various metal stoichiometries were obtained with Cu-poor and Zn-rich composition range ($0.70 \leq Cu/(Zn + Sn) \leq 0.85$ and

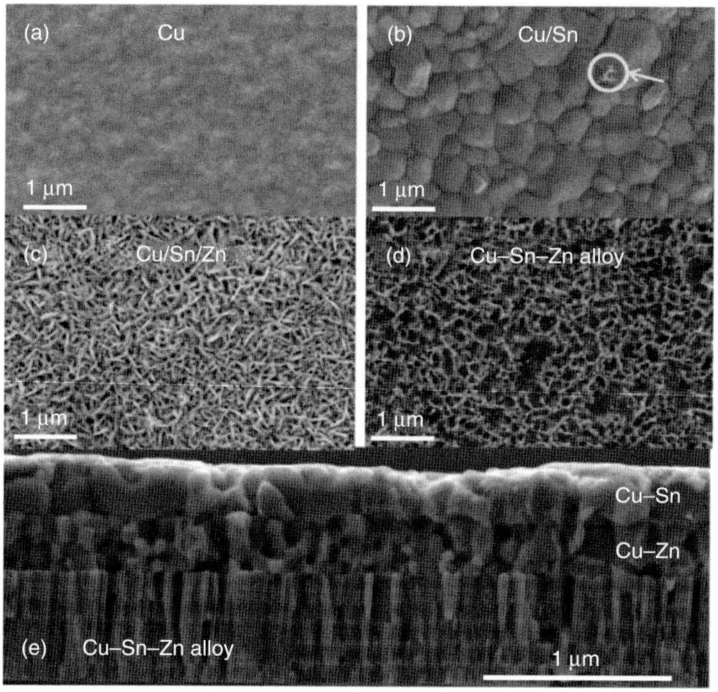

Figure 4.25 Surface scanning microscopy images of electrodeposited layers: (a) Cu layer deposited on SLG/Mo, (b) Sn layer on SLG/Mo/Cu, (c) Zn layer on SLG/Mo/Cu/Sn, (d) surface, and (e) cross section of the full stack precursor after alloying at 200 °C. Source: Vauche et al. 2016 [76]. Reproduced with the permission of John Wiley & Sons.

$1.00 \leq Zn/Sn \leq 1.25$, as measured by X-ray fluorescence (XRF)), for which the best device efficiencies of 7–9% have been reported to date [74–77].

Scanning electron microscopy images of the precursor metal stack surface as shown in Figure 4.25 reveal compact morphologies of the surfaces. The thickness of the complete precursor metallic stacks varies between 0.5 and 0.7 μm. The relative standard deviation (RSD) in thickness of each individual metallic layer was kept below ±6% over $15 \times 15\ cm^2$ substrates in order to obtain good precursor composition uniformity. Copper and zinc layers were deposited with good lateral uniformities over $15 \times 15\ cm^2$ (RSD < 4%), which is desirable in order to avoid deviations in composition. For tin, larger thickness variations were observed (RSD ≈ 5–6%).

The tin layer deposition was reported to be challenging also when using physical vapor deposition. As shown in Figure 4.25b, the tin layer forms small aggregates at the surface, probably because of the oxidation of Sn(II) to Sn(IV) in acidic tin plating baths [76, 78]. As the cause and impact of the aggregates was not fully understood, they were minimized by using a mostly fresh tin electrodeposition bath. In order to improve the uniformity and the inter-mixing of the elements, a low-temperature annealing was applied on the electrodeposited precursor [79]. A review on the challenges of using electrodeposition for kesterite thin film applications can be found elsewhere [80].

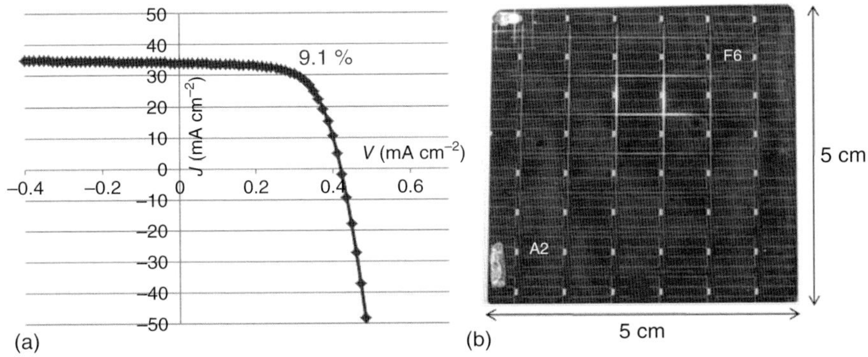

Figure 4.26 The *I*–*V* curve (a) and photographic image (b) of best efficiency cells (9.1%) from NEXCIS precursors with a deposited grid and without antireflective coating.

Tin losses were minimized by improving the uniformity of the absorber and the secondary phases, which were mainly segregated at the surface. If necessary, tin was also added during the annealing process. Using wet surface treatments for CZTSe, the secondary phases of ZnSe and Sn—Se can be removed by using $KMnO_4/H_2SO_4 + Na_2S + (NH_4)_2S$ [81, 82]. Finally, using an optimized CdS buffer layer and TCO with appropriate metal contact, the best CZTSe solar cell has achieved 9.1% power conversion efficiency. This set a new record for kesterite solar cells produced by electrodeposition [77] (Figure 4.26).

It has been observed that band tailing is a major contributing factor to the large open-circuit voltage (V_{oc}) deficit that is currently limiting the PV efficiency of Cu_2ZnSnS_4 (CZTS) PV devices. Spatial fluctuations in CZTS optical properties using low-temperature cathodoluminescence (CL) in a SEM were measured. Principal component analysis revealed three CL peaks whose relative intensity varied across domains of about 100 nm in size. The CL results were consistent with the presence of band tailing in CZTS [83].

4.4 From the Rotating Disk to the Paddle Cell as a Scale-up Platform

4.4.1 Introduction to Scale-up

Most of the electrodeposition reactions are diffusion controlled. This is particularly true for the elements that electroplate at high hydrogen overvoltage and for most of the metal alloys.

The majority of electroplating chemistries contain small amounts of organic additives or complexing agents that reduce the overvoltage and help to control the microstructure of the resulting deposit. Electrode reaction suppressors, accelerators, or inhibitors that act as levelers are added to the solution chemistry in order to achieve suitable deposit microstructures during electroplating. In order to produce uniform electroplating of the nanometer thin layers over a meter large planar surface, it is essential to achieve not only uniform current distribution but also to achieve very uniform and highly reproducible diffusion layer thickness over the entire surface area with uniformity at the same time nano- and macro-meter scales.

Up until 1969, the dominant methods of agitation used were air or nitrogen bubbling through shower heads located at the bottom of the tank with gas bubbles moving past the vertically suspended substrates with a mechanical movement of the part in the back and forth direction or by using recirculation of a large volume of solution and impinging the jet flow through moving nozzles at the surface being electroplated. These methods required considerable amount of energy for compressing and pumping large volumes of gas or for pumping large volumes of plating solution to achieve high-speed liquid impingement against the cathode being plated.

None of these methods, however, resulted in thin enough, uniform enough, and reproducible enough diffusion layer thickness at the cathode. Furthermore, none of these methods were readily scalable from smaller to larger plating tanks and from small to large plated surfaces.

4.4.2 Entirely New Solution Agitation Method

An entirely new concept of solution agitation with high-speed paddle moving close to the cathode resulted in uniform, reproducible, and scalable agitation method. Hence, uniform thickness films with uniform alloy composition could be achieved.

In order to solve this problem, after experimenting with many different conventional plating cells and agitation methods, Romankiw with his coworker has invented the paddle agitation method described first in the US Patent 3,652,442 [84]. This method achieved, for the first time, a very uniform NiFe alloy composition, essential for making possible the fabrication of thin film magnetic heads on 5″ × 5″ substrates.

This method uses a paddle consisting of two inverted triangles, which spans the entire width of the plating cell and travels from end to end of the plating cell at uniform separation of 1–3 mm away from the surface at a speed of at least 20 and preferably 120 $cm\ s^{-1}$ or higher. At such speeds and distances from the cathode, the paddle edge cuts into the diffusion layer. The solution is scooped up by the edge of the paddle, flows through the opening between the two triangles where it is mixed, and is returned back to the surface.

The flat surface of the lower part of the paddle, due to the sheer force, sets up in the diffusion layer eddy currents, which remain active up to the time when the paddle passes in the opposite direction. This maintains a very uniformly thin and highly reproducible diffusion layer at the surface. The frequent periodic reversal prevents a unidirectional growth of grains and nodules and the potential further complications that arise from such growth. Using the paddle as a means of agitation ensures uniform supply of metal ions and of various organic additives and a very uniform, reproducible, and constant removal of reaction by-products. In the case of reaction in which hydrogen is a by-product of the reaction, this method of agitation very quickly and efficiently removes the hydrogen bubbles. This method of agitation also ensures a very uniform hydrogen and potential overvoltage from spot to spot on the cathode.

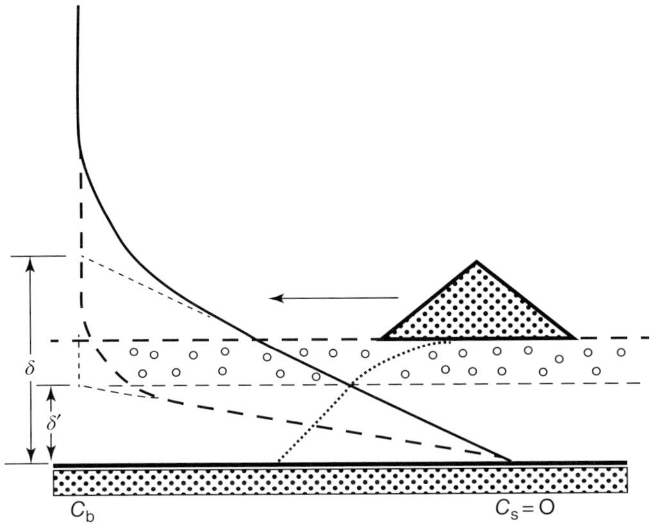

Figure 4.27 Paddle agitation mechanism [84].

Figure 4.27 pictorially shows the described mechanism of paddle agitation. The sketch in Figure 4.27 represents the observation made by video and photos during the die injection studies carried out in a specially built translucent paddle-plating cell [84]. The described mechanism was also confirmed by single-point-limiting current density studies carried out during the operation of the paddle equipped with capability to make single-point measurements.

This method became the standard electroplating equipment for magnetic alloys for thin film magnetic storage heads by the magnetic storage industry. This agitation method was subsequently scaled-up to 200 mm and for 300 mm diameter silicon wafers for the fabrication of solder bumps. Because of the entirely new approach to agitation, which was readily scalable, we have extended it very successfully first to 30 cm × 60 cm batch thin film solar panels and subsequently to 60 cm × 120 cm solar panels in a continuously moving line.

4.4.3 The Paddle Agitation Technique Is More Readily Scalable

Unlike many other techniques of agitation, which are very difficult to impossible to scale from small to large substrates, the paddle agitation method described above follows certain rules that permit to scale the agitation from small to larger dimensions regardless if the wafers are circular or square shaped. Scaling to rectangular shapes becomes more difficult but with good understanding of key concepts, it can be readily achieved.

The plating cell shown in Figure 4.28 described in the US Patent 3,652,442 [84] solves three other very important problems that exist in plating of magnetic thin film heads, variety of electronic devices, and of the solar devices. The first of these problems is elimination of edge effects in plating due to current spreading and crowding at the edges. The second problem is achieving good electrical contact

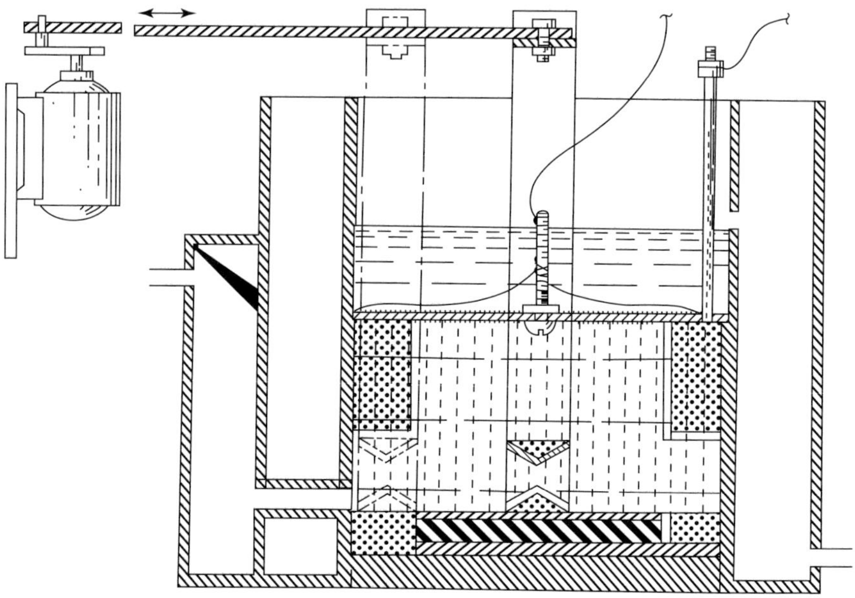

Figure 4.28 Paddle cell as described in the US Patent 3,652,442 [84].

to a thin seed layer on top of the wafer. The third problem is overcoming the resistance drop in the seed layer from the edge to the center of the wafer with the use of a separately powered current thief. The thief diverts some of the anode current to the center of the wafer and thereby compensates for the current drop in the seed layer between the edge and the center of the wafer.

By making the anode the same size as the cathode and surrounding the anode and the cathode by vertical walls, the US Patent 3,652,442 [84] solves most of the problems of the primary current distribution when the substrate is highly conductive.

As in most electronic devices and also in solar devices it is necessary to plate on a very thin seed layers of limited conductivity, the primary current falls off from the edge to the center of the wafer. We have demonstrated that by surrounding the wafer with a secondary cathode (the current thief) that is powered by a separate power supply, it is possible to compensate for the IR drop in the thin seed layers.

In the arrangement shown in Figure 4.29, part of the anode current is diverted toward the center of the glass wafer metalized with the very thin Mo/Cu seed layer discussed in EP 2,469,580 [87]. An excellent quality, unoxidized Mo/Cu seed surface was found to be essential for good uniformity of the plated metal thin films and for good adhesion of the solar cell to the substrate.

For the smaller parts and relatively conductive seed layers, the current thief is sufficient to achieve the desired result. For very large wafers with very resistive seed layers, it is necessary to use, in addition to the current thief, current deflectors. The current deflectors placed between the anode and the cathode, above the paddle, permit additional compensation for the current density distribution. The current deflectors along with the current thief provide more precise

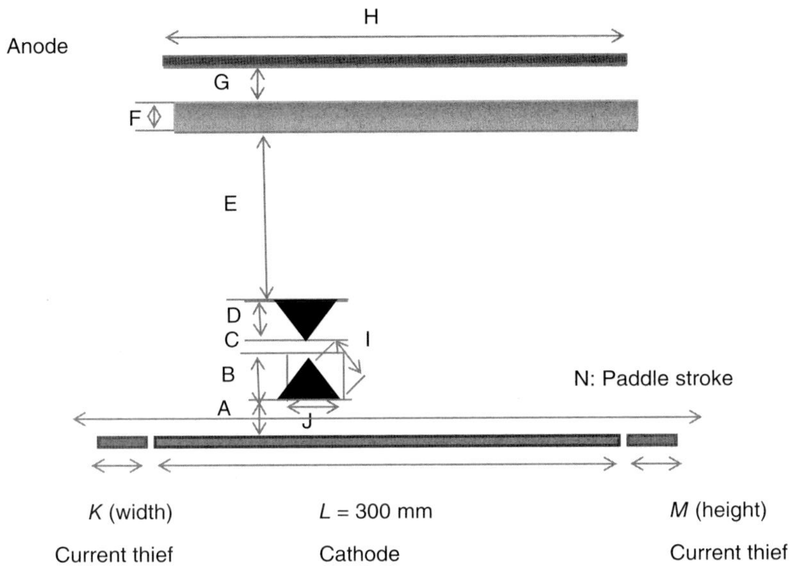

Figure 4.29 Schematic cross section of paddle cell [85, 86].

ability to compensate for the current drop in large samples with very thin seed layers, in which there is an appreciable IR drop from edge to center. The proper combination of current thief and current deflector are particularly very important when dealing with non-symmetrical or rectangular wafers. The proper configuration of the current thief and of the current deflector combination can be achieved by experimental means for each cell configuration and for each thin film metallization thickness on the sample being electroplated.

4.4.4 Electrical Contact Between the Thin Seed Layer and the Source of Current

In Figures 4.28 and 4.29, power to the cathode is supplied by a very thin metal lip overlapping the metalized surface by a couple of millimeters, thus providing contact to the entire periphery of the seed layer. The lip is less than 1 mm thick; therefore, it produces a minimum distortion of the current around the periphery of the sample. As a result, the plated substrates can have only 2–5 mm exclusion layer around the periphery of the sample. More recently, some authors proposed to substitute the lip contact with multiple sharp point contacts. At this point, there is no published comparative study for the two methods on identical substrates in identical plating solutions.

4.4.5 Previous Scale-up of the Paddle Cell

This agitating concept has been successfully scaled up in IBM at least four times from originally $5 \times 5\,\text{cm}^2$ samples to $15 \times 15\,\text{cm}^2$, to 20 cm, and to 30 cm diameter circle. Each time not only scaling the plating cell size and the paddles, using proprietary scaling, but also improving cathode holders, refining agitation method and the current thieving methods to handle much larger highly resistive substrates. Some of the improvements are described in the US Patent 5,312,532 [88] dual cell for plating 12.5 × 12.5 cm square substrates in a horizontal position and for plating 20 and 30 cm diameter wafers in a vertical position as described in the US Patent 5,516,412 [85].

These concepts have been used very successfully in manufacturing of thin film heads, semiconductor thin film package, plating of solder interconnects, and other electronic applications for many years with excellent results. Electronic companies are using paddle agitation method in manufacturing under IBM licensing agreement. At least three tool-manufacturing companies, Semitool acquired by Applied Materials, Ebara, and NEXX Systems, have come up with their own variant of the paddle-plating tool and are selling such tools for manufacturing of electronic devices. Other companies such as Furukawa Electronic Co. Ltd., Headway Technologies, Inc., Hitachi Global Storage, TDK Corp., and Tausetsu KK have received patents of their own version of the paddle-plating equipment for their own internal use.

Based on our experience with the paddle-plating tools, we have selected the paddle-plating tool as the best possible choice for plating on large-area glass/Mo samples to achieve the desired, agitation rate, current distribution and the desired run to run reproducibility, and eventual capability to scale-up for manufacturing large plates at high rate of throughput.

4.4.6 Scale-up of the Paddle Cell to 15 cm × 15 cm

An electrodeposition process for CIGS deposition with the stacked metal approach was developed with the paddle cell platform targeting 400 nm of indium on 300 nm of copper using Heliofab IN-390 chemistry from Enthone.

Low resistance and reliable electrical contact were found to be very critical for plating on 15 cm × 15 cm substrates in paddle cell. A thin film of indium plated using the Heliofab indium chemistry in a paddle cell at 20 mA cm^{-2} with only four pin contacts at the four corners of the substrate was plated. A very small area at the center of the substrate was plated with indium. In addition, not only indium was not plated at the four corners but also the copper layer was dissolved, at the corners. Figure 4.30 shows the impact of the current density on the appearance of the indium film. It is evident that a low current density results in more uniform plating, consistent with the impact of contact resistance and the substrate sheet resistance being very important. However, the indium film plated at lower current density has poor coalescence, namely, a poor microscale uniformity.

The plating performance was significantly improved by using a continuous contact at the four edges of the substrates. This can be achieved by a lip contact, ring-sealed continuous pin contacts, or a conducting contact tape combined with contact pins.

It was found that using a continuous conducting copper tape coupled with 12 pin contacts greatly improved the uniformity as compared with the four contact pins. However, a lip contact was used at NEXCIS. No significant difference

(a) (b) (c)

Figure 4.30 Indium plated in paddle cell with four pin contacts at current densities (a) 20 mA /cm^{-2}, (b) 10 mA cm^{-2}, and (c) 5 mA cm^{-2}.

(a) (b) (c)

Figure 4.31 Contact methods for 15 × 15 cm^2 substrates, (a) sketch and (b) the plated samples with copper tapes and with 12 contact pins and (c) sketch of ring sealed continuous pin contacts.

was observed between the two contact schemes. We also developed a ring-sealed continuous pin contacts (see Figure 4.31).

The within-wafer uniformity was measured by XRF based on 7 × 7, i.e. 49 points of measurements. The thief current was optimized for the best uniformity and a standard deviation below 3% was achieved. The conditions used with a horizontal paddle cell setup were a Pt anode, 20 × 18 cm^2, ~7 cm away from and parallel with the cathode, a solution pH in the range of 10 or 11, room temperature 20–21 °C, paddle agitation of 1 Hz, wafer current density of 20 mA cm^{-2}, and thief current density of 10 mA cm^{-2}.

The total target thickness of copper plus indium was 600–750 nm, and the Cu/In ratio between 1.4 and 1.8. More specifically, the targets were 270 nm Cu, 360 nm In and 1.65 Cu/In ratio. We found out that the thickness of the plated indium film when the indium solution pH was 10.2–10.5 decreased with plating time of 100 seconds or more. Increasing the solution pH to 11 enabled a constant indium plating rate for longer plating time.

Other parameters were also studied such as oxygen level in the solution, anode, thief, rinsing steps, pH of the rinsing/dipping water, and idle time in indium solution. Among them, pre-dipping of the plated Cu layer in a solution without metals and with the pH adjusted to the pH of the indium solution seemed to show some improvement. However, the increase in the pH to 11 greatly improved the reproducibility and the pre-dipping seemed to be unnecessary.

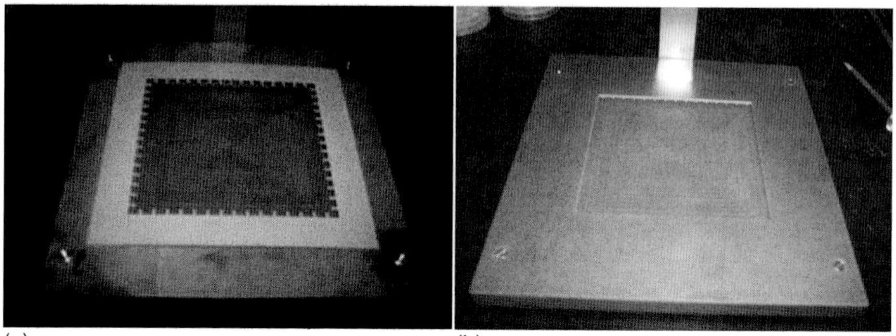

(a) (b)

Figure 4.32 (a) Teflon-coated continuous pin contact where electrical connection to the wafer is made from the back side. (b) A cathode holder after assembly.

Among them, the thief current, or the thief–cathode current density ratio needs to be optimized according to the plating setup. This ratio is strongly dependent on the area of the cathode, area and geometry of the thief, and the geometry of the plating cell. It also depends on the solution properties including concentration, conductivity, and pH. Therefore, in a continuous operation of the plating process, this thief–cathode current ratio may need to be adjusted to compensate for solution changes such as bath aging.

We demonstrated improvements in reproducibility of the resulting indium-plated thickness at solution pH 11. An improvement of the thickness σ was from 100 to 30 nm and the σ of Cu/In improved from 0.4 to 0.2. As a result, the 15 cm × 15 cm processed glass plates at pH 11 were within the target process window and had a tight overall thickness distribution. These adjustments were necessary for a high-throughput or even a continuous manufacturing process.

In addition, a paddle cell study was carried out for gallium electroplating on glass/Mo/Cu/In substrates. The paddle cell was used to scale-up the gallium electroplating to a 15 cm × 15 cm size substrate. A cathode holder that was designed with a continuous contact scheme (Figure 4.32) was used in this gallium plating study. The parameters monitored and maintained during the plating and post plating processes are room temperature at which plating was conducted, which varied from 20 to 22 °C, plating bath temperature that was varied: 19–21 °C, rinse water temperature: 19–21 °C, pH of the electrolyte was ~13, clean room temperature, when soft annealing was conducted at 18–20 °C, cathode to thief current density ratio was 1.76, paddle cell agitation was 1 Hz, and plating rate was 8.33 nm s^{-1}.

Figure 4.33 shows the microstructure of a 300 nm gallium film plated in a paddle cell using an alkaline basic IBM gallium chemistry at room temperature by applying 30 mA cm^{-2} for 36 seconds at 1 Hz paddle speed. The electroplated gallium thin films show a very uniform grain size ranging from 800 to 1000 nm. The film was quite smooth. It can be seen that the grain size of the as plated gallium is on the order of 0.8–1.0 μm. Electroplated 300 nm gallium on 15 cm × 15 cm glass/Mo/Cu/In substrates where gallium was plated had typical thickness uniformity of 1.56%.

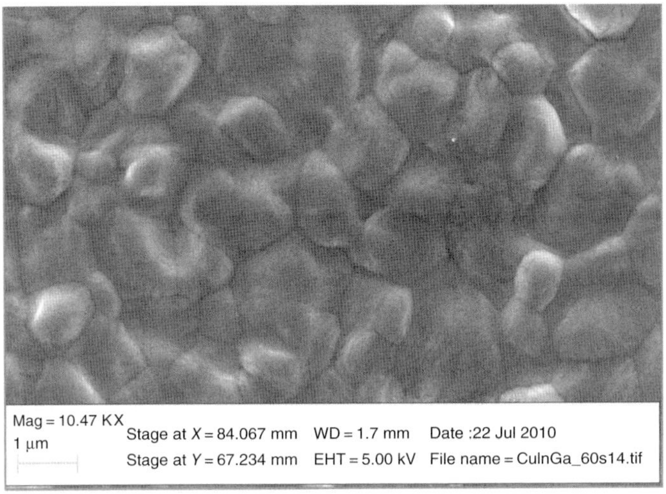

Figure 4.33 The microstructure of a 300 nm gallium film plated in a paddle cell using an alkaline gallium chemistry at room temperature and using 30 mA cm^{-2} current density for 36 seconds and 1 Hz paddle speed.

4.4.7 Scale-up of the Paddle Cell to 30 cm × 60 cm

The NEXCIS team with support from IBM started the 30 cm × 60 cm plating line and produced the first samples with copper and indium, which met the desired thickness uniformity. A 15 cm × 15 cm section, from the center of the 30 cm × 60 cm sample, was made into a solar cell device. The device yielded greater than 8% solar efficiency with a very narrow spread of values.

In order to improve the uniformity of the 30 cm × 60 cm solar cells, various aspects of electrochemistry and electrochemical engineering were implemented such as current deflectors and the experimental approach to using the combination of the current deflectors together with current thieves to achieve the most uniform thickness and the best reproducibility of parts in the 30 cm × 60 cm line. We have discussed the importance of the high solution pH (pH of 11) for the indium, and we installed a high pH solution pre-dip of parts after copper plating and before plating in the Enthone indium plating bath. In Figure 4.34, we show the thickness uniformity results from the 30 cm × 60 cm electroplated plates.

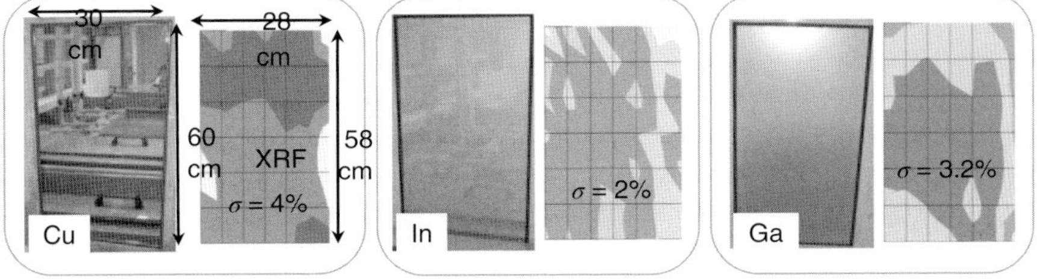

Figure 4.34 Thickness uniformity achieved on 30 cm × 60 cm after the optimization of the paddle cell parameters.

4.4.8 Improving Within-Wafer Uniformity, Reproducibility, and Demonstration of Scalability

The process flow and tools necessary to pre-clean the parts, plate copper, performing the rinsing and drying operation, plating of the indium and gallium, were designed. For the indium and gallium electroplating, the cathode was moved past an air knife to minimize indium and gallium solution drag out, and a cascade rinsing and drying were utilized.

4.4.8.1 Within-Wafer Uniformity

The plates were stored in an inert nitrogen cabinet before plating. The humidity of the storage cabinet was verified by an internal probe to be 0–3% of H_2O. During the usual storage time, 20 cycles of opening/closing of the cabinet were allowed. During such opening and closing, the relative humidity changed between 35% and 45% for several minutes. Copper plating was carried out at current density of 10 mA cm^{-2}, thief current density of 23 mA cm^{-2}, and a current ratio of 0.43. Indium was plated at a current density of 20 mA cm^{-2} and the same thief current density was used.

When using very large substrates, the within-wafer standard deviation depended mainly on the seed layer thickness uniformity and reproducibility over the entire wafer, which, in turn, affected the IR drop. It also depended on the seed layer surface condition such as surface roughness, pin holes, and presence of oxides or other adsorbed contaminants because of prolonged storage. The copper oxide forms very quickly in the presence of humidity in the air. The water tends to condense on the hydrophilic copper surface within minutes after removal from the sputtering system. Roughness of Mo and porosity of very thin hydrophobic Cu seed layer tend to enhance the water condensation. Condensation on rougher or more porous areas of surfaces of copper may take place at even lower relative humidity than 40% at room temperature. If the micropores in Cu reach all the way down to the Mo surface, local electrochemical cells form, which result in local corrosion micro pits. The corrosion cells manifest as white corrosion spots. The micro pits further aggravate the problem. In a PV device, such micro pits tend to enlarge during prolonged storage. This, in turn, may result in lowering the solar absorber efficiency. Therefore, it is essential to minimize the storage time of the copper seed layer sputtered wafers to an absolute minimum. Storage should be less than one to two hours in low-pressure nitrogen or in-house vacuum.

Most of the copper oxide that is formed during storage time is subsequently removed in the copper plating solution with a 10 or 20 seconds dwell time in the acidic copper plating solution before the current is switched on. As the copper seed layer is very thin, the dwell time needs to be extremely tightly controlled because while removing the oxide, copper is also removed. The copper seed layer was only 80 nm thick; therefore, there is a possibility that in some spots, particularly where there is thicker copper oxide, more copper is removed than in other areas. This could result in a local variation of the copper seed layer thickness and hence greatly affect the electroplated thickness uniformity. It is highly recommended that experiments be carried out to investigate the length of storage of PV parts. Otherwise, it is recommended that the storage time is kept to a maximum of two hours and that the storage time and the dwell time in the plating solution be tightly controlled before copper plating.

In addition to tight control of the storage time and the dwell time, one can additionally remedy the situation by increasing the thickness of the copper seed layer from the present 80–120 nm. This would partly alleviate the problem because with thicker seed layer, the IR drop per unit length would be smaller. The copper plating thickness distribution would become more forgiving.

It is important to note that this particular problem is very specific to the batch plating processes on very thin very reactive seed layers. In a continuous flow process operation, if the plating equipment and the process are properly designed, the problem will be greatly minimized.

From the data, it was concluded that the within-wafer uniformity is better for indium plating than for the copper plating. This underscores the fact that the IR drop here also plays an important role. With 320 nm of electroplated copper on top of sputtered Mo/Cu and 325 nm of plated copper, the IR drop of the Mo/Cu/Cu stack is almost an order of magnitude smaller; hence, the uniformity of current distribution is easier to achieve and indium plating, therefore, shows smaller standard deviation than copper plating.

Optimum within-wafer uniformity can be achieved either by mathematical modeling or by experimental means using experiments with the current thief and the current deflectors. It is important to note that the experiments with the current deflectors and the thief had to be repeated every time that the seed layer thickness has been changed. The plated indium film thickness depends on the thickness and the resistivity of the Mo/Cu film on which it is plated. Large thickness variation in the plated copper either within the wafer or from wafer to wafer will affect the thickness distribution map of indium even if the deflector shape and the wafer to thief current are kept the same.

4.4.8.2 Wafer-to-Wafer Reproducibility

The wafer-to-wafer reproducibility in a batch process depends mainly on having a reproducible surface condition, pretreatment of the surface before entering the plating solution, the pH of the rinsing water post plating, and the pH of the plating solution. Indium and gallium plating solutions from Enthone were used at pH between 10.5 and 11.5 and rinsing water pH of 6.

In the case of copper plating on a sputtered copper seed layer, the importance of quality of the copper surface, the storage time, storage condition, and other issues have been discussed before and will not be repeated here again. Sufficient to say that the thinner the copper seed layer, the more readily it oxidizes metal, and the larger the surface area, the more difficult it becomes to achieve a very small standard deviation.

For indium plating, there was a choice of either entering dry into the indium bath or skip the drying operation and go into the bath wet after the water rinsing operation. As the rinsing water is at pH 6 and the indium plating bath operates in the range between pH 10.5 and 11.5, even the smallest amount of the rinse water remaining on the surface will be carried over and will dilute the plating bath.

The dilution can make the plating bath composition drift out of the preferred plating range. Dipping wet parts in the plating solution will not only change the pH but will also take out of the optimum plating range all the other solution components. The dilution of the bulk of the plating bath is a slow process and will show up only over extended periods of time.

More important for wafer-to-wafer reproducibility is how to increase the initially low surface pH, increase the concentration of indium ions, and increase the levels of the complexing agents and the bath additives at the surface when entering the plating bath. To overcome this, we kept the cathode immersed in the plating bath for a predetermined time (dwell time) while agitating without the current on. The experiments showed that for the Enthone alkaline indium plating bath, the dwell time should be between 10 and 20 seconds. During this time, not only the water was replaced with the alkaline indium plating bath but also some of the copper was dissolved into the bath. The plating bath gradually acquired a bluish color. The mechanism of the exchange of the water with the alkaline plating solution cannot be explained without an in-depth scientific investigation. It was, however, possible to control the electroplating wafer-to-wafer reproducibility by tightly controlling every step between the rinsing operation and the switching on of the current when the part was already in the plating bath.

To partly mitigate the problem, a pre-dip of the plated copper in a water solution with pH adjusted to 10.5 or 11.0 (the pH at which the indium or gallium plating bath was operated). Ideally, the pre-dip should be made in an identical composition electrolyte to the one used for plating of the indium but without the indium ions in it. We used a sodium hydroxide solution at pH of 10.5 as a pre-dip in the experiments on 15 cm × 15 cm, and this proved to be effective improving the wafer-to-wafer plating reproducibility for the 30 cm × 60 cm.

Initiating the plating process without the pre-dip or without the established dwell time in the bath before switching the current on could result in a delayed nucleation, different nucleation density, and hence different and irreproducible film thickness and structure of the deposit.

4.5 Scaling-up to 60 cm × 120 cm from Tiny Electrodes to Meters

4.5.1 A 1 m^2 min^{-1} Continuous Industrial Scale

To assess the large-scale electrodeposition process, compatible with mass production, a prototype able to process sequentially copper, indium, and gallium at 60 cm × 120 cm scale was built at NEXCIS in partnership with an equipment supplier (Figure 4.35). Attention was paid to the key parts of the tool (conveyor belt, plating cells, rinse, and drying sections) to validate the proof of concept of the machine.

For substrate sizes greater than 30 cm × 60 cm, it was determined that the electrodeposition uniformity and roughness could be enhanced using a plurality of stirrers (paddles) according to the invention described in the US Patent 9,647,151 B2 [86].

Figure 4.35 General and detailed view of the NEXCIS 60 × 120 cm² plating line.

Figure 4.36 Top-down view of an electrolytic cell (CU) with a plurality of paddles AG1, …, AGN [86, 89].

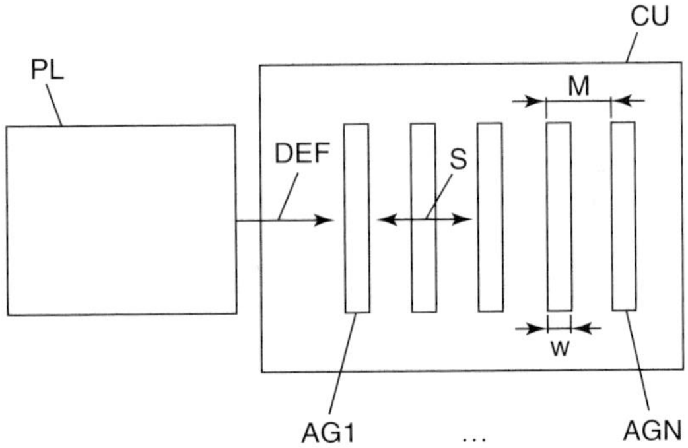

Tests were carried out for the electrodeposition of indium on copper. Plates with dimensions 60 cm × 120 cm were placed on a conveyor running at a speed between 0.5 and 2 m min^{-1}, for example in the region of 1 m min^{-1}, in an electrolytic bath (or "tank"), with the deposition side facing downward. As represented in Figure 4.36, the 120 cm of the plate PL is counted in the direction of the arrow DEF in the tank CU. The length of the electrolytic tank was 90 cm and the tank had several paddles AG1, …, AGN. Each paddle moves back-and-forth with an amplitude *S*, perpendicular to the major axis, and parallel to the direction of movement DEF of the plate. In particular, the back-and-forth movement of the paddles is in phase, having the same speed and the same amplitude.

It was deemed necessary to optimize the stirring parameters as the topographic (surface) and local thickness uniformity is closely dependent on these parameters. For example, marks, parallel to the major axes of the stirrers, may appear on the depositions. These marks are associated with the formation of local stationary turbulences, induced by each stirrer. An eddy generated by each stirrer interacts with the eddy generated by the other stirrers and may give rise to screening phenomena having an impact on the local thickness of an interaction "layer" between the material to be deposited in ionic form (for example In^{3+}) and particularly the electrons applied to promote electrolysis. This layer is referred to as the "diffusion barrier layer." It is further necessary to define a parameter for characterizing electrolytic efficiency, referred to as the "faradaic efficiency." The faradaic efficiency is the ratio of the amount of material that it is electrodeposited and the amount of expected electrodeposited material based on the charge corresponding to the current flowing between the electrodes during electrolysis.

Variations in the diffusion layer have an impact on the local topography of the electrodeposited layer, and consequently on the electrical properties. In order to reduce the variability in the diffusion layer and ascertain uniform transport of species to the cathode, it is necessary to select the paddle movement parameters carefully, for example, the movement amplitude S, the paddle speed, the width W of each paddle (measured, for example, at the base of a triangle), and the distance M between two successive paddles (center to center).

For example, the distance M was between 100 and 200 mm. Each stirrer displayed the double triangle geometry shown in Figure 4.29. The distance g between the paddles and the deposition electrode was chosen to be 2 mm. The movement amplitude S of each stirrer was greater than the distance M between two successive stirrers less one stirrer thickness W and less than twice the distance M between the stirrers. It was observed that if the amplitude was less than this range, the faradaic efficiency for the deposition of indium dropped and the appearance of the plate became "milky."

For a plate with dimensions of 60 cm × 120 cm, the results were obtained with eight or with four stirrers. The distance between two successive paddles was 100 mm when eight stirrers were used. The sweeping speed is $0.16\,\text{m}\,\text{s}^{-1}$. The speed of each paddle was greater than one movement amplitude S, per second, and preferentially less than or equal to $0.2\,\text{m}\,\text{s}^{-1}$.

With the use of four paddles, for a 60 cm × 120 cm plate, it was possible to obtain a coefficient of variation of the uniformity of indium thickness of less than 3% as measured by XRF. Thickness measurements were done with XRF at 140 measurement points uniformly distributed across the entire plate surface.

The faradaic efficiency for the deposition of indium obtained with four or eight stirrers compared to using a single stirrer was greater than 80%. When the faradaic efficiency for indium deposition was less than 80%, the appearance of the plate was milky, indicating poor removal of hydrogen bubbles that formed during electrolysis. The hydrogen evolution reaction competes with the indium deposition reaction and it is necessary to remove the hydrogen bubbles during electrodeposition of indium. Without effective stirring, this hydrogen is not removed from the surface, reducing the fraction of the electrode surface accessible for indium deposition.

Multiplying the number of paddles, Nb AG thus gives rise to an increase in the faradaic efficiency for the deposition of indium by 30% when four stirrers are used. Further increasing the number of stirrers per unit of tank length, a significant gain in lateral thickness uniformity may be obtained. The results achieved for indium were obtained with parameters in the same ranges for gallium and very similar parameters for copper.

The main challenge was controlling the lateral uniformity of the stack during the electrodeposition process. As described in other sections, with particular attention to bath control, flow uniformity, and electrical contacts. XRF on a SMX system from the Solar Metrology Company with 200 points of measurement all over each stack surface was used to measure the CIG uniformity (Figure 4.37).

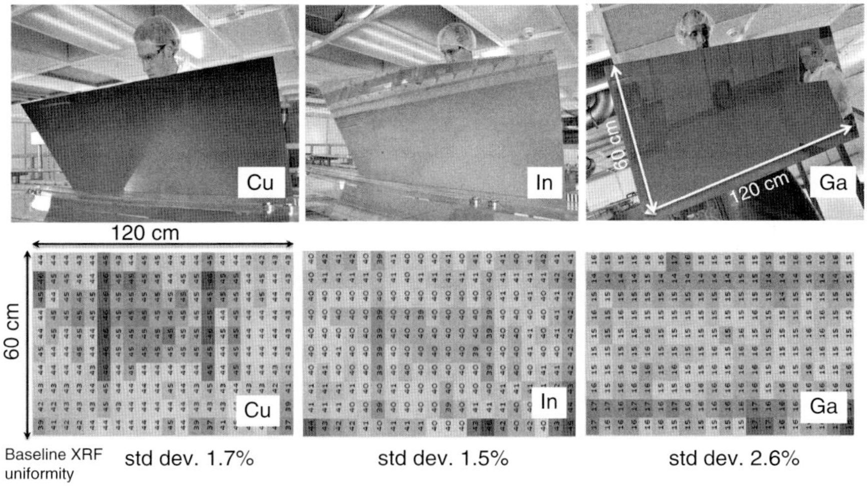

Figure 4.37 View of the Cu, In, and Ga electrodeposited layers and typical XRF mapping for each element.

The main parameters known to affect the CIGS properties are the Cu/(Ga + In) ratio (CGI), the Ga/(Ga + In) ratio (GGI), the chalcogen content, and more specifically the Se/(Cu + In + Ga) ratio (Se/M). Those three parameters were measured by energy dispersive XRF. The standard sampling consisted of 64 measurement points per $60 \times 120\ \text{cm}^2$ plate.

To assess the electrodeposition process robustness and capability, a statistical process analysis has first been performed on 524 plates produced from NEXCIS pre-industrial electrodeposition line, focusing on the parameters most relevant to prove the high reproducibility of the electrodeposition production step, namely the CGI, GGI, and thickness of the metallic precursor. The CGI, GGI ratios, and precursor thickness shown in Figure 4.37 were measured by XRF after electrodeposition on an optimized back-contact and before selenization.

The C_p and C_{pk} values calculated for all three indicators CGI, GGI, and thickness in Figures 4.38a–c are all above 1, meaning that the process is tightly controlled ($C_p > 1$) and well centered ($C_{pk} > 1$). This is the characteristic of a controllable process, which is mandatory for industrialization of CIGS modules. The SPC values were calculated based on a range of ±0.04 for CGI, ±0.015 for GGI, and ±35 nm for the precursor thickness. The ranges were chosen by producing plates over the full composition span and selecting the specifications so that the efficiency of cells made with composition at the boundary did not drop significantly in comparison with the results of plates in the middle of the specifie d range.

Out of the 524 electrodeposited plates, 100 were annealed under chalcogen atmosphere to produce CIGSSe absorber, and the histogram of Figure 4.39 proves that the Se/(Cu + In + Ga) ratio measured by XRF on all plates was kept very well in the specification range of ±0.05. The sulfur content could not be routinely and acceptably measured with the available tools, but new measurement methodologies are being developed to include this element in the future.

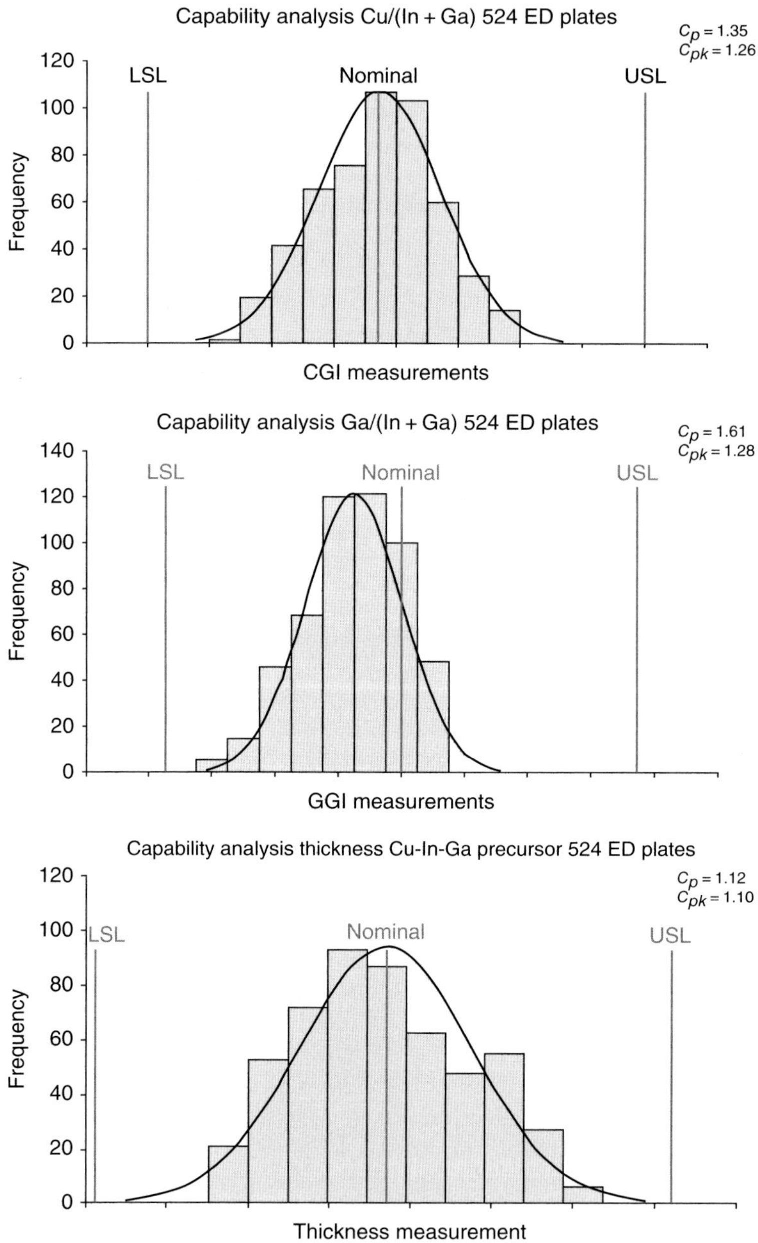

Figure 4.38 Capability analysis of (a) Cu/(In + Ga), (b) Ga/(In + Ga), and (c) Cu—In—Ga precursor total thickness after electrodeposition.

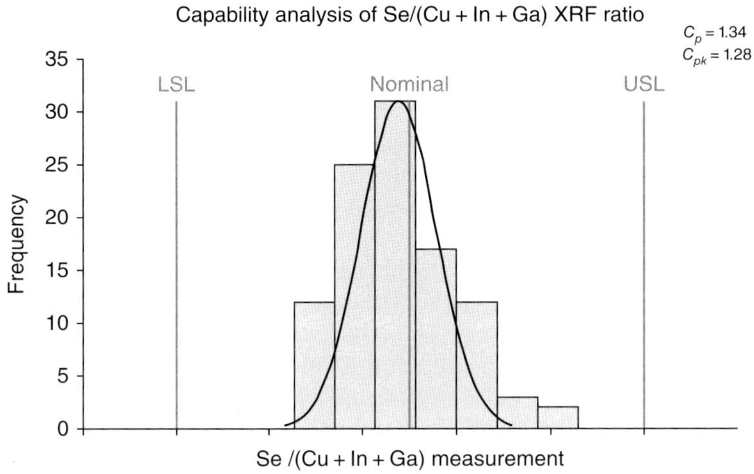

Figure 4.39 Capability analysis of the Se/(Cu + In + Ga) XRF ratio measured on CIGSSe absorbers after annealing.

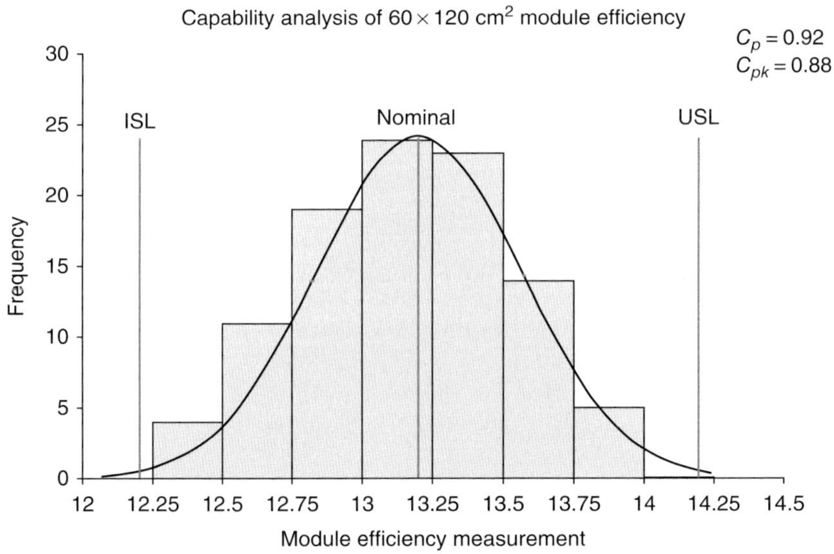

Figure 4.40 Capability analysis of the 60 × 120 cm² modules efficiency.

To complete the SPC analysis, the efficiency of one hundred 60 × 120 cm² standard modules was measured and the histogram of the efficiency (Figure 4.40) shows a broad distribution reaching both the upper and lower specification level. The C_p and C_{pk} values obtained for these module efficiencies with a range of ±1% were 0.92 and 0.88, respectively, very close but still slightly lower than unity, which indicates that considering the whole 100 plates, the process must be further optimized.

One obvious reason for this is that the module efficiency does not only depend on the precursor formation and selenization but also on other process steps such as CdS or TCO deposition, as well as patterning and cell interconnection for which further improvements can be implemented.

To further assess the process capability, the SPC analysis was restricted only to the 200 latest plating runs of the 524 plating runs produced within this study. In this case, due to continuous improvement of the control and maintenance of the electrodeposited process, C_p and C_{pk} values for the GGI and precursor thickness were calculated above 2, which confirms the capability of the electrodeposition process (C_p(GGI) = 2.43; C_{pk} (GGI) = 2.37; C_p (thickness) = 2.53; C_{pk} (thickness) = 2.37).

The capability values above 1.67 mean that the process is very well controlled inside its current specifications. Such results could even allow the specifications' range to be tightened, provided it is valuable for the product, for instance, by improving the modules' output power or by reducing the efficiency distribution scattering.

Out of the 100 plates considered for the efficiency histogram of modules (Figure 4.40), 43 were part of the 200 most recently electrodeposited plates. Forty-three pieces is a small statistical sample to conduct a proper statistical analysis but the average of those 43 modules reached 13.25% efficiency with a standard deviation of only 0.26%. The capability values calculated from those 43 modules gave a $C_p = 1.26$ and $C_{pk} = 1.19$ considerably better than the values obtained from the complete set of 100 modules. This confirms that the overall module electrodeposition-based manufacturing process has been further improved by these tightened and well-controlled electrodeposition requirements.

Higher C_p and C_{pk} values above 1.33 and up to 2 are often needed for established industrial processes producing very large volumes but, taking into account that the current $60 \times 120\,\text{cm}^2$ process has been optimized only from a few hundred plates, the Cp/Cpk values above 1 for modules efficiencies are already noteworthy results.

These results demonstrate that the electrodeposition-based CIGS formation process could be well suited to produce efficient commercial size 60 cm × 120 cm modules.

4.5.2 Bath Control

From a general point of view, electrodeposition solutions used in industrial conditions should ideally have long-term stability (when in use), can be easily replenished, have high resistance to reaction by-products and/or contaminants, allow the use of a high deposition rates >200 nm min^{-1} with faradic efficiency >90%, operate at low temperature < 50 °C, are re-used and recycled at the end of useful life, are of low cost, and most importantly have all bath non-toxic, non-carcinogenic, non-mutagenic bath components.

For PV applications, the plated film needs to be conformal and adherent to the substrate. Depending on the frequency of use (number of square meter plated per day), bath components consumption (drag out and deposited into the substrate),

Table 4.1 List of parameters to be monitored in a copper plating chemistry for thin film PV applications.

	Parameter	Method	Online (yes/no)	Offline frequency (daily/weekly)	Automatic dosing
Copper bath	Density	Density meter	No	D	
	Copper	AAS/ICP/titration	No	D	X
	Sulfuric acid	Titration	No	D	X
	Chloride	Titration	No	D	X
	Brightener, leveler, suppressor	CVS/HPLC	No	D	X
	Metallic impurities (Fe, Cr, …)	AAS/ICP	No	W	—
	Solution temperature (°C)	Thermometer	Yes	—	—
Copper plating	Solution flow ($l\,min^{-1}$)	Flow meter	Yes	—	—
	Wiper frequency ($cycle\,min^{-1}$)	Cycle meter	Yes	—	—
	Wiper amplitude (cm)	Odometer	Yes	—	—
	Current (A)	Amp meter	Yes	—	—
	Voltage (V)	Volt meter	Yes	—	—

and plating parameters, in-line or off-line bath monitoring was chosen. The list of parameters to be monitored depends on the bath composition and the specifications of the target application. As an example, the parameters in Table 4.1 can be monitored for copper plating for PV thin film applications.

Each bath needs to be carefully monitored to maintain a steady state for plating due to metal consumption on the substrate, drag out, evaporation, degradation of components such as organic additives, which can be oxidized or reduced, respectively, at the cathode or the anode.

Two main approaches are generally used for bath maintenance:

- *Bleed and Feed*: periodically, a part of the used bath is removed ("bleeded") and replaced by fresh electrolyte ("feeded"). This principle is often used in microelectronic industry for copper plating of on-chip interconnects. The bath life in this case is theoretically infinite. It can be a costly approach depending on the price of the bath.
- Replenishment of the consumed species based on real-time bath analysis. In this case, due to accumulation of reaction by-products or contaminants in the bath, periodic treatment with selective electrolysis, filtration, or treatment with active carbon may be appropriate.

A method to monitor the accelerator SPS and its breakdown components by an HPLC electrochemical detector have been developed. The method is described

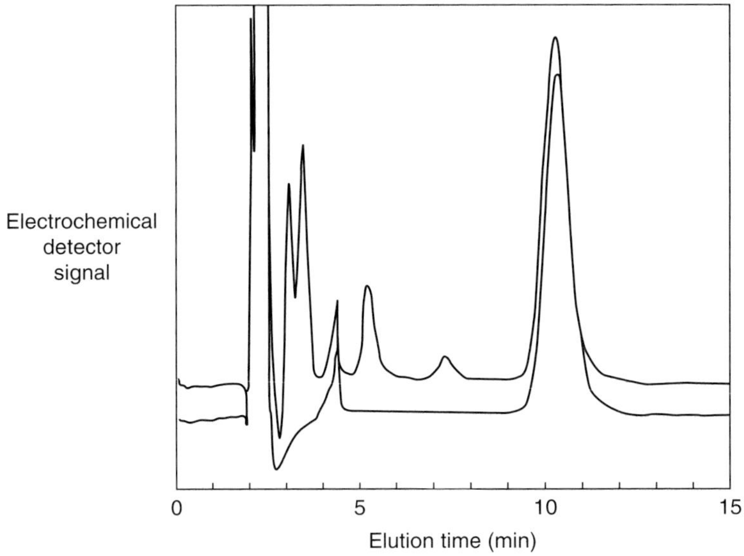

Figure 4.41 HPLC chromatograph showing a fresh bath with SPS only (peak at 10.5 minutes) and an aged bath with SPS, first (peak at 3 and 3.5 minutes), second (peak at 5.5 minutes), and third (peak at 7.5 minutes) oxidation by-products [25].

in US Patent 7,678,258 B2 [25] and it is used to measure SPS and the first and second oxidation by-products of SPS as shown in Figure 4.41.

The choice of anode (soluble or insoluble) depends strongly on the application.

4.5.2.1 Insoluble Anode

An insoluble anode should have good conductivity and high chemical and electrochemical stability. Anodic gas evolution reaction or oxidation of bath components can occur, leading to pH changes in the bulk electrolyte. In this case, the replenishment of electrodeposited metal must be done using metallic salts (such as sulfate and nitrate) or oxide.

4.5.2.2 Soluble Anode

The following parameters were considered:

- Uniform dissolution of the anode
- The faradaic efficiency of the anode and the cathode should be as close as possible.
- Anode should be of high purity in order to avoid contamination of soluble impurities.

A study of dwell time in indium solution was carried out to investigate the effect of the possible dissolution of copper into the indium chemistry. Indium electroplating solutions contain a chelating agent to improve the tolerance to metallic contamination. The chelating agent ties the metallic ions and brings the equilibrium potentials to more negative values to avoid the plating of the metallic species. The stability constant of the Cu^{2+} complex is estimated to be high. In other words, the dissolution of Cu into Cu^{2+} is expected to be enhanced by the chelating agent(s). More importantly, transfer of the cathode structure (frame and the substrate) inevitably brings small amount of copper solution into the indium plating solution. These effects were manifested by a gradual color change of the indium solution toward blue, due to the formation of Cu^{2+} complex. Copper concentration in the indium chemistry can be monitored by UV–Vis spectrum. A calibration of the Cu^{2+} concentration in the indium chemistry will be needed to perform the quantitative analysis. Another approach for the analysis of Cu^{2+} and other metallic contamination is using ICP.

The effect of copper ion in the indium solution was studied and Figure 4.42 shows the cyclic voltammogram of the indium solution with different amounts of copper added. No significant changes were observed until 25 mM Cu^{2+} was used. A current plateau was observed for 25 mM Cu^{2+}, corresponding to the Cu deposition. The plating of copper starts at a potential very close to indium plating, indicating that the complexing agent used in the indium plating solution is also a strong complexing agent for Cu^{2+}.

When the concentration of foreign metallic species is high enough to disrupt the indium plating process, measurements will be needed. Suggestions are shown in the flow diagram in Figure 4.43.

4.5.2.3 Bath Maintenance and Reproducibility and Steady-State Operation

Although many plating operations only analyzed for the most critical components in the plating solution and periodically replenished with the metal salts and the organic additives, this is not the best method of operation to achieve good reproducibility and re-use of the plating chemistry. Using such approach,

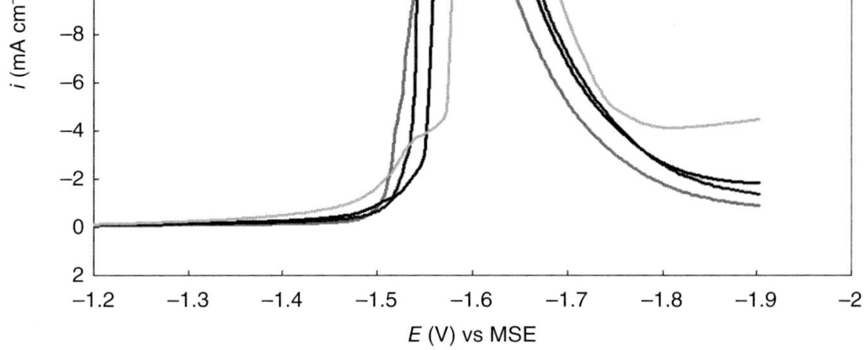

Figure 4.42 Voltammogram of Heliofab IN-390 chemistry with the addition of different amounts of Cu^{2+}.

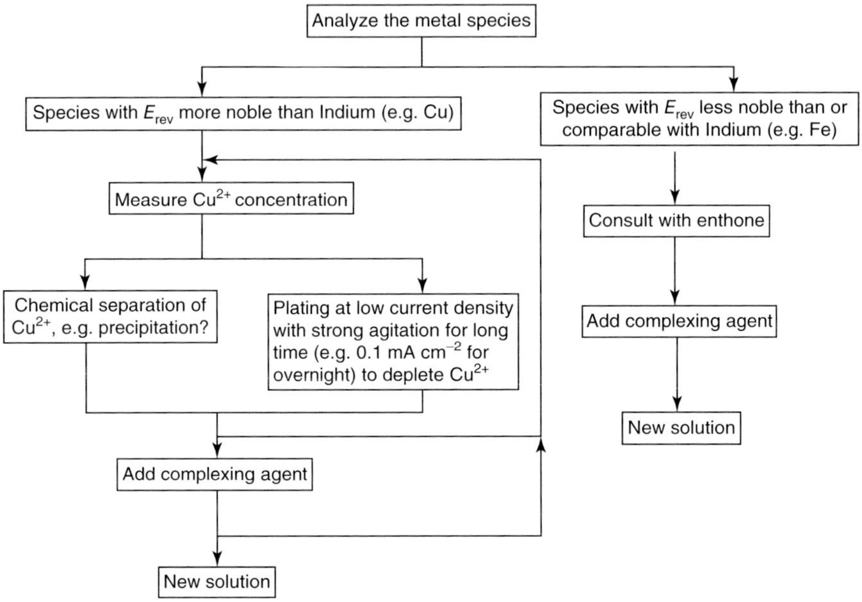

Figure 4.43 Process flow for analysis, monitoring, and control of the metallic contamination in the Enthone indium chemistry.

the anions added with the metal salts and the decomposition products of various additives used in modifying the plating operation keep building up with time. This results in the build-up of these species in the plating bath, change in the specific gravity, viscosity, film nucleation and structure, and most importantly the drop in the efficiency of the plating bath and of the film thickness.

A breakthrough was made by IBM in 1986 when a steady-state operation of the plating bath was introduced in the NiFe plating bath, which obviated the rapid aging of the bath. The process named as "Bleed and Feed operation" has demonstrated that it is possible to use only small volume of the plating bath and to continue using it almost for an infinite length of time. The approach is now used not only in the plating of the magnetic materials for the magnetic head applications but also for the maintenance of the copper, gold, tin, lead, and many other metals and metal alloys in the electronic industry. The process is described in the US Patent 5,352,350 [89].

The thickness of the indium films with 34 g l^{-1} of indium in the bath was approximately 500 nm. As the concentration of the indium in the bath was gradually depleted, despite the fact that the additives were kept constant, the thickness of the film gradually decreased down to approximately 425 nm when the metal ion concentration fell to approximately 29 g l^{-1}.

As the metal ion concentration in the plating bath was increased to between 30 and 33 g l^{-1} range, the film thickness seemed to have recovered to an average of 450–475 nm thickness range. As the concentration of the indium ion was varied in subsequent experiments, the indium film thickness seemed to have changed approximately proportionately to the concentration of the indium ion concentration on the plating bath. Based on the data that were collected, one may conclude that it should be possible to maintain the film thickness in the desired thickness range by keeping the additives constant and adjusting the indium ion concentration. One would have to operate the bath longer and to stress it much more in order to see the effect of the built up of the anions and the product of the decomposition of the organic additives.

The copper bath was started with an average $29\,g\,l^{-1}$ of copper in the bath, which gave 335 nm of thickness. As the cupric ion concentration fell to about $22\,g\,l^{-1}$, the plated copper thickness decreased to approximately 320 nm. Furthermore, as the copper ion concentration in the bath increased back to $30\,g\,l^{-1}$, the thickness of the films increased to 325 nm and eventually gradually reached 350 nm. The copper film thickness also shows a correlation with the ion concentration in the plating bath but not as strong as the indium plating.

Run-to-run reproducibility depends on the **maintenance** of the plating bath by measuring and adjusting the key bath parameters that affect thickness and if necessary for more precise control maintaining the bath chemistry in a steady-state condition. The steady-state condition can best be achieved using a "bleed and feed method" described in the US Patent 5,352,350 [90]. This method, when soluble anode is used, maintains all the additives and complexing agents in the "steady-state condition." Material balances of all the components are maintained.

- Bath control was achieved using ICP for Cu, In, and Ga, HPLC for the additives in the plating chemistries, and titration for chloride and sulfate.
- Aging of the bath and anode condition impacted drastically wafer-to-wafer and within-wafer uniformity.

We have also discussed the importance of drying of the substrate before entering the plating bath. A film of pH 7 DI water on the surface, however thin, when carried over into the plating bath of pH 11 may cause many different problems. One of these problems is lack of reproducibility of plating from wafer to wafer. The complexing agents of indium in the Enthone high pH plating bath, when contacted with the pH 7 water film on the surface of the copper, become ineffective. In such case, indium oxide may precipitate on the surface of copper and slow down the initiation of nucleation. This will also affect the film growth. In addition, the Enthone proprietary organic additives are selected to operate in the pH range of 10.5–11.5. The bath behavior will also be affected by pre-dipping in the same pH solution before plating. The alternate to pre-dipping in a high pH solution is to enter the indium plating solution with a dry copper surface. As the Enthone gallium plating solution is also operating in a highly alkaline pH range, everything discussed above for the Enthone indium plating solution applies also to the Enthone gallium plating solution.

4.6 Conclusions

We described our journey of scaling-up thin film solar cell technology using electrodeposition of metal stacks and annealing to create the final chalcopyrite p-type absorber layer. We overcame many technical challenges. The fundamentals of electrodeposition of copper, indium on copper, and gallium on indium, copper, and gold were studied using a rotating disk electrode and a rotating electrode of 2.5 cm × 2.5 cm (1 in.^{2}). Nucleation and growth phenomena as a function of solution chemistries for indium and gallium were investigated in depth. We also investigated gallium chemistries that work in the basic pH range. We further briefly looked at the co-deposition of CuInGa and concluded that it was very challenging due to the complexity of controlling the chemical composition of the bath and the compositional uniformity of the thin film precursor deposit. From these early studies, it became evident that a simpler single metal deposition approach was more manufacturable than a metal alloy deposition methodology.

Next, we scaled-up the electrodeposition method to 15 cm × 15 cm size soda-lime glass substrates. We had to overcome the poor adhesion of the solar absorber to the glass/Mo substrate, and we installed a thin sputtered copper seed layer that covered the molybdenum surface and which prevented formation of molybdenum oxides, thereby improving adhesion. Another significant issue was that the compositional uniformity of the precursor absorber layer was a function of the thickness uniformity of each metal layer. Controlling the uniformity and roughness of each metal layer with σ variation of less than 3% was done with the use of independently powered thief electrodes and strong agitation with a single paddle. The paddle cell was used as a platform to further scale-up the method to 30 cm × 60 cm and to 60 cm × 120 cm full industrial scale.

Scaling-up the paddle cell to 30 cm × 60 cm encountered new challenges in controlling the current distribution. For this purpose, we used current thieves and deflectors and successfully controlled the resistance drop across the seed layer by making the seed layer thicker and by providing a continuous current contact across the plate. Because the copper plating chemistry was acidic and the indium and gallium chemistries were alkaline, special provisions to transition the parts from an acidic to an alkaline environment had to be made. Finally, scaling-up to the 60 cm × 120 cm size was accomplished by designing a new multi-paddle cell system, which was able to not only provide optimum agitation across the large plate but also improve the faradaic efficiency of electrodeposition and achieve a very smooth and nanometer uniform thin film deposit.

This concept of scale-up can be duplicated to create many small continuous low-cost production lines in many places around the globe. We hope that this wonderful journey of scaling-up the thin film solar cell technology will result in fueling the renewable energy economy by producing thin film solar cells for many different applications, including electric cars, green buildings, electronic devices, and sensors for the IoT.

Acknowledgments

We would like to express our deep and sincere appreciation to all the people who took part in the NEXCIS technology development. We would like to extend special thanks to the NEXCIS team, the suppliers, and the industrial, academic, and institutional partners. We would also take the opportunity to acknowledge the

IBM colleagues who contributed to key parts of the work described within this book chapter. As this book chapter is mainly focused on electrodeposition, we would like to extend our special thanks to Salvador Jaime Ferrer and Philippe de Gasquet from the NEXCIS team. NEXCIS was a spin off of IRDEP (Institut de Recherche et Développement sur l'Energie Photovoltaïque) launched in 2009. IRDEP was a joint laboratory between CNRS, EDF and Chimie Paristech. Since 2018 IRDEP has been integrated into the Institut Photovoltaïque Ile de France (IPVF) created in 2013.

References

1 Renewable Energy Policy Network for the 21st Century (2016). Renewables 2016 – Global Status Report.
2 K. Yoshikawa, H. Kawasaki, W. Yoshida, T. Irie, K. Konishi, K. Nakano, T. Uto, D. Adashi, M. Kanematsu, H. Uzu, K. Yamamoto, *Nat. Energy* 2 (Art. No.: 17032) (2017).
3 Jackson, P., Wuerz, R., Hariskos, D. et al. (2016). *Phys. Status Solidi RRL* 10 (8): 583–586. doi: 10.1002/pssr.201600199.
4 Deligianni, H., Ahmed, S., and Romankiw, L.T. (2011). *Interface* 49–55.
5 Andricacos, P.C., Uzoh, C., Dukovic, J.O. et al. (1998). *IBM J. Res. Dev.* 42: 567–574.
6 Vereecken, P.M., Binstead, R.A., Deligianni, H., and Andricacos, P.C. (2005). *IBM J. Res. Dev.* 49 (1): 3–18.
7 Li, X., Drews, T.O., Rusli, E. et al. (2007). *J. Electrochem. Soc.* 154: D230.
8 Rusli, E., Xue, F., Drews, T.O. et al. (2007). *J. Electrochem. Soc.* 154: D584.
9 Qin, Y., Li, X., Xue, F. et al. (2008). *J. Electrochem. Soc.* 155 (3): D223–D233.
10 Yokoi, M., Konishi, S., and Hayashi, T. (1983). *Denki Kagaku* 51: 460–464.
11 Healy, J.P., Pletcher, D., and Goodenough, M. (1992). *J. Electroanal. Chem.* 338: 155–165.
12 Kelly, J.J. and West, A.C. (1998). *J. Electrochem. Soc.* 145: 3472–3476.
13 Hayase, M., Taketani, M., Aizawa, K. et al. (2002). *Electrochem. Solid-State Lett.* 5: C99–C101.
14 Hayase, M., Taketani, M., Hatsuzawa, T., and Hayabusa, K. (2003). *Electrochem. Solid-State Lett.* 6: C92–C95.
15 Fabricius, G., Kontturi, K., and Sundholm, G. (1996). *J. Appl. Electrochem.* 26: 1179–1183.
16 Kelly, J.J., Tian, C., and West, A.C. (1999). *J. Electrochem. Soc.* 146: 2540–2545.
17 Healy, J.P., Pletcher, D., and Goodenough, M. (1992). *J. Electroanal. Chem.* 338: 167–177.
18 Healy, J.P., Pletcher, D., and Goodenough, M. (1992). *J. Electroanal. Chem.* 338: 179–187.
19 Kim, J.J., Kim, S.-K., and Kim, Y.S. (2003). *J. Electroanal. Chem.* 542: 61–66.
20 Wong, S. and Porter, M.D. (2000). *J. Electroanal. Chem.* 485: 135–143.
21 Chon, S. and Paik, W. (2001). *Phys. Chem. Chem. Phys.* 3: 3405–3410.
22 Zhukauskaite, N.A. and Malinauskas, A.A. (1989). *Prot. Met.* 25: 132–136.

23 Frank, A. and Bard, A.J. (2003). *J. Electrochem. Soc.* 150: C244–C250.

24 Bard, A.J. and Faulkner, L.R. (2001). *Electrochemical Methods: Fundamentals and Applications*, Chapter 9, 2nde, 352. New York: Wiley.

25 Andricacos, P., Chung, D., Deligianni, H., Fluegel, J., Kwietniak, K., Locke, P., Restaino, D., Seo, S.-C., Vereecken, P., and Walton, E. (2010). Void-free damascene copper deposition process and means of monitoring thereof. US Patent 7,678,258 B2, issued 16 March 2010.

26 Hung, C.-C., Lee, W.-H., Hu, S.-Y. et al. (2008). *J. Electrochem. Soc.* 155 (5): H329–H333.

27 Lovrecek, B. and Markovac, V. (1962). *J. Electrochem. Soc.* 109: 727.

28 Markovac, V. and Lovrecek, B. (1965). *J. Electrochem. Soc.* 112: 520.

29 Markovac, V. and Lovrecek, B. (1966). *J. Electrochem. Soc.* 113: 838.

30 Valderrama, R.C., Miranda-Hernándeza, M., Sebastiana, P.J., and Ocampo, A.L. (2008). *Electrochim. Acta* 53: 3714.

31 Huang, Q., Reuter, K., Amhed, S. et al. (2011). *J. Electrochem. Soc.* 158 (2): D57–D61.

32 Calvo-Barrio, L., Perez-Rodriguez, A., Alvarez-Garcia, J., et al. (2001). *Vacuum* 63: 315–321.

33 Grand, P.-P., Jaime-Ferrer, S., Broussillou, C., Reuter, K. B., Huang, Q. (2011). Production of a multi-layer structure for photovoltaic uses with perfected electrolysis conditions. WO2011086327 A1/PCT published 21 July 2011.

34 de la Breteque, P. (1964). *Ind. Eng. Chem.* 56 (6): 54.

35 Wade, K. and Banister, A.J. (1975). *The Chemistry of Aluminum, Gallium, Indium, and Thallium, Texts in Inorganic Chemistry*, vol. 12. Pergamon Press.

36 Tengi, L., Lv, Y.-G., Lui, J., and Zhou, Y.-X. (2005). *Forsch. Ingenieurwes.* 70 (4): 243–251.

37 Pendey, P.K., Sahu, S.N., and Chandra, S. (ed.) (1996). *Handbook of Semiconductor Electrodeposition*, vol. 206. New York: Marcel Dekker.

38 Chandra, S. and Khare, N. (1987). *Semicond. Sci. Technol.* 2: 214–219.

39 Contreras, M.A., Egaas, B., Ramanathan, K. et al. (1999). *Prog. Photovolt.* 7 (4): 311–316.

40 Wieting, R. (1999). *AIP Conf. Proc.* 462: 3–8.

41 Burgess, R.M., Chen, W.S., Devaney, W.E. et al. (1988, 1988). *The Conference Record of the Twentieth IEEE Photovoltaic Specialists Conference, 1988*, vol. 2, 909. IEEE.

42 Jasenek, A., Rau, U., Weinert, K. et al. (2001). *Thin Solid Films* 387 (1–2): 228–230.

43 Chen, W.S., Stewart, J.M., Stanbery, B.J. et al. (1987). *IEEE Photovoltaic Specialists Conference, 19th, New Orleans, LA, May 4–8, 1987, Proceedings (A88-34226 13-44)*, 1445–1447.

44 Takahashi, T., Yamane, T., Minamino, Y., and Kimura, T. (1989). *J. Mater. Sci. Lett.* 8 (8): 882–884.

45 Marinkovic, Z. and Simic, V. (1992). *Thin Solid Films* 217: 26–30.

46 Simic, V. and Marinkovic, Z. (1976). *Thin Solid Films* 34: 179–183.

47 Nakahara, S. and Kinsbron, E. (1984). *Thin Solid Films* 113: 15–26.

48 Purwins, M., Enderle, R., Schmid, M. et al. (1976). *Thin Solid Films* 515: 5895–5898.

49 Crisp, R.S., Henry, W.G., Schroeder, P.A., and Wilson, R.W. (1966). *J. Chem. Eng. Data* 11 (4): 556.

50 Kampmann, A., Rechid, J., Raitzig, A. et al. (2003). *Mater. Res. Soc. Symp. Proc.* 763: B8.5.1.

51 Oda, Y., Minemoto, T., Takakura, H., and Hamakawa, Y. (2007). *Mater. Res. Soc. Symp. Proc.* 1012: Y12–Y39.

52 Ganchev, M., Kois, J., Kaelin, M. et al. (2006). *Thin Solid Films* 511–512: 325–327.

53 Dharmadasa, I.M., Chaure, N.B., Tolan, G.J., and Samantilleke, A.P. (2007). *J. Electrochem. Soc.* 154 (6): H466–H471.

54 Hanket, G., Shafarman, W., and Birkmire, R. (2006). *Proc. WCPEC-4* 560.

55 Ahmed, S., Reuter, K.B., Huang, Q. et al. (2012). *J. Electrochem. Soc.* 159 (2): D129–D134.

56 Ahmed, S. and Deligianni, H. (2013). Gallium electrodeposition processes and chemistries. US Patent 8,545,689 B2, issued on 1 October 2013.

57 Ahmed, S., Deligianni, H., Huang, Q., Reuter, K. B., Romankiw, L. T., Vaidyanathan, R. (2016). Electrodeposition methods of gallium and gallium alloy films and related photovoltaic structures. US Patent 9,401,443 B2 issued on 26 July 2016.

58 Taunier, S., Sicx-Kurdi, J., Grand, P.-P. et al. (2005). *Thin Solid Films* 480–481: 526–531.

59 Chassaing, E., Roussel, O., Ramdani, O. et al. (2007). *ECS Trans. Electrochem. Soc.* 6 (2): 577–585.

60 Ramdani, O., Guillemoles, J.F., Lincot, D. et al. (2007). *Thin Solid Films* 515 (15): 5909–5912.

61 Saucedo, E., Izquierdo-Roca, V., Parissi, C.M. et al. (2009). *Thin Solid Films* 517 (7): 2268–2271.

62 Izquierdo-Roca, V., Alvarez-García, J., Calvo-Barrio, L. et al. (2008). *Surf. Interface Anal.* 40 (3–4): 798–801.

63 Saucedo, E., Ruiz, C.M., Chassaing, E. et al. (2010). *Thin Solid Films* 518 (14): 3674–3679.

64 Delbos, S. (2008). Turbulence controlled Cu–In–Se electrodeposition for solar cell production. Ph.D. thesis. Pierre et Marie Curie University - Paris 6 and Institute for Hydromechanics Karlsruhe University.

65 Delbos, S., Chassaing, E., Grand, P.P. et al. (2011). *Advanced Topics in Mass Transfer*, 43–60. InTech.

66 Fernandez, A.M. and Bhattacharya, R.N. (2005). *Thin Solid Films* 474: 10–13.

67 Ramdani, O. (2007). Cu(In,Ga)Se_2 thin film electrodeposition for photovoltaic conversion. Ph.D. Thesis. Pierre et Marie Curie University - Paris 6.

68 Duchatelet, A., Sidali, T., Loones, N. et al. (2013). *Sol. Energy Mater. Sol. Cells* 119: 241.

69 Duchatelet, A., Savidand, G., Loones, N. et al. (2014). *J. Electrochem. Soc.* 161 (7): D1.

70 Kois, J., Ganchev, M., Kaelin, M. et al. (2008). *Thin Solid Films* 516: 5948–5952.

71 Ribeaucourt, L., Chassaing, E., Savidand, G., and Lincot, D. (2010). *Thin Solid Films* 519: 7241.

72 Ribeaucourt, L., Savidand, G., Lincot, D., and Chassaing, E. (2011). *Electrochim. Acta* 56: 6628.

73 Broussillou, C., Viscogliosi, C., Rogee, A., Angle, S., Grand, P.-P., Bodnar, S., Debauche, C., Allary, J.L., Bertrand, B., Guillou, C., Parissi, L., Coletti, S. (2015). 2015 IEEE 42nd Photovoltaic Specialist Conference (PVSC), New Orleans, LA: IEEE, DOI: 10.1109/PVSC.2015.7356224.

74 Ahmed, S., Reuter, K.B., Gunawan, O. et al. (2012). *Adv. Energy Mater.* 2 (2): 253.

75 Guo, L., Zhu, Y., Gunawan, O. et al. (2014). *Prog. Photovoltaics Res. Appl.* 22 (1): 58.

76 Vauche, L., Risch, L., Sánchez, Y. et al. (2016). *Prog. Photovoltaics Res. Appl.* 24 (1): 38–51.

77 Vauche, L. (2015). *Process Development and Scale-up for Low-cost High-efficiency Kesterite Thin Film Photovoltaics*. Ph.D. Thesis. Marseille, France: Université d'Aix-Marseille.

78 Crossay, A. (2016). *Electrodeposition and Selenization of Metallic Thin Films for Kesterite Solar Cell Application*. Ph.D. Thesis. Luxembourg: University of Luxembourg.

79 Berg, D.M., Izquierdo-Roca, V., Perez-Rodriguez, A. et al. (2014). Simplified formation process for Cu_2ZnSnS_4-based solar cells. *Thin Solid Films* 573: 148–158.

80 Colombara, D., Crossay, A., Vauche, L. et al. (2015). *Phys. Status Solidi A* 212 (1): 88–102.

81 López-Marino, S., Sánchez, Y., Placidi, M. et al. (2013). *Chem. Eur. J.* 19 (44): 14814–14822.

82 Xie, H., Sánchez, Y., López-Marino, S. et al. (2014). *ACS Appl. Mater. Interfaces* 6 (15): 12744–12751.

83 Mendis, B.G., Taylor, A.A., Guennou, M. et al. (2018). *Sol. Energy Mater. Sol. Cells* 174C: 65–76.

84 Powers, J. V. and Romankiw, L.T. (1972). Electroplating cell including means to agitate the electrolyte laminar flow. US Patent 3,652,442, issued 28 March 1972.

85 Andricacos, P.C., Berridge, K.G., Dukovic, J. O., Flotta, M., Ordonez, J., Poweleit, H. R., Richter, J. S., Romankiw, L.T., Schick, O. R., Spera, F., Wong, K.-H. (1996). Vertical paddle cell. US Patent 5,516,412, issued 14 May 1996.

86 Grand, P.-P., Jaime, S., De Gasquet, P., Deligianni, H., Romankiw, L.T., Vaidyanathan, R., Huang, Q., Ahmed, S. (2017). Checking of the stoichiometry of I–III–VI layers for use in photovoltaic using improved electrolysis conditions. US Patent 9,647,151 B2, issued 9 May 2017.

87 Grand, P.-P., Jaime Ferrer, J.S., Roche, E., Deligianni, H., Vaidyanathan, R., Reuter, K.B., Huang, Q., Romankiw, L., Mason, M., Zupanski-Nielsen, D.S. (2012). Improved interface between a I–III–VI material layer and a molybdenum substrate. EP2469580A1, Bulletin 2012/26.

88 Andricacos, P.C., Branger, M., Browne, R.M., Dukovic, J. O., Fu, B.W.B., Hitzfeld, R. W., Flotta, M., McKenna, D., Romankiw, L.T., Sahami, S. (1994). Multi-compartment electroplating cell. US Patent 5,312,532, issued 17 May 1994.

89 Deligianni, H. and Romankiw, L.T. (2016). Design of a $60 \times 120\,cm^2$ prototype electrodeposition cell for processing of thin film solar panels. US Patent 9,362,440 B2, issued 7 June 2016.
90 Andricacos, P.C., Dukovic, J.O., Romankiw, L.T. (1994). Method of controlling chemical species concentration. US Patent 5,352,350 B2, issued 4 October 1994.

5

Thin-film Head and the Innovator's Dilemma

Keishi Ohashi

Waseda University, Research Organization for Nano & Life Innovation, 513 Waseda-tsurumaki-cho, Shinjuku, Tokyo 162-0041, Japan

5.1 Introduction

Information systems are designed to store huge amounts of data and are networked to each other in an easy-to-use format such that they provide the means to create new value. Data storage devices cover various data capacities ranging from very large data centers to tiny mobile memories. All the data in storage systems are stored in digital format for electronic use. The memory hierarchy of these systems consists of multiple memory levels characterized by their respective access speeds and storage capacities [1]. That is, the combination of high-speed memory with low-cost data recording has increased the performance/cost ratio of storage systems. Magnetic recording is the key enabling technology for low-cost recorders. Among the various magnetic recording devices, the hard disk drive (HDD) has been the most commonly used storage device since the late 1950s [2]. The magnetic head, which writes data to and reads data from the magnetic medium, is one of the most important components in a magnetic recording device. Its technological evolution as a result of ongoing innovation has contributed greatly to reduce the cost per megabyte of a HDD from $10 000 (1957) to ¢1 (2014) [3].

Thin-film head technology introduced microfabrication technology to the magnetic recording head. This technology reduced the size of a magnetic head core to less than one millionth in volume of that of a conventional ferrite head core and brought about extremely lower inductance and, consequently, a faster signal response, which has become indispensable to data storage access. In terms of manufacturing technology, from a historical viewpoint, it is necessary to recognize that the thin-film head was originally developed by IBM [4] and to acknowledge the role of the ingenious engineer Romankiw [5]. This innovation introduced a precise fabrication technology, especially for micromagnetic devices by combining electroplating with microphotolithography. Related insulator technologies were also developed by him and his colleagues [4, 5]. Furthermore, the through-mask Cu plating technique had finally been developed such that it became possible to produce interconnections on a

Electrochemical Engineering: The Path from Discovery to Product,
First Edition. Edited by Richard C. Alkire, Philip N. Bartlett, and Marc T. Koper.

large-scale integration (LSI) chip by using an electroplating technology known as Cu damascene [6].

From the viewpoint of business, however, it has been argued that the thin-film head was simply a sustaining technology rather than a disruptive innovation [7]. This chapter attempts to reposition thin-film head technology in the HDD business sector from the author's point of view and also introduces the research contributions his group has made to this field.

5.2 Thin-film Head Technology

5.2.1 Magnetic Properties for HDD

Ferromagnetic materials, such as those used in HDDs, have three engineering applications [8]. Soft magnetic materials are used for magnetic cores. Their magnetization direction and the associated magnetic flux easily respond to an applied magnetic field. The sensitivity to the external field is expressed by the magnetic permeability. Hard magnetic materials are used for permanent magnets in motors and actuators for HDDs. Usually, their magnetization does not change even when an external field is applied. Semi-hard magnetic materials are used for magnetic recording media. Their magnetization changes when the applied field is sufficiently strong to switch the direction of magnetization.

The coercivity (H_c) is defined as the field capable of reversing half of the magnetic moments or as the field when the net magnetization is zero. Soft magnetic materials are characterized by small or almost zero H_c. The write heads in HDDs should generate a stronger field than the H_c of recording media. Thus, the saturation magnetization (M_s) of a write head should be accordingly high.

5.2.2 Permalloy

Permalloy (Ni–Fe alloy) is the most well-known soft magnetic material. Soft magnetic properties, such as low H_c and high permeability, are structure sensitive in that the composition, shape and size, defects, and internal stress of the alloy affect them greatly and the fabrication method should be carefully explored. Electroplated permalloy films show several different properties compared with evaporated permalloy films and bulk permalloys with the same composition. Electroplated permalloy films usually incorporate impurities such as hydrogen (H_2), sulfur (S), and associated vacancies of the order of a few percent. These impurities could produce lower activation energy components for anisotropy change during annealing. This is an inherent property of electroplated permalloy films, resulting in magnetic properties that differ slightly from those of other permalloys.

Several in-depth studies on the soft magnetic properties of permalloy films and their thermal stability have been conducted in universities in Japan as well as in universities in the United States and Europe. The author started his graduate research in the Uchiyama Laboratory at Nagoya University, Japan, in 1977, where magnetic thin films were extensively investigated. The laboratory was equipped with an ultrahigh-vacuum evaporation machine with a rotating field magnetometer in a vacuum chamber, which measured the anisotropic field of a magnetic film without breaking the vacuum after deposition. The machine was constructed to investigate low activation energy components usually found in an electroplated permalloy [9]. The experiments revealed that the permalloy films evaporated onto a substrate at room temperature did contain the low activation energy component when the film was not exposed to the air. This result was useful to overcome the groundless superstition that electroplated films were of low quality and different from physically deposited materials. It had already been confirmed that an appropriate annealing procedure ensured that electroplated permalloy films had the requisite long-term stability [10].

However, unfortunately, the perception gradually arose that research on permalloy films was becoming outdated in electronics. This was because the plated wire memory technology, which had been intended to replace ferrite core memory with permalloy film, had already lost the development race with semiconductor memory in the early 1970s. At the time, permalloy research efforts in universities shifted their attention to magnetic bubble memories followed by magneto-optical (MO) recording. Therefore, the author decided to pursue research on amorphous rare-earth/transition-metal films intended to be used in bubble memories and MO recording toward his Master's thesis.

At around that time, Romankiw et al. at IBM succeeded in developing permalloy electroplating for thin-film heads [5]. In Japan, vacuum deposition was more popular than electroplating with a few exceptions [11]. One of the reasons why they preferred sputtering could be that the main target for them was magnetic bubble memory and beyond. The magneto-resistive (MR) read sensor for bubble memories requires relatively thin (<100 nm) films, and a highly ambitious project such as that involving Bloch line memory, which uses a lineal substructure of spin configurations in a domain wall, was expected to increase the capacity of magnetic bubble memory phenomenally. The latter was a large collaborative research project between Kyusyu University and NEC [12]. They concluded that extremely fine patterning tools were available only for the rapidly progressing state-of-the-art semiconductor dry process. The complicated mechanism of using major and minor loops similar to those employed in bubble memories introduced another risk in the form of a long access time.

Many researchers investigating soft magnetic films in Japan had been interested in higher M_s materials because of their application in video cameras, which have a large market. Permalloys did not attract these researchers whose main research interest was toward Fe-based amorphous and nanocrystalline films using the aforementioned dry process. Fe- (and Co-) based films with higher M_s have relatively large crystalline anisotropy and resulting large H_c, which can be reduced if the crystalline size is sufficiently small. One of the well-known nano-crystalline soft magnetic films was Fe-Ta-N ($M_s = 1.5–1.6$ T) [13].

5.2.3 Thin-film Head

In 1979, the author joined NEC. He was engaged in two jobs. The first one was to replace Ni–Zn ferrite (saturation magnetization $M_s = 0.4$ T) with Mn–Zn ferrite ($M_s = 0.5$ T) for a magnetic head core to double the capacity of an HDD with maximum data capacity at that time. This was technical improvement work typical of that period in the Japanese computer industry. He managed to introduce low-resistivity Mn–Zn ferrite, originally developed for video recording, for use in an HDD head by changing the design of the core shape and coil-winding method.

The second but more important job was to work with the team members responsible for thin-film head development. In 1979, IBM announced the 3370 disk drives, which employed a thin-film head with electroplated permalloy (Ni–Fe) cores and Cu coils. At first, the policy was to continue the development work toward thin-film heads at NEC's Central Research Laboratory by using LSI-type processing as mentioned in [5]. Subsequently, the decision was made to set up a new small team in the Peripherals Division to develop a thin-film head following the announcement of the IBM 3370. There were only three young engineers in the development team. None of these members even had a title.

Thin-film heads were expected to replace bulk ferrite heads soon because of their potential productivity improvement from the labor-intensive coil-winding process to the photolithography-based batch fabrication process. Contrary to expectation, however, the replacement required some time. The continuing effort in Japan paid off as the metal-in-gap (MIG) head became one of the developments to disrupt the rise of thin-film head in the market. Reduction of the volume of the ferrite part in the MIG head succeeded to decrease the inductance sufficiently to enable their use for HDDs, and MIG head technology survived. Typical metals for the MIG head were sendust (Fe–Al–Si), Fe-Ta-C, and Fe-Ta-N. Fe-Ta-N heads for video cameras were mass-produced at NEC Kansai, Ltd., for years [14]. Although there was an effort to introduce Fe-Ta-N in a thin-film head for commercial HDDs, the low-temperature requirement in the fabrication process (<300 °C) prevented its ability from being demonstrated at that time.

Researchers in the Central Research Laboratory provided us with many must-read documents. The most important ones were those produced by Romankiw as listed in [5], which include papers, conference proceedings, patents, and technical documents describing non-patented inventions. The well-known "hard-baked resist as insulation" technology was described in the IBM Technical Disclosure Bulletin [15].

The primary purpose of our thin-film head development team was to be a fast follower of IBM. It may sound strange that, in this complicated situation, although novel ideas were positively introduced in the area of conventional technology, the senior engineers wanted to be a follower in the area of cutting-edge technology. Sometimes, rumors spread in the industry without any reliable data to support them. One of the examples is found in the shape of magnetic cores that was found in a drawing and a micrograph disclosed by IBM. Surprisingly, the thin-film head of the 3370 had a heart-shaped core with rounded corners. Soon several head researchers tried to construct a theory suggesting that a rounded core theoretically creates smooth magnetic domain wall motion and eliminates domain noise.

However, to their surprise, the next-generation double-layer coil thin-film head for the IBM 3380 K had a home-base-shaped magnetic core with sharp corners. Naturally, no papers or patents relating to the domain wall motion on the effect of rounded corners were published since then. Thus, many thin-film head engineers at that time had been engaged in following up specific technologies pioneered by IBM.

In the early 1980s, thin-film head engineers other than those at IBM discussed and exchanged their limited information. It was frequently observed that many engineers believed that they individually had found the same secret to manufacturing a good thin-film head. For instances, many head manufacturers adopted a slightly negative magnetostriction magnetic core to obtain a stable output signal. It had been described in a patent [16] as the combined effect of mechanical stress and the inverse effect of magnetostriction. The effect was so remarkable that whoever witnessed, it was disposed to explain the mechanism in their particular way. The author was one of them [17], and his involvement will be described in the subsequent Section 5.2.4. Nevertheless, he tried to develop thin-film heads on a rigidly theoretical basis.

Experiments to increase the yield of the prototype thin-film head were conducted intensively. At first, most of the activities were simply confirmatory studies in pursuit of the results obtained by IBM. Fortunately, the Engineering Manager responsible for developing all the magnetic heads, including ferrite heads and thin-film heads for HDDs, multi-track heads for tape drives, and floppy heads, said "I can't understand thin-film head technology. Anyway, I will get the budget. I trust you. Do your best." After some initial difficulties with high cost due to the low yield, a new small plant started to supply the first mass-produced thin-film heads in Japan in 1982 [18].

Then, the develop team members were dispatched to NEC Ibaraki, Ltd., to take up a temporary assignment to build a mass production line for thin-film heads, where thin-film disks were fabricated by using electroless plating [19] and multilayer ceramic substrates with electroplated fine lines for supercomputers [20] had already started volume production. The first combination of thin-film heads with thin-film disks was realized as the 10.5-in. HDD GEMMY (gigabyte capacity magnetic memory), of which the basic design was conducted by NTT [21]. Three computer manufacturers, NEC, Hitachi, and Fujitsu, built GEMMY with different component technologies. NEC successfully introduced thin-film technology for both heads [22] and disks [23]. The permalloy cores and Cu coils in the heads were electro-plated and the Co-P recording media were electroless-plated. The technology achieved the highest recording density as a commercial HDD for a while and, therefore, paved the way for HDDs with key magnetic devices containing all thin-film technology.

5.2.4 Magnetic Domain Noise

In magnetic materials, the orientations of magnetic moments usually divide into multiple magnetic domains. The boundaries of these domains are known as domain walls with finite thickness in which the direction of the magnetic moments gradually change. The typical thickness of a domain wall is between

1 nm and 1 μm [8, p. 413]. The motion of a domain wall is not always smooth but is associated with discontinuities that cause Barkhausen noise [24]. When a magnetic body is large, the relative volume of each of the domains is usually small and each discontinuity may be negligible. Conversely, when a magnetic body is sufficiently smaller than the size of the domain wall, there is no Barkhausen noise due to domain wall motion. In the case of a magnetic body of intermediate size, the magnetization divides into a few or several domains and the domain noise associated with domain wall discontinuity would be prominent. Unfortunately, a thin-film head core is of an intermediate size.

The size of each of the magnetic domains and their configuration are strongly affected by the shape of the magnetic body as a design matter, and by the magnetic anisotropy energy as a process matter. The magnetic anisotropy has two main origins. One is magnetocrystalline anisotropy energy, which is determined by composition and atomic arrangement such as directional order [8, p. 300]. It is also modified by the induced anisotropy generated by annealing under a magnetic field and by mechanical stress, which deforms the lattice.

Several researchers proposed a "good" domain configuration to obtain a stable waveform in the read signal of a thin-film head. A typical model is shown in Figure 5.1, in which the thin-film head core has hexagonal central domains with triangular edge-closure domains. Such a domain configuration appears when the direction of the magnetic anisotropy is perpendicular to the signal flux from the disk and in the direction horizontal to the paper.

The closure domains had been suspected to cause Barkhausen noise in the thin-film head [25]. Thus, the closure domains were eliminated by fabricating laminated magnetic cores with thin nonmagnetic layers [26, 27]. The author theoretically investigated the magnetization dynamics of domain configurations with edge-closure domains. The motion of the edge domain wall couples with the rotation of the magnetization in the central domains, and it was thought that the domain wall would move faster than predicted by classical wall motion theory. It was concluded that when the edge-closure domains are sufficiently smaller

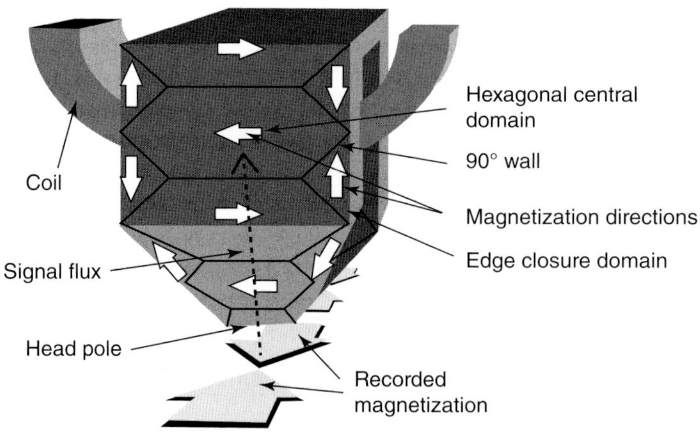

Figure 5.1 Domain configuration model for thin-film head core.

than the central domains, the flux change through the entire core is only slightly affected [28, 29]. Experimental observation of magnetic flux conduction confirmed this fast flux conduction through the central domains [30, 31]. In conclusion, a complicated laminate structure is not necessarily required to suppress the Barkhausen noise. Instead, induction of horizontal magnetic anisotropy to shrink or diminish the closure domains is essential to obtain a stable read signal.

Observation of the domain configuration during the fabrication process revealed that the configuration was changed by several processes [17]. Even when a domain configuration appeared ideal before overcoating and machining, as shown in Figure 5.1, the head often revealed different domain configurations after the final fabrication process. The author observed the domain configurations throughout the wafer process and the slider process. At that time, a high-resolution observation was only possible by using the powder-pattern (Bitter) method [8, p. 388]. The final and the most important change in the domain configuration was observed after slicing the wafers into head sliders. In this process, the magnetic cores are elongated downward because application of the overcoat to the thin-film head usually required compressive stress to avoid chipping during machining. Figure 5.2 shows the change in the domain pattern due to the elongation, as observed with permalloy dummy cores on the overcoat [17]. The edge domains of the slightly negative magnetostriction permalloy core (82% Ni and 18% Fe) disappeared after slicing. In contrast to this, domain walls parallel to the signal flux, which are suspected to cause Barkhausen noise, appeared in the slightly positive magnetostriction permalloy core (<81% Ni).

The experiments leading to the above results were conducted in 1981 and NEC succeeded in improving the yield to ship the first HDD incorporating a thin-film head in Japan [18]. The obtained results were kept highly confidential at that time and were prohibited from being presented at academic meetings for almost 10 years. However, it was a nonsensical action as already described in Section 5.2.3 because, at that time, an IBM patent already disclosed the technology in which negative magnetostriction was used [16]. In discussions between magnetic head engineers other than those at IBM, they realized that there were many similar original findings that were kept secret in their company or organization.

During these early days, however, a few engineers remained unconvinced that the chosen approach was correct and decided not to follow it. All the proposed models were based on the belief that the induced magnetic anisotropy occurred in the plane of the magnetic core and was perpendicular to the magnetic flux from the disk. Some engineers, however, found that a more stable read output was sometimes obtained with much more negative magnetostriction heads (more than 84% Ni). The configuration of the magnetic domain in these cores seemed to differ from that in the conventional model. The magnetization direction could be perpendicular to the core as the Ni film sometimes showed, and the magnetization change may not have required domain wall motion. Actually, these heads were fabricated by one of the head suppliers in Japan and their read signal instability was small. This could be a good example showing the strength of a manufacturer who is not an armchair theorist.

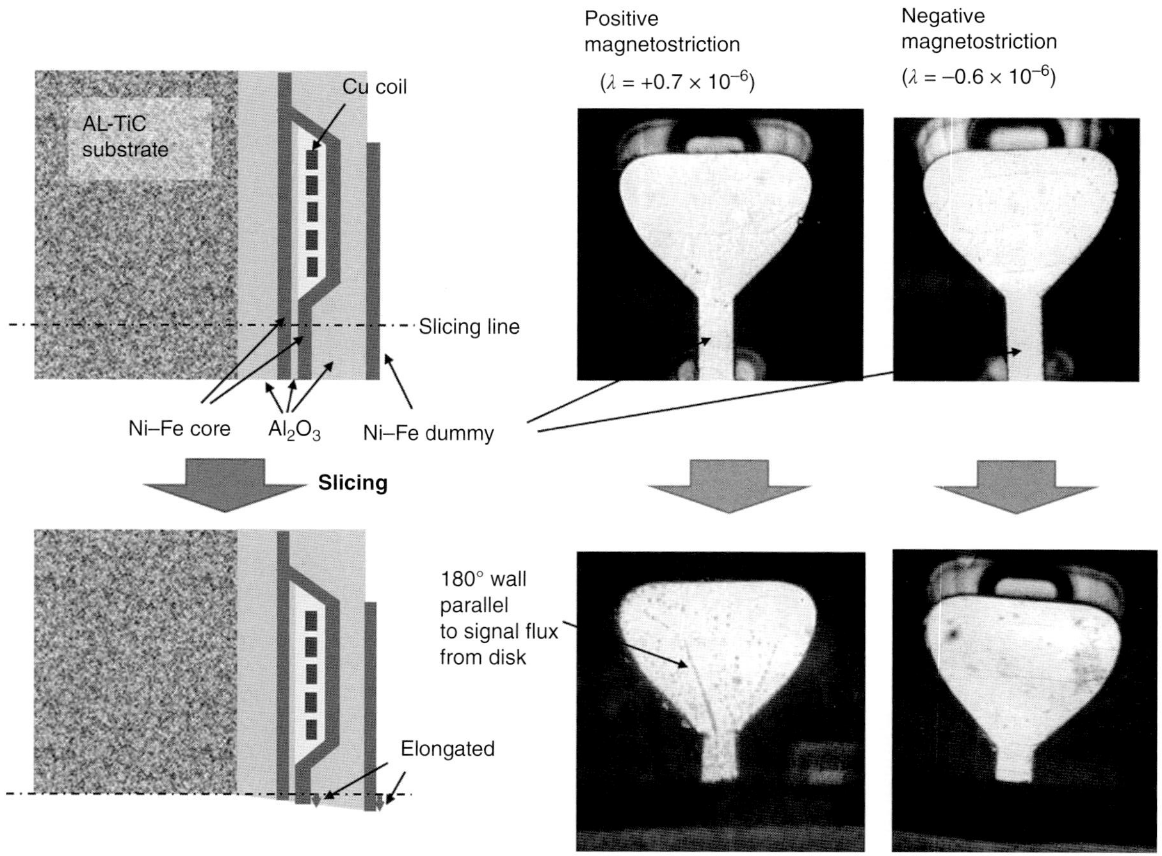

Figure 5.2 Domain configurations before and after slicing the head wafers [18]. The domain wall in the core with positive magnetostriction after slicing is a 180° wall as the magnetization directions on either side of the wall have opposite directions, up and down along the vertical direction of the paper. Domain changes in the dummy core are expected to reflect the domain changes in Al_2O_3, which cannot be observed by using the powder-pattern method.

5.3 Data Storage Business in Japan

5.3.1 Magnetic Thin-films for HDD in the 1980s

In the 1980s, Japanese HDD manufacturers decided to challenge the U.S. manufacturers by taking the initiative away from them based on their strong capability in testing and in redesigning improved products, especially by making small product modifications [32]. NEC's version of GEMMY was the first HDD incorporating both thin-film media and a thin-film head and delivered the highest performance at the time. The author presented the thin-film head technology applied to GEMMY at the 1987 ECS Fall Meeting [22] and participated in a panel discussion on the "future of electrochemistry in the electronics industry" [33] with Tetsuya Osaka (Waseda University), Lubomyr T. Romankiw (IBM), and Ion Croll (IBM).

It had been an open secret that electroless-plated media had gradually been losing the advantages it held over rapidly developing sputtered media around that time. When large 14- and 10.5-in. disks were dominant in the computer industry, the investment in manufacturing facilities for electroless-plated media was much smaller than that of sputtered media, thereby reducing the manufacturing cost of the former. However, as the demand for smaller HDDs for use in desktop and mobile personal computers (PCs) was so strong that the disk size continued to shrink from 5.25- to 3.5-in. and then to 2.5-in., which made it possible to divert the silicon wafer fab vacuum deposition machines of the previous generation to thin-film disk fab. Thus, the manufacturing cost of electroless-plated media was overtaken by that of sputtered media.

A more critical problem with the technology presented itself. Thin-film media introduced a new read-error source known as transition noise, which was generated by irregular magnetization transitions between written bit "domains." After cumulative efforts by sputtered media researchers, the transition noise was suppressed by reducing intergranular exchange coupling [34]. Research into noise reduction with electroless-plated media was consequently delayed. Osaka reported electroless-plated perpendicular magnetic recording media [35], which could avoid transition noise. However, we had to wait for 20 years until the commercialization of perpendicular recording HDD. It was regrettable that the electroless-plating facility at NEC Ibaraki was closed in the early 1990s.

Osaka believes in collaborative research and reasoned that the slow improvement in plated disks was a consequence of NEC's secretive manner and that this emphasizes the importance of open research in high-tech business. Actually, it has been recognized that, relative to US companies, the links between scientific research and invention are weaker in Japan [36]. Croll and Romankiw kindly provided the author with additional advice that simple methods, especially in bath composition, and control techniques usually lead to improved results when the technology is transferred from research to production and would also be useful in developing new materials. Later, these thoughts influenced the development of thin-film heads at NEC.

5.3.2 Use of Optics

A magnetic recording medium exhibits hysteresis in the magnetization response against an applied magnetic field or write head field. The magnetic domain pattern recorded using the hysteresis has two opposite directions that are expected to be stable for a long period. However, as the magnetization direction of smaller grains is more susceptible to thermal fluctuation, data missing with higher density thin-film media became a critical issue in the 1990s [37]. This thermal decay was accelerated by the introduction of thin-film media with a strong demagnetizing field in the magnetization transition area.

MO recording offered more stable magnetic data than hard disk recording [38, 39]. This stability was achieved by strong perpendicular anisotropy and low magnetization of the amorphous rare-earth/transition-metal film media, which substantially reduced the demagnetizing field. Magneto-optical drives (MODs), which were developed and commercialized intensely, especially in Japan, maintained higher recording density than was achieved by HDDs in the 1980s–1990s. However, the development work to increase the recording density of MODs was slowed by the diffraction limit of light. From a business perspective, the backward compatibility with legacy media put the MODs in a disadvantaged position with respect to increasing the recording capacity. MODs could, therefore, not establish any firm market competing against HDDs before the latter approached the thermal fluctuation limit.

Near-field optical recording by using a surface-plasmon antenna was investigated [40, 41]. It would be necessary to combine an optical technology with magnetic recording such as in heat-assisted magnetic recording (HAMR) [42]. The primary difference between HAMR and MO recording is that the former uses a MR sensor as a read head. The superiority of an MR sensor as the head element for reading a weak signal from a tiny bit will be discussed in Sections 5.4.3 and 5.4.4. Nevertheless, in spite of the potential advantage of the combination of magnetic recording with optical technology, the sudden decision by the US companies, especially by IBM, to focus on HDDs seems to have been the right one in the end.

5.3.3 High-Moment Head Core Material

Attempts to prevent the thermal decay of magnetic data in HDDs led to three approaches being tried: (1) increase the effective volume per magnetic bit, (2) reduce the demagnetizing field, and (3) increase the coercivity (H_c) of the recording medium. Similar to MO recording, perpendicular recording [43, 44] provides a larger effective volume with a bit area and also reduces the demagnetizing field. The last approach (3) requires a write head that generates a stronger magnetic field than ever. More specifically, a higher moment magnetic core for the write head has been indispensable for a high-H_c medium for thermally stable high-density recording.

To enable writing in a high-H_c medium, two approaches were tried to develop high-M_s soft magnetic films in NEC. The first involved a sputtered Fe-Ta-N that was already used for video recording [13]. The other was electroplated films.

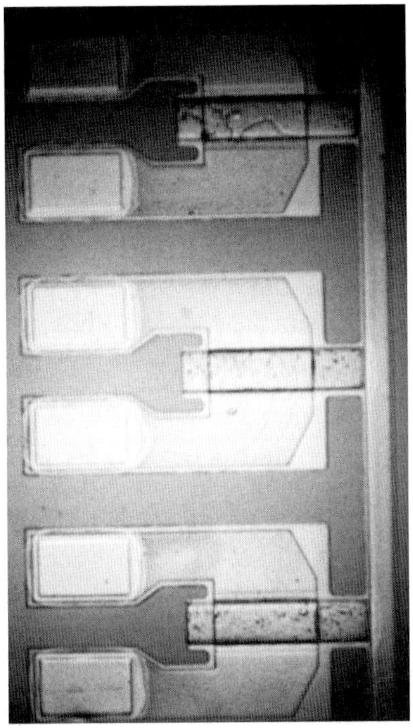

Figure 5.3 The "Electroplated Head" developed in the Peripherals Development Division at NEC (1974). The head core material was electroplated $Ni_{45}Fe_{55}$. The 1-turn Cu coils were electrodeposited from a pyrophosphate bath. Source: Courtesy of Shizuo Hanagata.

In 1992, Osaka proposed to the author a collaborative research project to develop new high-M_s soft magnetic films by using electrodeposition. Several papers on core materials for a higher M_s head for thin-film heads had already been published. For example, the first prototype thin-film head in NEC named "Electroplated Head" (Figure 5.3) used a high-M_s electroplated permalloy (45% Ni and 55% Fe) [11, 22].

In spite of these trials, no one succeeded in producing the high-M_s materials that would have satisfied the requirements for a thin-film head application: (i) good soft magnetic properties such as high permeability, low-H_c, and small magnetostriction λ; (ii) corrosion resistance and a small aging effect; (iii) low-temperature fabrication process (<300 °C); and (iv) low fabrication cost. The "Electroplated Head" showed noticeable Barkhausen noise, perhaps due to uncontrolled domain structures originated from large positive magnetostriction. In the history of thin-film heads, the electroplated permalloy with slightly negative magnetostriction was firmly believed to be the only material satisfying all the requirements for commercial thin-film heads. Their M_s was limited to 0.9–1.0 T.

The entrenched idea of using permalloy (82% Ni and 18% Fe) for the head core material, however, changed after the introduction of the MR element as a separate read sensor that was employed in combination with an inductive write element [45]. This inductive-write and MR-read dual-element head held advantages for the design of the write head in that no strict requirement in terms of sensitivity to a weak magnetic field from the disk existed. It permitted the use of magnetic films with slightly larger coercivity and larger magnetostriction for the write core of the MR head than those for a conventional thin-film head. Consequently, some MR heads were produced using an electroplated permalloy (45% Ni and 55% Fe) with a high-M_s of 1.6 T [46].

The first electroplated high-M_s material in our program was Fe-P. The electroplating bath contained sodium hypophosphite as an additive, which was familiar to electrochemists because of its use in electroless plating for disk media. The electroplated Fe-P film showed good soft magnetism with relatively higher M_s of 1.4 T by adjusting the P content to obtain fine Fe grains [47]. A problem occurred, however, after the first experimental trial in which Fe-P plating was used at a small production plant at NEC when using the bath composition developed by Waseda University. The permalloy films plated the following day had a slightly different composition even when the permalloy bath was controlled as usual. The abnormal state continued for several days. The production team leader suspected the bath tank was contaminated by the experiment. Additionally, he forbade us to use the plating system for the experiment. This event reminded the author of the advice from Croll and Romankiw not to use too many additives.

Osaka next decided to attempt to develop a new material that could be used in real products by targeting Co–Ni–Fe ternary alloys. The additives used in the initial experiment were almost the same as those used in conventional permalloy plating, namely sulfur-containing additives such as saccharin sodium as a stress reducer [48]. The stress reduction was explained by the inclusion theory, in which the C—S bonding in sulfur-containing additives breaks to release S. The sulfur reduces the strong tensile stress induced in films in which no such additives are used.

The outcome of using a new Co–Ni–Fe bath by changing the additives from standard saccharin sodium [49], through thiourea [50], to sulfur-free additives [51–53], has already been well described [54]. The first important finding was that the f.c.c.–b.c.c. phase boundary in the Co–Ni–Fe ternary system shifted toward the lower M_s (~1.7 T) region when the deposits were electroplated from the bath containing thiourea. The grain size of the films near the f.c.c.–b.c.c. phase boundary was as small as 10–20 nm; therefore, they had a sufficiently good soft magnetic property for a write head with an MR read element. The films electroplated from the bath with thiourea contained a larger amount of S. This seemed to suggest that the phase boundary would shift toward the higher M_s region by eliminating sulfur-containing additives, and this proved to be true. Thus, we obtained a film with a good soft magnetic property ($H_c < 160$ kA m^{-1} (2 Oe)) and higher M_s than ever (2.0–2.1 T) by electroplating Co–Ni–Fe from the bath without any sulfur-containing additives except for the surfactant [53].

Initially, the problem was that the films were delaminated owing to a relatively large film stress of 600–900 MPa. However, the accumulation of stress is prevented by using a thinner film and by delineating the film to a small pattern. This was achieved by the addition of two high-M_s Co–Ni–Fe thin films to the gap side of upper and lower permalloy poles [55]. This structure is similar to that of the MIG head where the leakage of magnetic flux from thin higher-M_s film is confined in an adjacent thick lower M_s layer. Later, this double-layer core was replaced by an ultra-small Co–Ni–Fe core. The removal of the sulfur-containing additives from the plating bath also improved the pitting corrosion resistance of the film [56, 57].

5.3.4 High-M_s Write Heads

After the demonstration of the superior writing performance of the Co–Ni–Fe head, however, several HDD engineers, especially those who had been engaged in high-performance HDD head design, were concerned that the Co–Ni–Fe head would not operate at high recording frequencies because of the anticipated large eddy current loss due to the low resistivity of the Co–Ni–Fe film. Actually, our $Co_{65}Fe_{22}Ni_{13}$ film has relatively low electrical resistivity of 18×10^{-8} Ω·m compared to that of a conventional permalloy $Ni_{82}Fe_{18}$ film of 25×10^{-8} Ω·m and a higher M_s permalloy $Ni_{45}Fe_{55}$ film of 50×10^{-8} Ω·m.

When a magnetic field is applied to an electrically conductive magnetic film, an eddy current is induced in the conductor, especially near the surface, thereby preventing the field to penetrate the conductor. The penetration depth of a magnetic field is known as the skin depth. When the thickness is sufficiently small, the skin depth is proportional to the square root of the resistivity and the inverse of the square root of the frequency. However, according to this simple approximation, the required core thickness at high write frequencies (>100 MHz) is very thin (<1 μm), which results in flux leakage from the core and lowers the write field strength.

Despite this simple pessimistic estimation, the eddy current was significantly reduced by shrinking the length and the width of the core, even though the thickness exceeded 1 μm [58]. Figure 5.4 shows the results of a 3-D calculation and

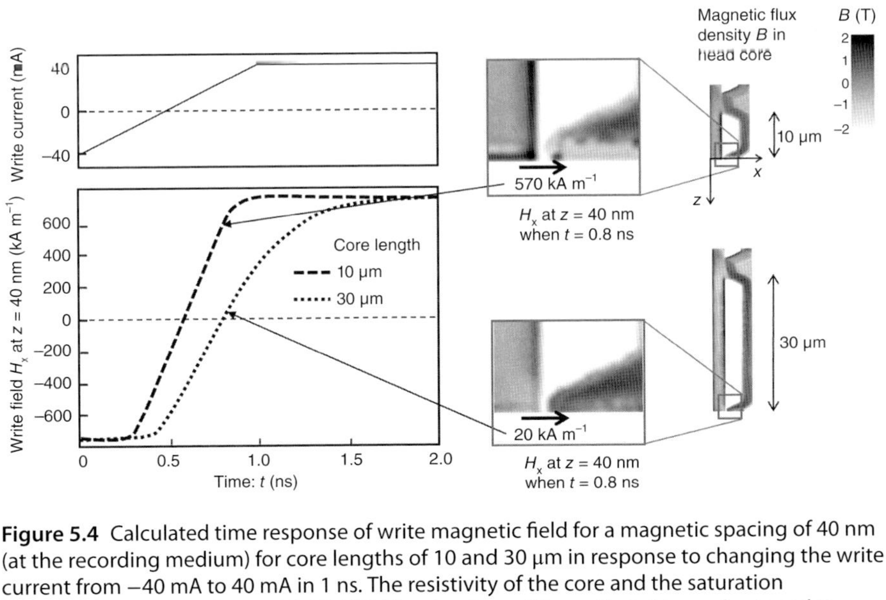

Figure 5.4 Calculated time response of write magnetic field for a magnetic spacing of 40 nm (at the recording medium) for core lengths of 10 and 30 μm in response to changing the write current from −40 mA to 40 mA in 1 ns. The resistivity of the core and the saturation magnetization is 0.2 μΩm and 2.0 T [58]. The rise time of the write field was shortened to <1.0 ns by focusing the magnetic flux.

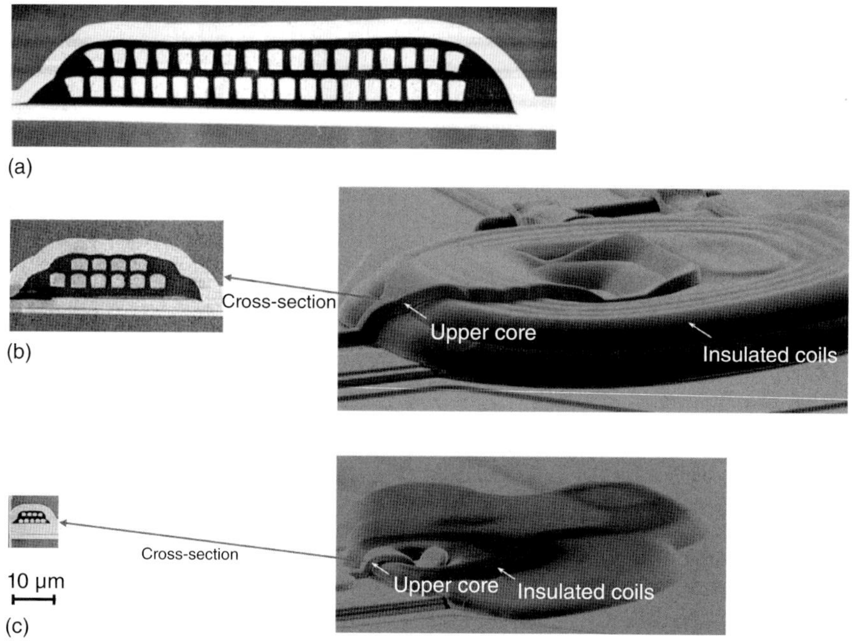

Figure 5.5 Cross-sectional view of thin-film heads with different core lengths: (a) 120 μm, (b) 50 μm, and (c) 10 μm. It significantly reduced the eddy current loss at high frequency. (a) Thirty-six turn coil thin-film head (1991). (b) Ten turn coil GMR head (1999). (c) Nine turn coil TMR head (2000).

shows that by shrinking the core length, the flux changes fast enough because of the reduction of the eddy current. This result persuaded the high-performance HDD engineers, after which high-M_s materials such as electroplated Co–Ni–Fe films with the smaller core design were employed in subsequent HDDs. Figure 5.5 depicts the history of core shrinkage of heads developed at NEC.

5.4 The Innovator's Dilemma

5.4.1 Thin-film Head is not Disruptive

In the latter part of the twentieth century, Japanese companies were overwhelmed with criticism that they had simply been copying the technologies invented in the United States. On the other hand, the strong yen gradually weakened the competitiveness of companies in the electronics industry in Japan after the Plaza Accord in 1985. In the late 1980s, Japanese executives noted that their HDDs were not as price-competitive as they had been [59]. NEC decided to cease designing and manufacturing its own HDDs despite several prominent original technologies having been produced.

Then, the author chose to start investigating new technologies for new business other than HDD. At around that time, he found a description not to be passed without notice in *The Innovator's Dilemma* [7], a best-selling business book written by Clayton M. Christensen, a professor at Harvard Business School. He proposed the concept of "disruptive technology," which brings a value proposition very different from any ever proposed by existing customers. The leading companies in a conventional market concentrate their investment on "sustaining technology" for customers in the existing market. Sound decisions by great managers in a great company sometimes fail because they always hear the voice of the most important existing company and cannot concentrate their investment on disruptive technology. This is the innovator's dilemma.

With reference to thin-film heads, Christensen states that "because of the risk involved in the technology's development and its potential importance to the industry, the trade press began speculating in the late 1970s about which competitor would lead with thin-film heads" [7, p. 123]. The author of this chapter was pleased to find the familiar term "thin-film head." He and many other HDD engineers had recognized that the thin-film head was one of the most important watershed technologies affecting the pace at which HDD makers were able to increase the recording density. He was pleased to find his result in Figure 6.1 of the book as the fourth company in the world that had adopted thin-film technology.

However, soon, he was bewildered that Christensen positioned the thin-film head as a typical example of a sustaining technology in the existing market. He says, "Yet, it turned out, whether a firm led or followed in this innovation did not make a substantial difference in its competitive position" [7, p. 123]. He proceeded to argue that "there is no evidence that any of the leaders in developing and adopting sustaining technologies developed a discernible competitive advantage over the followers"... and "there is no relationship between leadership and followership in thin-film heads and any subsequent technological edge" [7, p. 124].

During these years, ferrite heads succeeded in catching up with thin-film heads by introducing MIG technology, which consisted of adding two high-M_s metal films between two ferrite cores and a magnetic gap. The less-expensive MIG heads provided almost the same write performance (strong write field) as permalloy thin-film heads. At IBM, where the HDDs were for high-end use, thin-film heads were very useful because they have low inductance and high-speed response compared with MIG heads. However, in the low-end emerging personal computer market, MIG heads provided sufficient performance until MR head technology was introduced. At NEC, however, as its early PC business achieved fair success, the HDD business strategy could not be focused and the resource power was divided into high-end enterprise, PC desktop, and mobile markets. NEC's leadership in thin-film technology did not promote effective use for competitive advantage, resulting in no significant gain in market share [7, p. 124].

5.4.2 Small HDD

According to Christensen, the disruptive aspect of HDD technology was "the architectural innovations that shrunk the size of the drives" [7, p. 15]. Smaller diameter disks enabled significant power reduction by reducing windage loss [60]. In terms of the fluid mechanics, the power loss in the laminar flow close to the disk is proportional to the fourth power of the disk diameter [61, p. 66]. Thus, the wind loss in 2.5-in. HDDs decreases to 0.1% of that in 14-in. HDDs. This effect was more pronounced for high-performance HDDs with higher revolutions per minute (RPM) to reduce the latency time. Another advantage of the small form factor is the significant improvement in shock resistance. The shock power is proportional to the mass and square of the velocity and continued to decrease with the shrinkage in the disk size. From the viewpoint of logistics for manufacturing, the cost of air shipping smaller (<3.5 in.) HDDs became negligible compared to the total cost of the product as an additional benefit [62].

A number of HDD manufacturers including NEC did seriously consider the smaller form factor and tried to develop new markets with existing HDD technologies. However, their existing customers did not welcome the small HDDs because their performance considerably changed for the worse by reducing the disk size [7]. Actually, a larger disk always provides a higher effective recording area, thereby resulting in higher bit cost devices. Did the manufacturers who remained with conventional large HDDs lack futuristic vision? Was the market growth for the smaller HDDs a confirmed one at that time?

5.4.3 MR Head

Before the introduction of a MR element, the signal flux from a disk was read by an inductive head element composed of a soft magnetic core and a coil. The inductive head was used for both write and read operations. The read voltage is proportional to the rate of flux change and therefore to the relative velocity between the head and the disk. Shrinking the size of hard disks meant that the signal output voltage decreased, which inevitably lowered the performance (in terms of data capacity) of the HDDs. However, the magnitude of the read signal output voltage of the MR head did not depend on the velocity [63]. This difference had its origins in the different energy sources that were used to generate the read output. The energy of the inductive read head was supplied by the motion of disks, whereas that of the MR head was supplied by the sense current, which does not depend on the velocity.

Within a short period, this velocity-insensitive property of the MR heads became one of the main driving forces for increasing the data capacity of small HDDs. We can consider the MR head technology as a key for disruptive small HDD innovation. Thin-film head technology is regarded as an ancestor of MR heads, as they would not have appeared without the revolution in the production engineering of thin-film heads. We must remember that the MR head of an HDD was developed as an extension of the dual-element thin-film head [64]. It replaced the inductive read element of the dual inductive head with a simple MR read element.

The process to fabricate a thin-film head is not the same as the silicon LSI fabrication process. The former is a batch fabrication process in which thousands of simple passive devices for HDDs are fabricated on a wafer. Meanwhile, an LSI is produced by connecting many semiconductor elements to produce a large system. One of the advantages of batch fabrication is that it permits rejection of only defective devices rather than rejecting a whole wafer. This advantage is believed to have assisted the transition from inductive thin-film heads to MR heads. In contrast to this, other functional integrated devices, such as magnetic random-access memories (MRAMs) [65], have been taking a longer time to become mainstream in electronics. We can see a similar example in the development of optical interconnects where many small optical parts are integrated to produce an interconnect system [66].

The technical advantage of MR heads was not obvious during the 1980s. As the read output of an inductive thin-film head is proportional to the relative velocity between the head and the medium (Figure 5.6), the combination of a large disk and a head with additional coil turns for a higher read output voltage could provide cost-effective HDDs. In 1991, IBM introduced the first MR head for commercial HDDs [67]. The effort to increase the coil turns was limited by the bandwidth of the read circuit. Furthermore, the aforementioned mechanical disadvantages associated with a large disk accelerated the uptake of MR heads.

However, the effect of the improvement by the MR head was limited, as seen from Figure 5.6. Moreover, in the small-sized storage market, the nearest rival, flash memory, started to appear in the market in 2000 in the form of the USB flash drive.

5.4.4 GMR Head

Small form factor HDDs succeeded in leaving the flash drive behind with the introduction of the giant magneto-resistance (GMR) head in 1997. The sensitivity of the read head was much increased by replacing the MR head with a GMR head, which enabled higher capacity small HDDs to be realized.

Increasing the recording density of HDDs is limited by the deterioration of the bit error rate (BER) associated with the smaller read output signal from a smaller

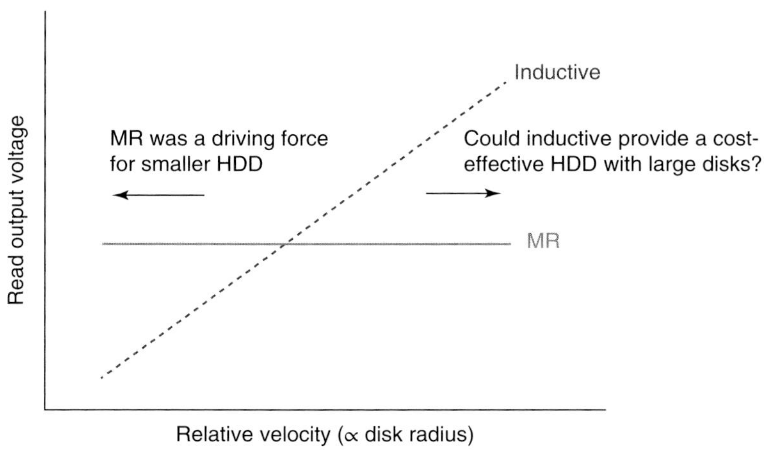

Figure 5.6 Read output voltage of heads vs relative velocity between head and disk.

bit pattern. The BER is improved by increasing the signal-to-noise ratio (SNR) of the read signal, which increases with the MR ratio, the maximum difference in resistance over the minimum resistance when magnetic fields are applied to the read element. The original MR head used the anisotropic magneto-resistance (AMR) of magnetic materials such as soft magnetic Ni–Fe. The origin of the AMR is the increase in the collision frequency of the conduction electrons when they travel parallel to the magnetization. The MR ratio of the AMR heads was about 1%.

The GMR effect was discovered with the creation of ferromagnetic/nonmagnetic multilayers in 1988 [68, 69]. Its origin is the spin-dependent scattering of the electrons, which increased the observed MR ratio to more than 30%, suggesting the possibility to obtain a read element that generates output voltage that is more than a magnitude higher. However, the problem was that the original GMR multilayer structure required a large magnetic field of more than 10 kOe (800 kA m^{-1}) to magnetically saturate the GMR multilayers to obtain such a large MR ratio, whereas the conventional AMR using a soft magnetic Ni–Fe film saturated with a much weaker field of ~10 Oe (0.8 kA m^{-1}).

In an attempt to solve this problem, Shinjo and Yamamoto examined a multilayer structure consisting of Co/Cu/Ni–Fe/Cu [70]. The magnetization direction of the high coercivity semi-hard Co film changed little but that of soft magnetic Ni–Fe flipped under a weak magnetic field (within ±50 Oe), resulting in a GMR effect with an MR ratio of 9.9%. In 1992, Yamamoto asked the author's magnetic head development team at NEC to build a practical GMR head. Our initial attempt to fabricate a head resulted in a low percentage of completeness. The author suspected that such a device would become unstable if a strong field is accidentally applied. To develop a practical GMR head, IBM invented a spin-valve structure consisting of a nonmagnetic metal interlayer sandwiched between ferromagnetic film pairs; the magnetization of one of the films was free to rotate (soft magnetic), whereas the magnetization of the other film was pinned by an antiferromagnetic film, which had been investigated to suppress the Barkhausen noise in an AMR head [71]. They succeeded to ship the first commercial HDD implementing GMR heads in 1997. The performance of the GMR head was excellent compared to that of the AMR head as shown in Figure 5.7, and the combination

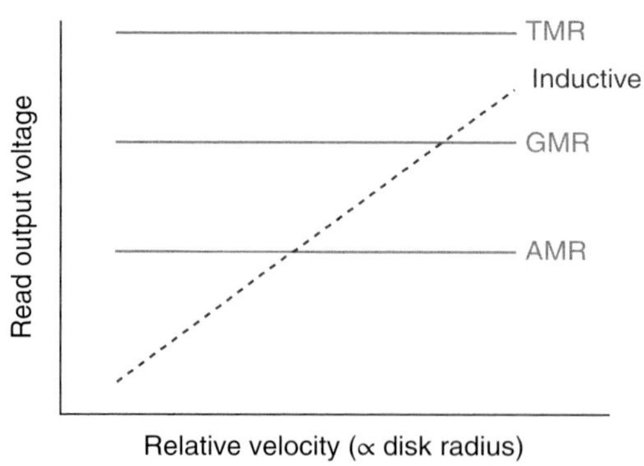

Figure 5.7 Inductive read lost decisively by increasing the read output voltage by the evolution of MR technology from AMR to GMR and then to TMR.

with small form factor HDDs opened new markets such as mobile applications and video files. This development provided the final trap to the inductive read head and the fortunes of the large HDD. At this time, the money did not flow to the small form factor HDD manufacturers but tended to gather at the GMR head manufacturers because "the companies that are positioned at a spot in a value chain where performance is not yet good enough will capture the profit" [72].

5.5 TMR Head

5.5.1 Infinite MR Ratio

Several candidates vied for the position of successor of the GMR head in 1998. Tunneling magneto resistance (TMR) was particularly attractive because the MR ratio could be infinite (∞) in a simple model. The possibility to provide an incommensurable increase in the SNR of the read signal voltage, which could lead to continuing the capacity increase of HDDs, had been discussed. TMR also attracted the attention of the solid-state memory industry. After the introduction of magnetic wire memory in computer and telephone switching, magnetic devices such as MRAMs no longer attracted attention except for a small magnetic bubble memory business [73]. One of the main reasons for this absence of interest came from the small on–off ratio of the resistance in MRAMs. The expected large on–off ratio of TMR seemed to raise MRAMs from the ashes.

5.5.2 Suspicions Surrounding the TMR Head

Many people, nevertheless, believed that the TMR read head was unpractical because of its large electrical resistance. In a TMR head, electrons are transmitted through the insulator barrier by tunneling, whereas in a GMR head, they travel through metal films. The resistance of a tunnel junction is usually characterized by the resistance–area product ($R{\cdot}A$). The $R{\cdot}A$ is the resistance of a junction with an area of 1 μm^2, which roughly corresponds to that of a read head with a track width of about 1 μm required for 10 Gbits/in.^{2} recording, as was the case around the year 2000. A typical $R{\cdot}A$ value of a TMR junction was about $10^8\ \Omega\ \mu m^2$ when the tunneling barrier was produced by plasma oxidation of aluminum [74]. In comparison, the estimated resistance of a TMR head was 100 MΩ, whereas the resistance of a typical GMR head for 10 Gbits/in.^{2} was about 10 Ω. Read heads would not operate at such a high resistance. Or, even if they could operate, the very large amount of shot noise generated by a device with such a high resistance would impede the reading data signal. To make matters worse, HDDs with a much higher recording density for 100 Gbits/in.^{2} or higher need a smaller junction area for TMR, which strongly suggested that the SNR of a TMR head would never exceed that of future GMR heads.

In addition, another potential demerit for a spin-valve type TMR head came from the suspicion that a thin tunnel barrier exposed on a head slider facing the disk surface may be problematic. The barrier might be smeared out during the slider-lapping process to produce a short circuit between both sides of the barrier. Furthermore, damage due to electric discharge might easily result from such a narrow electrical gap. To avoid these potential problems, flux guide-type TMR heads, in which the tunnel junction occupied a recessed position, were proposed and investigated [75, 76].

5.5.3 Low-Resistance TMR Head

A simple circuit simulation can be used to predict that the SNR is approximately proportional to the [MR ratio]/[square root of the head element resistance]. This suggested the importance of low resistance as well as a high MR ratio for achieving a lower bit error rate in higher density HDD applications. Thus, the SNR would be worse as the track width is narrowed. This discussion brought the pessimistic conclusion that TMR head development for future high-density HDDs was unpromising [77]. Besides the high resistivity, anxiety caused by the potential electrical short circuiting and the electrical discharge were discussed, suggesting the need for new techniques.

Under these negative circumstances, the research team of Hisanao Tsuge at NEC reported a low-resistance TMR junction fabricated by the in situ natural oxidation (ISNO) of a thin Al film [78]. In his process, a thin vacuum-deposited aluminum film was slowly oxidized by oxygen without breaking the vacuum to form a thin tunneling barrier. In their first paper, the $R{\cdot}A$ was reported to be 1.5 kΩ μm^2, which was soon decreased to less than 100 Ω μm^2, but it remained too high for narrow track HDD applications.

The author's device research team and Tsuge's materials research team started to collaborate to develop a practical TMR head with superior performance to the GMR heads prevailing in those days. The first TMR head that was fabricated by using ISNO had a resistance of 33 Ω [79]. Figure 5.8 shows photographs of this TMR head during the wafer fabrication process. Figure 5.9 shows one of the first TMR heads and its low-noise read output signal. The read output of the first TMR head was almost the same as that expected from test elements at a wafer level. This result dispelled the anxiety about deterioration arising through the head production process. The structure of the first TMR head was similar to that of the spin-valve GMR head except that the sense current flowed perpendicular to the plane of the element. One of the advantages of this current perpendicular to the plane (CPP) type of the TMR element is that the separation between shields, which responds to the spatial resolution of the read element, is less than half for the TMR head than for the GMR head.

Further improvement was achieved by developing a fabrication technique for producing a smooth barrier surface. Fujikata et al. developed a method to improve the roughness by introducing O_2 surfactant gas on the Ta-seed-layer under a TMR stack [80]. The $R{\cdot}A$ was decreased to 6 Ω μm^2 and the TMR ratio was increased to 15%, which would provide a recording density of more than 40 Gbit/in.^{2} [81].

These spin-valve-type TMR heads were manufactured on a GMR head production line. Moreover, the fabrication process from the wafer to the final head test was almost the same as those of conventional GMR heads. More important was

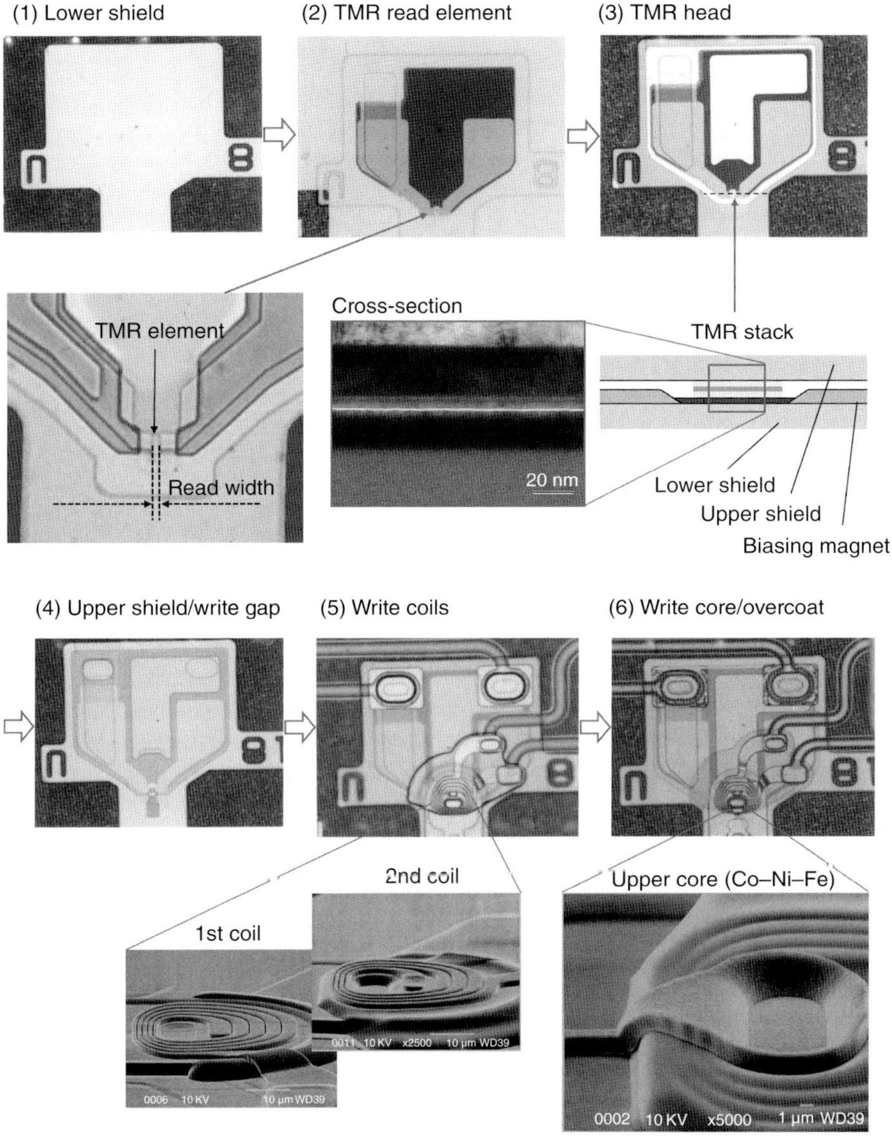

Figure 5.8 Photographs of the first practical TMR head. Left: Spin-valve-type TMR head element immediately after the last process of the read head had finished. Right: Combined head structure of a Co–Ni–Fe write head on a TMR element [79].

the compatibility for the head customers (HDD manufacturers) who had already been using GMR heads. The transition from GMR to TMR was smooth, similar to the previous shift from AMR to GMR [82] because TMR heads can be used as an almost upwardly compatible device to GMR heads by the customers except that it has higher signal output. No notable damage as a consequence of electric short circuitry or by electric discharge was observed for carefully fabricated

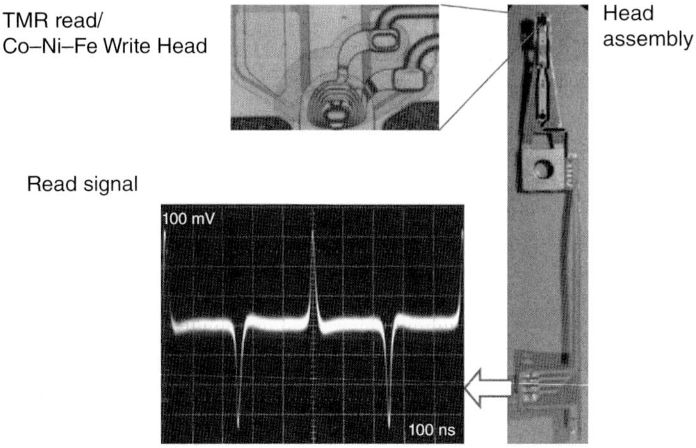

Figure 5.9 TMR head assembly (left) and read signal of a hard disk from the TMR head (right) [79].

TMR heads. We succeeded in paving the way to offer TMR heads as standard key components for HDDs alongside GMR heads.

Our team showed that TMR is the most promising technology for the post-GMR era. In previous transitions between critical head technologies, such as those from ferrite to thin film, thin film to AMR, and AMR to GMR, almost all the HDD manufacturers and the head manufacturers followed the IBM recipe. The TMR head was the first technology in which a recipe other than that of IBM was employed.

5.5.4 MGO: The Final Push

The performance of our first TMR heads in 2000 was almost the same as or slightly higher than that of state-of-the-art GMR heads. However, future read heads for HDDs with a higher recording density would require a higher MR ratio because a smaller TMR junction has a higher $R{\cdot}A$.

In 2001, two papers predicted that an Fe/MgO/Fe(001) tunneling junction offers an extremely large TMR by coherent tunneling [83, 84]. Contrary to this result, early experiments with an MgO barrier revealed a much lower MR ratio than the prediction. One of the causes of the disagreement was speculated to be the use of a polycrystalline electrode and amorphous MgO [85]. Yuasa et al. in Japan reported an MR ratio of ~180% at room temperature (RT) in a fully epitaxial Fe/MgO/Fe TMR junction [86]. They prepared a single-crystal Fe(001)/MgO(001)/Fe(001) film using molecular beam epitaxy (MBE). In that same period, Parkin et al. at IBM demonstrated an MR ratio of ~220% at RT with a sputter-deposited polycrystalline TMR junction grown on an amorphous underlayer, but with highly oriented (100) MgO [87]. The latter suggested the possibility of using an MgO barrier to manufacture a TMR head with a high MR ratio TMR head.

Tsunekawa's team at ANELVA had developed a sputtering machine for depositing a smooth surface nano-oxide layer (NOL) in a GMR read head. They found the machine capable of providing an excellent TMR junction with a low $R{\cdot}A$ of 2.4 Ω μm^2 and high MR ratio of 138%, which promised to provide a read head technology for HDDs with higher density of more than 100 Gbit/in.^{2}, by magnetron sputtering using an MgO target to achieve a highly textured MgO [001] film [88]. Many head manufacturers subsequently introduced the machines to manufacture TMR heads. Finally, Seagate succeeded in producing the first commercial HDDs using a TMR head in 2004 [89], and soon all the read heads in HDDs became TMR that were fabricated by using an ANELVA's machine.

The exponential increase in the recording density of HDDs due to the radical and discontinuous innovations of MR to TMR was above that of semiconductor memory for about 20 years, during which time the use of solid-state drives (SSDs) for computers has been prevented.

5.5.5 Exploring New Markets

After NEC ended their HDD development program, they continued to supply floppy-disk drives (FDDs) for many years. The author started a new program to develop a TMR head for high-density FDDs. Prototype TMR heads originally designed for HDDs were reassembled as TMR heads for this purpose. A Zip 100-type flexible disk and a higher performance version of that supplied by a Japanese floppy disk manufacturer were used for the evaluation.

The evaluated error rate suggested that it could support a very high capacity of more than 1 GBytes/disk, which is 10 times higher than that of conventional Zip drives [90]. One of the problems with removable disk drives is the thermal asperity (TA), i.e. the noise generated when the read element hits the medium. In our experiment, the TA due to the TMR head was smaller than that of the AMR head, which could be explained by the smaller temperature coefficient of resistivity with TMR. These results supported the possibility to explore a new market for very high capacity removable disks. However, the technology trend of the times stopped this project. The high-density FDD business itself came to an end soon. These forms of removable storage were almost completely replaced with USB flash drives by around 2010.

5.6 Discussion

Christensen discussed small form factor HDD as a typical disruptive innovation [7]. However, the small form factor HDD by itself is not a sufficient condition for being a disruptive technology. One of the strongest technical driving forces in creating the new market for small HDDs was MR head technology. The signal read output of an MR head is independent of disk velocity or disk size, which formed the high-performance base of small HDDs. Even AMR head technology might not have been sufficient to enable HDDs to compete with semiconductor flash memories after 2000. It was the successive novel magnetic sensor development of GMR and TMR, neither of which was even predicted in physics textbooks before the 1970s, which had resulted in small HDDs becoming the main storage device in computers.

Thin-film head technology was regarded as a typical sustaining technology in the existing market in *The Innovator's Dilemma* [7]. From the viewpoint of manufacturing engineering, an MR head is a kind of thin-film head that is built by using film deposition and etching combined with photolithography or micro/nano-lithography rather than by machining bulk materials. Thin-film head technology requires careful treatment of the substrate surface and sometimes creates unforeseen excellent performance. The nanocrystalline Co–Ni–Fe film electrodeposited from a bath without sulfur was one of these successes in electrodeposition [51]. A similar technology was also developed in vacuum deposition known as the "surfactant" technology for GMR [91]. Tsunekawa's machine [88] for producing smooth TMR junctions, which had been used by all the TMR head manufacturers, would be seen as an extension of these technologies. Without the initial challenge to the thin-film head and its core idea of controlling the interface at the atomic level, the low-cost mass production of ultrahigh-performance magnetic heads for small HDDs would never have been realized.

Integrated magnetic circuits were proposed in the 1960s as a batch fabrication technology for magnetic heads [92]. In contrast to the rapid growth of the semiconductor IC market, commercialization of the integrated magnetic circuit required almost 20 years to enjoy success as a thin-film head [93]. There were many trials that showed no promise to succeed. It would be natural to conclude that the challenges of creating both market needs and technological innovation seeds are necessary for business.

Acknowledgments

The author is grateful toward Prof. Tetsuya Osaka and Prof. Richard Alkire for recommending and encouraging this chapter. He highly appreciates the contribution of Dr. Hiroshi Gokan, who led the collaborative research between Dr. Hisanao Tsuge and the author to develop the first practical TMR head, and who read this manuscript and gave considerable advice to improve it. He would also like to thank Dr. Koji Tsunekawa of ANELVA Corporation for the comments on the development of the magnetron-sputtering machine for MgO film of TMR heads. The author would like to thank many people in NEC especially the three engineers who took up the challenge to start the thin-film head business, Mr. Masaru Ito, Mr. Kozo Iijima, and Mr. Takao Matsubara. He also thanks Mr. Yutaka Ikeuchi and Mr. Makoto Watanabe for supporting the development of the thin-film head; and Mr. Takao Maruyama, Mr. Tsutomu Ishi, Dr. Nobuyuki Ishiwata, Dr. Masafumi Nakada, Prof. Mikiko Saito, Mr. Fujio Maeda, Dr. Junichi Fujikata, Prof. Madoka Takai, and Dr. Hiroaki Honjo for research discussions.

References

1 Von Neumann, J. (1945). *First Draft of a Report on the EDVAC*. Moore School of Electrical Engineering, University of Pennsylvania.
2 Daniel, E.D., Mee, C.D., and Clark, M.H. (1988). *Magnetic Recording, the First 100 Years*. New York: IEEE Press.
3 The Storage Engine (2016) *Computer History Museum*. http://www.computerhistory.org/storageengine/ (accessed 30 November, 2016).
4 Croll, I. (2005). *Oral history panel on advanced thin film disk head. Computer History Museum* X3362: 2006.
5 Romankiw, L.T. and Krongelb, S. (2015). The path from invention to product for the magnetic thin film head. In: *Electrochemical Engineering Across Scales: From Molecules to Processes* (ed. R.C. Alkire, P.N. Bartlett and J. Lipkowski). Wiley-VCH Verlag GmbH & Co. KGaA.
6 Andricacos, P.C., Uzoh, C., Dokovic, J.O. et al. (1998). Damascene copper electroplating for chip interconnections. *IBM Journal of Research and Development* 42 (5): 567–572.
7 Christensen, C.M. (1997). *The Innovator's Dilemma: When New Technologies Cause Great Firms to Fail*. New York: HarperCollins Publishers, Inc.
8 Chikazumi, S. (1997). *Physics of Ferromagnetism*, 2e, 600. Oxford: Oxford University Press Inc.
9 Takayasu, M., Uchiyama, S., Takahashi, K., and Fujii, T. (1974). Isothermal annealing of zero-magnetostrictive permalloy films II experimental. *IEEE Transactions on Magnetics* 10 (3): 552–555.
10 Anderson, R.L., Gangulee, A., and Romankiw, L.T. (1973). Annealing behavior of electroplated permalloy thin films. *Journal of Electronic Materials* 2 (2): 161–176.
11 Mitsuhashi, S., Hanagata, S., Takaba, T., Konayashi, T, and Komiyama, Y. (1974). Electrodeposited head. Annual Conference of IECE Japan, S17-8, In Japanese.
12 Hidaka, Y., Matsuyama, K., and Konishi, S. (1983). Experimental confirmation of fundamental functions for a novel Bloch line memory. *IEEE Transactions on Magnetics* 19 (5): 1841–1843.
13 Ishiwata, N., Wakabayashi, C., and Urai, H. (1991). Soft magnetism of high-nitrogen-concentration FeTaN films. *Journal of Applied Physics* 69 (8): 5616–5618.
14 Okumura, T., Osaka, A., Ishiwata, N. et al. (1993). High frequency read/write characteristics for laminated Fe-Ta-N heads. *IEEE Transactions on Magnetics* 29 (6): 3843–3845.
15 Romankiw, L.T. (1980). Use of AZ-type photoresist as insulation in multiturn thin film recording heads. *IBM Technical Disclosure Bulletin* 23: 2584.
16 Hempstead, R.D. and Money, J.B. (1979). Thin film head having negative magnetostriction. US Patent 4, 242,710, filed 29 January 1979 and issued 30 December 1980.

17 Ohashi, K., Ito, M., and Maruyama, T. (1990). Effect of plated NiFe composition on domain configuration. In: *Magnetic Materials, Processes, and Devices*, vol. 90-8 (ed. L.T. Romankiw and D.A. Herman Jr.,) 247–258. Pennington, NJ: The Electrochemical Society.

18 N7761 and N7765 Hard Disk Units (2016). Computer Museum. Information Processing Society of Japan, http://museum.ipsj.or.jp/en/computer/device/magnetic_disk/0047.html (accessed 3 September 2016).

19 Suganuma, Y., Tanaka, H., Yanagisawa, M. et al. (1982). Production process and high density recording characteristics of plated disks. *IEEE Transactions on Magnetics* 18 (6): 1215–1220.

20 Watari, T. and Murano, H. (1985). Packaging technology for the NEC SX supercomputer. *IEEE Transactions on Components, Hybrids, and Manufacturing Technology* 8 (4): 462–467.

21 Mitsuya, Y. and Takanami, S. (1987). Technologies for high recording density in large-capacity fast-access magnetic disk storage. *IEEE Transactions on Magnetics* 23 (5): 2674–2679.

22 Ohashi, K., Ito, M., and Watanabe, M. (1988). Application of electroplating to thin film heads: an overview. In: *Proceedings of the Symposium on Electrochemical Technology in Electronics*, vol. 88-23 (ed. L.T. Romankiw and T. Osaka), 525–542. Pennington, NJ: The Electrochemical Society.

23 Goto, F. and Osaka, T. (1989). The application of electroless plating to high-density hard disks. *Journal of the Surface Finishing Society of Japan* 40 (1): 46–47. In Japanese.

24 Becker, R. and Döring, W. (1939). *Ferromagnetismus*, 178. Berlin: Julius Springer Verlag.

25 Jones, R.E. Jr. (1979). Domain effects in the thin film head. *IEEE Transactions on Magnetics* 15 (6): 1619–1621.

26 Lazzari, J.P. and Melnick, I. (1971). Integrated magnetic recording heads. *IEEE Transactions on Magnetics* 7 (1): 146.

27 Herman, D.A. Jr., Argyle, B.E., Trouilloud, P.L. et al. (1988). Study of field-driven wall-configuration conversions for laminated permalloy in the easy-axis state. *Journal of Applied Physics* 63 (8): 4036–4038.

28 Ohashi, K. (1985). Mechanism of 90° wall motion in thin film heads. *IEEE Transaction on Magnetics* 21 (5): 1581–1583.

29 Ohashi, K. (1994). Magnetization dynamics in thin film heads. *Journal of Magnetism and Magnetic Materials* 134 (2–3): 262–267.

30 Re, M. and Kryder, M. (1984). Magneto-optic investigation of thin film recording heads. *Journal of Applied Physics* 55 (6): 2245–2247.

31 Kasiraj, P., Shelby, R.M., Best, J.S., and Horne, D.E. (1986). Magnetic domain imaging with a scanning Kerr effect microscope. *IEEE Transactions on Magnetics* 22 (5): 837–839.

32 Imai, K., Nonaka, I., and Takeuchi, H. (1985). Managing the product development process: how Japanese companies learn and unlearn. In: *The Uneasy Alliance: Managing the Productivity-Technology Dilemma* (ed. K.B. Clark, R.H. Hays and C. Lorenz), 337–376. Boston: Harvard Business School Press.

33 Romankiw, L.T. and Osaka, T. (1988). Preface. In: *Proceedings of the Symposium on Electrochemical Technology in Electronics*, vol. 88-23. Pennington, NJ: The Electrochemical Society.
34 Arnoldussen, T.C. (1995). Film media. In: *Magnetic Recording Technology*, 2e, vol. 4 (ed. C.D. Mee and E.D. Daniel), 4.7–4.19. New York: McGraw-Hill.
35 Osaka, T., Kasai, N., Koiwa, I. et al. (1982). A preparation of perpendicular magnetic recording media by an electroless plating method. *Journal of the Electrochemical Society* 130 (3): 568–571.
36 Okimoto, D. and Nishi, Y. (1994). R&D organization in Japanese and American semiconductor firms. In: *The Japanese Firm: Sources of Competitive Strength* (ed. M. Aoki and R. Dore), 178–208. Oxford: Oxford University Press.
37 Lu, P.-L. and Charap, S.H. (1994). Thermal instability at 10 $Gbit/in^2$ magnetic recording. *IEEE Transactions on Magnetics* 30 (6): 4230–4232.
38 Ohashi, K., Takagi, H., Tsunashima, S. et al. (1979). Magnetic aftereffect due to domain wall motion in amorphous TbFe sputtered films. *Journal of Applied Physics* 50 (B3): 1611–1613.
39 Mansuripur, M. (1995). *The Physical Principles of Magneto-Optical Recording*. Cambridge: Cambridge University Press.
40 Fujikata, J., Ishi, T., Yokota, H. et al. (2004). Surface plasmon enhancement effect and its application to near-field optical recording. *Transaction of the Magnetics Society of Japan* 4 (4–2): 255–259.
41 Ohashi, K., Fujikata, J., Ishi, T. et al. (2004). Enhancement of near-field light by surface plasmons. *Journal of the Magnetics Society of Japan* 28 (6): 774–780. In Japanese.
42 Rottmayer, R.E., Batra, S., Buechel, D. et al. (2006). Heat-assisted magnetic recording. *IEEE Transactions on Magnetics* 42 (10): 2417–2412.
43 Iwasaki, S.-I. (1980). Perpendicular magnetic recording. *IEEE Transactions on Magnetics* 16 (1): 71–76.
44 Honda, N., Ouchi, K., and Iwasaki, S.-I. (2002). Design consideration of ultrahigh-density perpendicular magnetic recording media. *IEEE Transactions on Magnetics* 38 (4): 1615–1621.
45 Bajorek, C.A., Krongelb, S., Romankiw, L.T., and Thompson, D.A. (1975). An integrated magnetoresistive read, inductive write high density recording head. In: *AIP Conference Proceedings of Magnetism and Magnetic Materials-1974*, vol. 24, 548–549.
46 Robertson, N., Hu, H.L., and Tsang, C. (1997). High performance write head using NiFe 45/55. *IEEE Transactions on Magnetics* 33 (5): 2818–2820.
47 Osaka, T., Takai, M., Nakamura, A. et al. (1994). Preparation of electrodeposited Fe-P films and their soft magnetic properties. *Journal of the Magnetics Society of Japan* 18 (S1): 187–190.
48 Wolf, I.W. (1962). Electrodeposition of magnetic materials. *Journal of Applied Physics*, Supplement to 33 (3): 1152–1159.
49 Nakamura, A., Takai, M., Hayashi, K., and Osaka, T. (1996, 1996). Preparation and magnetic properties of CoNiFe thin film by electrodeposition. *Journal of the Surface Finishing Society of Japan* 47 (11): 934–938.

50 Takai, M., Hayashi, K., Aoyagi, M., and Osaka, T. (1997). Electrochemical preparation of soft magnetic CoNiFeS film with high saturation magnetic flux density and high resistivity. *Journal of the Electrochemical Society.* 144 (7): L203–L204.

51 Osaka, T., Takai, M., Hayashi, K. et al. (1998). New soft magnetic CoNiFe plated films with high $B_s = 2.0–2.1$ T. *IEEE Transactions on Magnetics* 34 (4): 1432–1434.

52 Ohashi, K., Yasue, Y., Saito, M. et al. (1998). Newly developed inductive write head with electroplated CoNiFe film. *IEEE Transactions on Magnetics* 34 (4): 1462–1464.

53 Osaka, T., Takai, M., Hayashi, K. et al. (1998). A soft magnetic CoNiFe film with high saturation magnetic flux density and low coercivity. *Nature* 392 (6678): 796–798.

54 Osaka, T., Asahi, T., Kawaji, J., and Yokoshima, T. (2005). Development of high-performance magnetic thin film for high-density magnetic recording. *Electrochimica Acta* 50: 4576–4585.

55 Ohashi, K., Morita, N., Tsuda, T., and Nonaka, Y. (1999). Write performance of heads with a 2.1-Tesla CoNiFe pole. *IEEE Transactions on Magnetics* 35 (4): 2538–2540.

56 Osaka, T., Takai, M., Sogawa, Y. et al. (1999). Influence of crystalline structure and sulfur inclusion on corrosion properties of electrodeposited CoNiFe soft magnetic films. *Journal of the Electrochemical Society* 146 (6): 2092–2096.

57 Saito, M., Yamada, K., Ohashi, K. et al. (1999). Corrosion properties of electroplated CoNiFe films. *Journal of the Electrochemical Society* 146 (8): 2845–2848.

58 Nonaka, Y., Honjo, H., Toba, T. et al. (2000). Co-Ni-Fe write heads with a 10-μm yoke length for high-speed recording. *IEEE Transactions on Magnetics* 36 (5): 2514–2516.

59 McKendrick, D.G., Doner, R.F., and Haggard, S. (2000). *From Silicon Valley to Singapore: Location and Competitive Advantage in the Hard Disk Drive Industry*, 104–105. Stanford, CA: Stanford University Press.

60 Shirle, N. and Lieu, D.K. (1996). History and trends in the development of motorized spindles for hard disk drives. *IEEE Transactions on Magnetics* 32 (3): 1703–1708.

61 Schlichting, H. and Gersten, K. (2000). *Boundary Layer Theory*, 8th revised and enlarged edition, 124. Berlin, Heidelberg: Springer-Verlag.

62 McKendrick, D.G., Doner, R.F., and Haggard, S. (2000). *From Silicon Valley to Singapore: Location and Competitive Advantage in the Hard Disk Drive Industry*, 48. Stanford, CA: Stanford University Press.

63 Hunt, R.P. (1971). A magnetoresistive readout transducer. *IEEE Transactions on Magnetics* 7 (1): 150–154.

64 Romankiw, L.T. (1974). Integrated magnetoresistive read, inductive write, batch fabricated magnetic head. US Patent 3, 908,194, filed 19 August 1974 and issued 23 September 1975.

65 Stamp, R.L., Breitkreutz, S., Åkerman, J. et al. (2014). The 2014 magnetism roadmap. *Journal of Physics D* 47 (33): 33301.

66 Optical PCB interconnects, niche or mainstream? *ISSCC 2012*, Evening Discussion Session, ES4, 20 February 2012. http://ieeexplore.ieee.org/stamp/stamp.jsp?arnumber=6177111 (accessed 18 January 2017).
67 Tsang, C., Chen, M.-M., Yogi, T., and Ju, K. (1990). Gigabit density recording using dual-element MR/inductive heads on thin-film disks. *IEEE Transactions on Magnetics* 26 (5): 1689–1693.
68 Baibich, M.N., Broto, J.M., Fert, A. et al. (1988). Giant magnetoresistance of (001)Fe/(001)Cr magnetic superlatices. *Physical Review Letters* 61 (21): 2472–2475.
69 Binasch, G., Grünberg, P., Saurenbach, F., and Zinn, W. (1989). Enhanced magnetoresistance in layered magnetic structures with antiferromagnetic interlayer exchange. *Physical Review B* 39 (7): 4828–4830.
70 Shinjo, T. and Yamamoto, H. (1990). Large magnetoresistance of field-induced giant ferromagnetic multilayers. *Journal of The Physical Society of Japan* 59 (9): 3061–3064.
71 Dieny, B., Speriosu, V.S., Parkin, S.S.P. et al. (1991). Giant magnetoresistance in soft ferromagnetic multilayers. *Physical Review B* 43 (1): 1297–1300.
72 Christensen, C.M. and Raynor, M.E. (2013). *The Innovator's Solution: Creating and Sustaining Successful Growth*. Boston: Harvard Business School Publishing.
73 Eschenfelder, A.H. (1981). *Magnetic Bubble Technology*, 2nd corr. and updated. Berlin, Heidelberg: Springer.
74 Moodera, J.S., Kinder, L.R., Wong, T.M., and Meservey, R. (1995). Large magnetoresistance at room temperature in ferromagnetic thin film tunnel junctions. *Physical Review Letters* 74 (16): 3273–3276.
75 Coehoorn, R., Cumpson, S.R., Ruigrok, J.J.M., and Hidding, P. (1999). The electrical and magnetic response of yoke-type read heads based on a magnetic tunnel junction. *IEEE Transactions on Magnetics* 35 (5): 2586–2588.
76 Shimazawa, K., Redon, O., Kasahara, N. et al. (2000). Evaluation of front flux guide-type magnetic runnel junction heads. *IEEE Transactions on Magnetics* 36 (5): 2542–2544.
77 Mallary, M.L. (1999). Theoretical signal to noise comparison between GMR and magnetic tunnel junction sensors, in Intermag 99 Digest, BP-11.
78 Tsuge, H. and Mitsuzuka, T. (1997). Magnetic tunnel junctions with in situ naturally-oxidized tunnel barrier. *Applied Physics Letters* 71 (22): 3296–3298.
79 Ohashi, K., Hayashi, K., Nagahara, K. et al. (2000). Low-resistance tunnel magnetoresistive head. *IEEE Transactions on Magnetics* 36 (5): 2549–2553.
80 Fujikata, J., Ishi, T., Mori, S. et al. (2001). Low resistance magnetic tunneling junctions and their interface structures. *Journal of Applied Physics* 89 (11): 7558–7560.
81 Ishihara, K., Nakada, M., Fukami, E. et al. (2001). Read performance of tunneling magneto-resistive heads. *IEEE Transactions on Magnetics* 37 (4): 1687–1690.
82 Dendrick, J. and Kraemer, L. (2015). Who captures value from science-based innovation? The distribution of benefits from GMR in the hard disk drive industry. *Research Policy* 44: 1615–1628.

83 Butler, W.H., Zhang, X.-G., and Schulthess, T.C. (2001). Spin-dependent tunneling conductance of Fe|MgO|Fe sandwiches. *Physical Review B* 63: 054416.

84 Mathon, J. and Umerski, A. (2001). Theory of tunneling magnetoresistance of an epitaxial Fe/MgO/Fe(001) junction. *Physical Review B* 63: 220403(R).

85 Swagten, H.J.M. (2008). Spin-dependent tunneling in magnetic junctions. In: *Handbook of Magnetic Materials*, vol. 17 (ed. K.H.J. Buschow), 1–121. Amsterdam, The Netherlands: North-Holland.

86 Yuasa, S., Nagahama, T., Fukushima, A. et al. (2004). Giant room-temperature magnetoresistance in single-crystal Fe/MgO/Fe magnetic tunnel junctions. *Nature Materials* 3: 868–871.

87 Parkin, S.S.P., Kaiser, C., Panchula, A. et al. (2004). Giant tunneling magnetoresistance at room temperature with MgO (100) tunnel barriers. *Nature Materials* 3: 862–867.

88 Tsunekawa, K., Djayaprawira, D.D., Yuasa, S. et al. (2006). Huge magnetoresistance and low junction resistance in magnetic tunnel junctions with crystalline MgO barrier. *IEEE Transactions on Magnetics* 42 (2): 103–107.

89 Mao, S., Chen, Y., Liu, F. et al. (2006). Commercial TMR heads for hard disk drives: characterization and extendibility at 300 Gbit/in^2. *IEEE Transactions on Magnetics* 42 (2): 97–102.

90 Ohashi, K., Sato, A., Ishihara, K. et al. (2001). Thermal asperity of TMR heads for removable disk drives. *IEEE Transactions on Magnetics* 37 (4): 1919–1921.

91 Yang, D.X., Shashishekar, B., Chopra, H.D., Chen, P.J., and Egelhoff, W.F. Jr. (2001) Atomic engineering of spin valves using Ag as a surfactant, *Journal of Applied Physics*, 89 (11), 7121–7123.

92 Gregg, D.P. (1961) Deposited film transducing apparatus and method of producing the apparatus. US Patent 3, 344,237, filed 19 April 1961 and issued 26 September 1967.

93 Romankiw, L.T. and Thompson, D.A. (1979) Thin film inductive transducer. US Patent 4, 295,173, filed 18 October 1979 and issued 13 October 1981.

6

Development of Fully-Continuous Electrokinetic Dewatering of Phosphatic Clay Suspensions

Rui Kong, Arthur Dizon, Saeed Moghaddam, and Mark E. Orazem

University of Florida, Department of Chemical Engineering, PO Box 116005, Gainesville, FL 32611-6005, USA

Our group at the University of Florida has been working with Mosaic, a company that mines phosphate rock from nearly 200 000 acres of Mosaic-owned land in Central Florida to develop faster ways to separate water from clay suspensions generated as a waste stream in beneficiation of phosphate ores. The search for enhanced solids–liquid separation was motivated by the need to reduce both the land area dedicated to storage of phosphatic clay suspensions and the water consumed in the mining operation. The objective of this chapter is to illustrate the sequential development of a continuous electrokinetic separation process.

The present volume provides examples of experimental and/or computational methods that have been used to facilitate the task of moving atomistic-scale discoveries and understanding toward well-engineered products and processes based on electrochemical phenomena. The work presented in this chapter could be described as empiricism guided by the understanding gained through a constitutive relationship yielding the increase in solids content as a function of the applied electric field and elapsed time. Sophisticated mathematical models showed the deficiency of a batch application and motivated the use of continuous processing. While electron microscopy and surface analyses provided insight into the structure of the clay matrix, the design of successive prototypes relied on intuition, insights gained from previous prototypes, and interactions with our sponsor.

6.1 Introduction

The present work addresses the need for enhanced separation of phosphatic clay suspensions generated by the production of fertilizer from phosphatic ore deposits. The source of phosphatic clay suspensions, the scope of industrial practice, and associated challenges for separation of the suspensions are presented in this section.

Electrochemical Engineering: The Path from Discovery to Product,
First Edition. Edited by Richard C. Alkire, Philip N. Bartlett, and Marc T. Koper.

6.1.1 Phosphatic Clay Suspensions

A dilute 2-3 wt% suspension of phosphatic clay is produced as a waste stream in the beneficiation of phosphate ore. A typical Florida phosphate mining operation produces more than 100 000 gallons/min of phosphatic clay. The suspensions are pumped into large impoundments called clay settling areas (CSA) in which separation is achieved by hindered settling and self-consolidation. As it settles, the supernatant water is recycled for use in the beneficiation plant. A top crust is formed after a few years, but the clay beneath the crust has a large water content and a pseudo-plastic character that limits the amount of weight the settling area can support. As much as 25–50 years are needed to reach a solids content of 40 wt%. CSA cover about 30–40% of the mined land. Uses for the land are limited by the properties of the clay that leave the settling areas unstable. The Florida Department of Environmental Protection passed a law in 1982 that mandated the restoration of mined land and supported research in methods that would expedite the dewatering of phosphatic clay.

Methods that employ electrokinetic phenomena have been the subject of significant research as a possible means of accelerating the dewatering process. Reuss observed the movement of clay particles under the influence of an applied electric field [1], an early application of the Volta cell reported in 1800 [2]. Previous research to develop batch electrokinetic dewatering processes resulted in the determination that the electrokinetic dewatering was not economically feasible. The objective of our work was to overcome the economic hurdle through the development of continuous electrokinetic dewatering.

6.1.2 Industrial Scope

Phosphorus is an essential nutrient used in fertilizer. The U.S. Geological Survey reports [3] that in 2017, the United States was the third largest producer of phosphate. China and Morocco and Western Sahara were the first and second largest producers, respectively. The current global production of phosphate is shown in Figure 6.1. The U.S. Geological Survey [3] suggests that, given current consumption patterns, the distribution of phosphate mining will shift to locations where phosphate reserves are in abundance. Currently, the top three phosphate-producing countries account for 75% of the current production. In contrast, Morocco and Western Sahara hold 75% of the global phosphate reserves, whereas the U.S. and China account for only 6% of the global reserve. In order to support the global demand for phosphate, which is only expected to increase [4], phosphate mining must move to North Africa.

Phosphate mining has a long history in Florida starting in the late 19th century in the Peace River region near Arcadia, FL. Through industrial development, the capacity of Florida phosphate mining increased due to the introduction of electric drag-line excavators replacing the initial manual shovel and handpick mining. The increased mining was accompanied with increased generation of waste stream such as phosphatic clay. Phosphate mining occurs in seven locations in Florida, and Florida mines accounted for 65% of domestically mined phosphate [26]. The accompanying environmental impacts, such as CSAs, demand on local water, and land used for mining, are a concern for the industry [27]. Over 100 000 acres of CSAs are present in Florida [28].

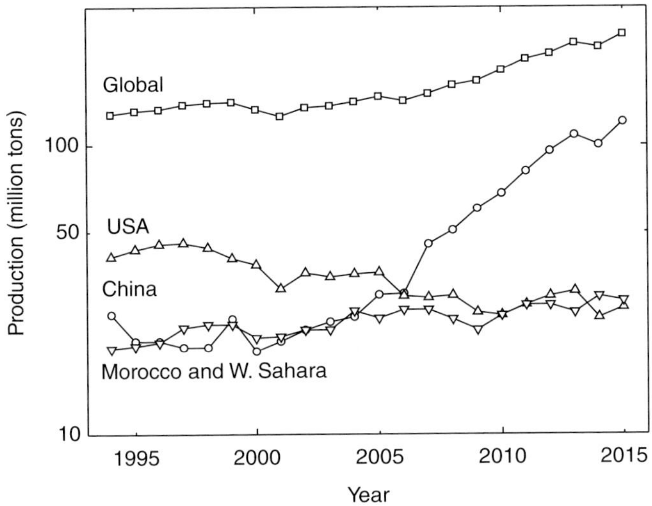

Figure 6.1 Phosphate production in major counties from 1994 to 2016. All data from U.S. Geological Survey, Mineral Commodity Summaries, Phosphate Rock [3, 5–23, 25].

Phosphate fertilizers are manufactured from phosphoric acid that is produced from mining phosphate deposits that contain clay minerals [29]. Phosphate mining is a strip-mining process in which drag lines expose the 10–20 ft of phosphate-rich matrix, which lies below 15–20 ft of overburden [30]. The phosphate-rich matrix consists of phosphate ore, clay, and sand in approximately equal proportions [31]. A flotation process is used to separate the desired phosphate ore from the mined matrix, leaving behind the sand and clay as waste streams. The phosphate ore is further processed to phosphoric acid, the sand is reinserted into the open-pit mines, and the phosphatic clay is transferred to the CSAs to begin the long dewatering process by gravity settling.

As a result of industry growth and mining technology development, the area allocated to CSAs increased. The problem of growing CSAs was identified by Tyler and Waggaman who suggested that phosphatic slimes could be further processed to remove remaining phosphate and decrease the volume of effluent destined for CSAs [32]. Continued concern resulted in the intervention of the Florida government in the Phosphate Land Reclamation Act 28.378(2) that aimed to address the CSAs by mandating the restoration of mined lands and the development of technologies to expedite dewatering.

6.1.3 Why is Separation of Water from Clay Difficult?

The difficulty of sedimentation and consolidation of phosphatic clay can be attributed to the particle–water interactions and clay particle geometry. The water surrounding particles can be categorized into four layers. The closest layer of water is the water of hydration, which is characterized as being chemically bound to the molecules of the clay particle and can only be removed by thermal means. The next layer consists of vicinal water, which is loosely bound to the surface of the clay particle. The interstitial water is the next layer that is bound into the agglomerated structures of the particle. This layer of water can be

Figure 6.2 Scanning electron microscope image of phosphatic clay suspension before dewatering. Source: Image taken by Lai and coworkers [35].

released if the agglomerated structure is destroyed. The last layer is free water, which is the bulk water of the suspension that is characterized by free movement and minimal interactions with the particles [33].

The clay minerals comprising phosphatic clays consist of layers of alumina and silica. Due to lattice defects and ion exchange, the particles develop an overall negative charge that contributes to the strong adherence of hydrating water and the binding of vicinal water. The layered molecular structure results in flat particles that are on average 10 μm in length and less than a 1 μm in thickness with a high surface area of 786 $m^2\ g^{-1}$ [34]. The particle charge and shape create a highly tortuous clay fabric that impedes settling and consolidation as observed in the scanning electron microscopy image of 2–3 wt% phosphatic clay shown in Figure 6.2.

6.2 Current Methods

The search for improved dewatering methods has led to an array of methods, including flocculation, which is employed mainly in water clarification applications, and various mechanical dewatering methods, which are used for clay thickening.

6.2.1 Flocculation

Flocculation is a common liquid–solid separation method utilized in water treatment, paper, and mining operations. Sedimentation is achieved by agglomerating the particles through the addition of flocculating agents. Flocculating agents are often high-molecular-weight molecules such as polyethylene oxide and polyacrylamide [36]. The agglomerates called flocs can then be mechanically separated from the dispersing liquid [37]. Recent work on sludge dewatering using ultra-high-molecular-weight flocculants like methyl acrylate-based flocculants shows the continuing interest in the method [38].

The effect of flocculating agents, specifically Unique Molecular Architecture flocculant, have been observed on phosphatic clay [39]. Currently in the phosphate mining, flocculation is used as the first phase of dewatering phosphatic clay and can produce a 10 wt% sludge from the initial 2–3 wt% slurry. Flocculation serves to accelerate the initial sedimentation, but cannot produce cake with high solids content.

6.2.2 Mechanical Dewatering

The most common methods for solid–liquid separation use mechanical energy, and use of mechanical methods is widespread in processing waste water sludge. Mechanical methods primarily remove the water in the free and interstitial layers. A review of these methods on activated sludge was presented by Vaxelaire and Cézac [40].

Mechanical methods such as filter presses and centrifuges apply mechanical forces to cause dewatering by the removal of free water [41–43]. The extent of dewatering by mechanical methods is limited because the free water is removed, leaving behind the more closely bound layers of water. Mechanical dewatering is often supplemented with methods that utilize electricity, thermal energy, magnetism, and acoustics. An example is the electrokinetic belt filter that has a conductive filter medium [44]. Thermal drying also has been added to assist in dewatering the discharge sludge from centrifuge processes [45]. Magnetic and acoustic techniques are also under development [46].

6.2.3 Electrokinetic Separation

The movement of clay particles under the influence of an electric field was reported by Reuss [1]. This observation could be explained by the mathematical concept of a double-layer [47]. In 1942, a thorough evaluation of electrokinetic phenomena was performed using kaolin [48].

The process of electrokinetic separation of clay and water can be divided into two main categories: electrophoresis and electro-osmosis. Electrophoresis is the movement of a charged particle in an electrolyte; whereas, electro-osmosis is the movement of an electrolyte in a charged channel. In both systems, the solid is charged, creating an electrical double layer. In the case of phosphatic clays, the negative charge of the particles attracts and adsorbs anions to the surface and repulse cations. Just outside the adsorbed layer, a diffuse region of charge exists where the electrolyte is partly movable. When an electric field is applied, a coulombic force acts on the fluid in the diffuse region and causes movement. In dilute clay suspensions, electrophoresis is the mechanism in which the negatively charged particle moves toward the anode. When the clay fabric forms a tortuous porous media, known as a cake, the coulombic forces act on the diffuse region, causing movement of the fluid toward the cathode. In the case of electrokinetic dewatering, electrophoresis is the primary mechanism by which a cake is formed by settling, whereas electro-osmosis is the mechanism by which additional water in the cake may be removed.

Phosphatic clay represents a difficult medium for traditional dewatering methods. Flocculation is only applicable for initial sedimentation up to a solid content of about 10 wt%. In processing clay for land remediation, the addition of flocculants is not desired. Mechanical methods alone can achieve significant dewatering, but use of mechanical methods to achieve solids content greater than 20 wt% is cost prohibitive. Electrokinetic dewatering has the potential to become an efficient and economically viable method.

The use of electrokinetic phenomena in geotechnical applications has been studied extensively in recent decades. For example, heavy metal contaminants in soils can be leached utilizing electro-osmosis. Electrokinetic remediation of contaminated soil has been employed for removal of zinc, nickel [49], copper, chromium, and arsenic [50, 51]. Electro-osmosis has been used to deposit additives into low-permeability clay contaminated with chlorinated solvents [52].

Interest in the development of electrokinetics for dewatering is also widespread, as it may be applicable to many types of mine effluent or tailings. In addition to phosphate mining, examples include Canadian oil sands [53, 54], African diamond mines [55], glass–sand plant tailings in Turkey [56], and various metal- and petroleum-mining operations in South Africa and Western Australia [57, 58]. Clarification using electrophoresis has been studied on effluent from coal washing [59]. The work in coal washing resulted in the creation of a prototype that could run as a batch or semi-continuous operation [60] as well as the development of a numerical model for electrophoresis [61].

Shang [62] demonstrated that electrokinetic methods could be used to dewater phosphatic clays to yield material suitable for use of a soil cover. Shang used a bench-top cell to explore the dewatering of Florida phosphatic, Canadian brown, and Canadian gray clays under the influence of an electric field [63]. Geometric configuration, current density, and electrode materials were found to influence electrokinetic dewatering [64]. The efficiency of electrokinetic dewatering can be optimized with increased understanding relevant factors. Incremental developments to improve the applicability of electrokinetic phenomena have sustained continued work. Conductive polymers were proposed as electrode materials suitable for electrokinetic dewatering of sewage sludge [65]. Other modifications to existing dewatering technologies include the rotating anode [66], the electrokinetic belt filter press [44, 67, 68], the electrokinetic bag filter [69], and in situ electrokinetic consolidation [57].

6.3 Development of Dewatering Technologies for Phosphatic Clays

Fourie [70] noted that, while the concept of dewatering soft clays has been used with some success, "difficulties with corrosion of electrodes, loss of contact between soil and electrodes, excessive energy consumption, and logistical problems of water collection and management have precluded the widespread use of the technology." Our early work showed that use of a batch process was not feasible; hence, a sequence of prototypes was used to develop the electrokinetic dewatering process from batch to semi-continuous to fully continuous designs. The compiled learnings from each prototype led to design elements that were integrated into subsequent designs. This section illustrates the sequential nature of the process development.

6.3.1 Lab-scale Batch Dewatering

The initial investigation into electrokinetic dewatering of phosphatic clay used a lab-scale cell on clay that originated from the Four Corners region in Central Florida [71]. The bench-top cell design was based on the cell design by Shang [62]. The cell was constructed from an acrylic tube with acrylic plungers securing iridium-oxide-coated titanium electrodes to sandwich the clay suspension contained within the tube. The cell was oriented vertically with the cathode as the upper electrode and the anode as the bottom electrode. Under application of an electric field, clarified water collected at the upper (cathode) electrode and solids accumulated on the anode. The electric field was controlled by the gap between the cathode and anode and by the magnitude of the applied potential. The experiments were performed using electric fields ranging from 0.08 to 5 V cm^{-1}. The duration of the experiments extended up to 170 hours to ensure that the full extent of possible dewatering was achieved at a specific electric field.

The results from the batch experiments showed that an electric field can induce separation of water from central Florida phosphatic clay suspensions. A proof-of-concept experiment, performed at a cell potential of 80 V (or 4 V cm^{-1}), showed that, in an elapsed time of nine hours, a uniform 9 wt% solids suspension could be converted to clarified water and a lump with solids content approaching 35 wt%. A photograph of the bench-top cell loaded initially with a 9 wt% solids suspension is shown in Figure 6.3a. After an elapsed time of nine hours, the layer of clarified water was decanted, leaving the thickened lump of clay as shown in Figure 6.3b consisting of 33.5 wt% solids.

A series of experiments were performed to identify a constitutive relationship to relate the increase in solids content to operating time and the applied electric field. These experiments were designed under the working hypothesis that a small electric field applied for a long time would yield the same increase in solids content as a large electric field applied for a short time. The results presented in Figure 6.4 showed that, for $Et < 20$ V h cm^{-1}, the increase in solids content was proportional to the product of the electric field and elapsed time; whereas, the limiting solids content could be expressed as a function of applied electric field.

The solid lines presented in Figure 6.4 are taken from an interpolation formula given as [71]

$$\Delta w_c = [(0.77tE)^{-n} + (7.1\log_{10}(E) + 16.5)^{-n}]^{-1/n} \tag{6.1}$$

where n is a dimensionless parameter that controls the transition from short-time to long-time behavior, E has units of V cm^{-1}, and t has units of hours. The parameter n had a fitted value equal to 5.

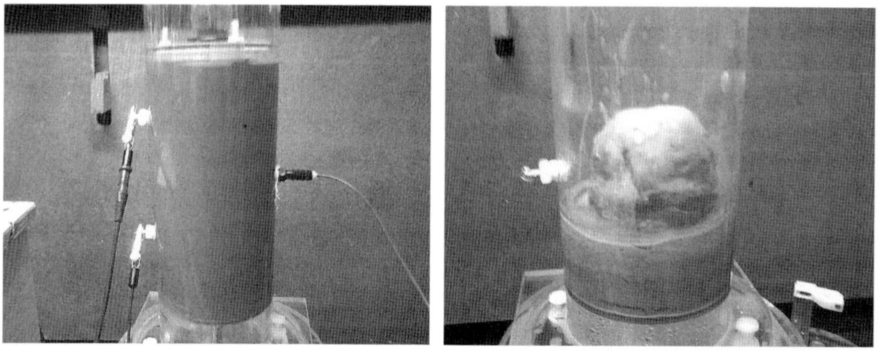

(a) (b)

Figure 6.3 Photographs of the bench-top batch dewatering system: (a) charged initially with a 9 wt% solids suspension and (b) after a duration of nine hours under an 4 V cm^{-1} applied electric field. The layer of clarified water was decanted, leaving the thickened lump of a thickened lump of clay consisting of 33.5 wt% solids. Source: Taken from [71, 72].

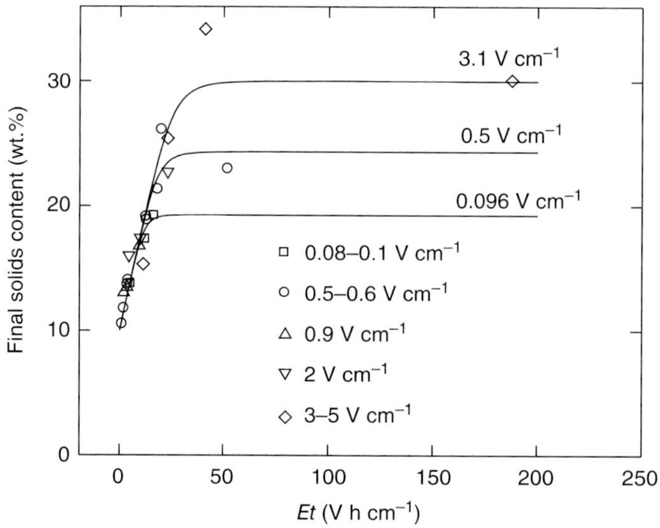

Figure 6.4 Final solids content as a function of the product of electric field and elapsed time with applied electric field as a parameter. The initial solids content was 10 wt%. The solid lines are taken from Eq. (6.1). Source: Data taken from [71, 72].

Simulations based on the constitutive relationship were used to evaluate the power demands for electrokinetic dewatering of CSA [73]. The tailings impounded in the CSAs were assumed to have an initial 10 wt% solids content. The dimensions of the CSA were assumed to be 1 mile (1.6 km) wide by 1 mile (1.6 km) long and 40 ft (12.2 m) deep. Two different electrode configurations were explored. For the horizontal orientation, cylindrical cathodes were placed near the surface of the simulated CSA and cylindrical anodes were placed near the bottom. The cylindrical electrodes were assumed to be 6 in. (15.2 cm) in diameter and placed 6 ft (1.8 m) apart. The electrode centerlines were placed

1 ft (0.305 m) below the air/CSA interface and 1 ft (0.305 m) above the CSA bottom, yielding a 38 ft (11.6 m) spacing between the electrode centerlines. The electrodes were one mile (1.6 km) long, consistent with a one-square-mile (2.56 km^2) CSA. They were assumed to be spaced equally along the entire CSA width of one mile (1.6 km).

The second configuration employed electrodes oriented vertically, with cathodes and anodes placed diagonally from one another. The electrodes were represented as 3 in. (7.6 cm) diameter cylinders, 39.5 ft (12 m) long. The electrodes were 10 ft (3.05 m) apart, with a 14 ft (4.31 m) diagonal spacing. A separator or filter cloth is required for vertically oriented cathodes to filter the clay from the water entering the inner diameter of the cathode. Such a system has been proposed by Fourie et al. [57] for dewatering mine tailings and by a Canadian company [74] for application to oil–sand tailings.

The simulation results are presented in Table 6.1 [73]. The simulation results suggest that, while the electrical requirement for batch dewatering a CSA is reasonable, the application of a significant electric field over a short time results in an unacceptable power requirement. Thus, a batch dewatering operation is not economical.

The methods presented in this work illustrate the manner in which a constitutive equation relating the change in solids content to time and applied electric field can be used to perform an economic assessment of electrokinetic dewatering. The observation that the electric field controls both the maximum solids content achievable and the time required to reach it severely constrains the application of electrokinetic dewatering strategies to an entire CSA. The approach, however, may be feasible if implemented as part of a continuous or semi-batch plant process.

Table 6.1 Results of power and energy calculations for dewatering of simulated one-square-mile clay-settling area.

Configuration	Horizontal simulation	Vertical simulation	Bench-top experiment
Number of anodes	880	69 696	1
Number of cathodes	880	69 696	1
Average electric field (Vcm^{-1})	1.2	1.2	1.1
Cell potential, (V)	1 445	858	20
Current (A)/electrode	34 617	586	0.04
Current required (A)	4.09×10^7	3.05×10^7	—
Time required (h)	19	35	9
Power required (MW)	44 000	35 000	12 000
Energy (kWh)	8.5×10^8	1.1×10^9	2.8×10^8
Water removed (kg)	2.02×10^{10}	2.02×10^{10}	0.57
Energy requirement ($Wh\,kg_{H_2O}^{-1}$)	43.6	53.7	13.9

Solids weight percent was increased from 10 to 25 wt%.
Source: Taken from [73].

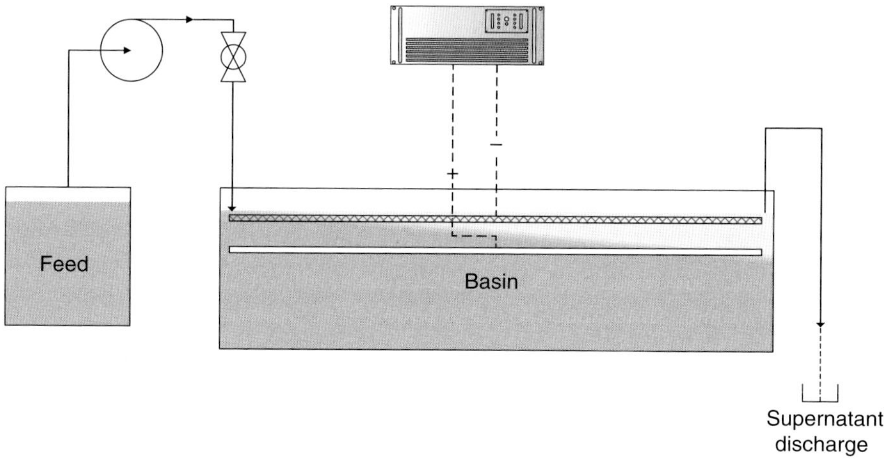

Figure 6.5 Schematic representation of the semi-continuous prototype used to recover clear supernatant water. Source: Kong and Orazem 2014 [75]. Reproduced with permission of Elsevier.

6.3.2 Semi-continuous Operation to Recover Clear Supernatant

The first step in developing a continuous electrokinetic dewatering process was to devise a system to remove the supernatant water continuously in a system that stored the separated solids. A schematic representation of this design is shown in Figure 6.5 [75]. A large basin was fitted with two horizontally oriented electrodes. The top and bottom electrodes served as cathode and anode, respectively, that maintained the flow of supernatant toward the cathode and clay toward the anode. The electrodes were made of corrosion-resistant iridium-oxide-coated expanded titanium. The feed clay was added to the basin to maintain a liquid level above the cathode while the supernatant was removed using an overflow weir. The large basin was intended to accumulate solids.

The efficacy of producing low-turbidity water using electrokinetics was evaluated by measuring the supernatant turbidity as a function of elapsed time. The supernatant turbidity is presented in Figure 6.6 as a function of dimensionless time, where elapsed time is scaled by the residence time τ. The residence time was calculated as the effective volume between the electrodes divided by the volumetric feed rate. Samples were collected every thirty minutes, and the turbidity was measured immediately after sampling and again after 24 hours. The turbidity readings at the time of sampling are presented in Figure 6.6 as squares, and the corresponding values after settling for an elapsed time of 24 hours are represented by circles. The supernatant turbidity decreased sharply after the start of the experiment and reached a steady value after approximately four hours ($t/\tau = 0.3$). The turbidity readings remained in the lower range until $t/\tau = 1.8$ hours, after which the turbidity increased. The time frame between $t/\tau = 0.3$ and $t/\tau = 1.8$ was assumed to be a period of pseudo-steady-state operation. The increase in turbidity after an elapsed time of $t/\tau = 1.8$ was observed when the solid storage capacity in the tank was exceeded and gas evolution at the anodes entrained solids-rich water that was mixed with the supernatant.

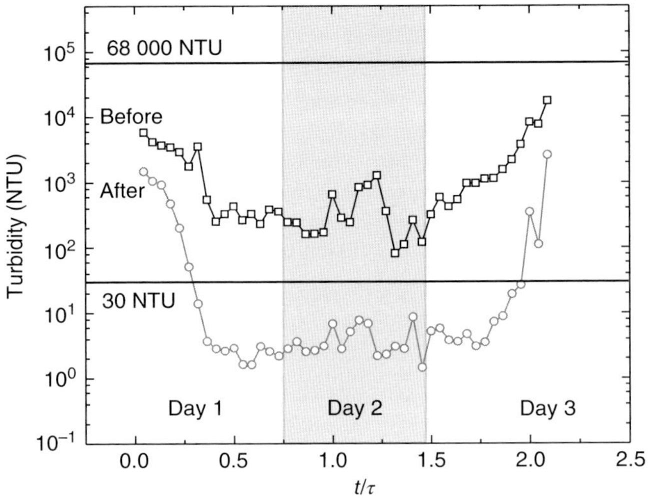

Figure 6.6 Turbidity of supernatant liquid as a function of scaled elapsed time for an electric field of 3 V cm^{-1} (300 V m^{-1}) and a feed flow rate of 40 ml min^{-1} (6.67×10^{-7} m^3 s^{-1}, $\tau = 11$ h): □ measured immediately after the sample was collected and ○ measured 24 hours after the sample was collected. The electric field and flow were discontinued overnight. Source: Kong and Orazem 2014 [75]. Reproduced with permission of Elsevier.

Figure 6.7 Representative images showing the supernatant samples 24 hours after collection. Source: Kong and Orazem 2014 [75]. Reproduced with permission of Elsevier.

A representative series of images is presented in Figure 6.7 for supernatant samples that were allowed to settle for more that 24 hours after collection. The turbidity was high for the first five samples of Day 1, corresponding to the first 2.5 hours of operation, and for the last samples of Day 3. These images are in good agreement with the turbidity reported in Figure 6.6. A thin layer of solids is visible

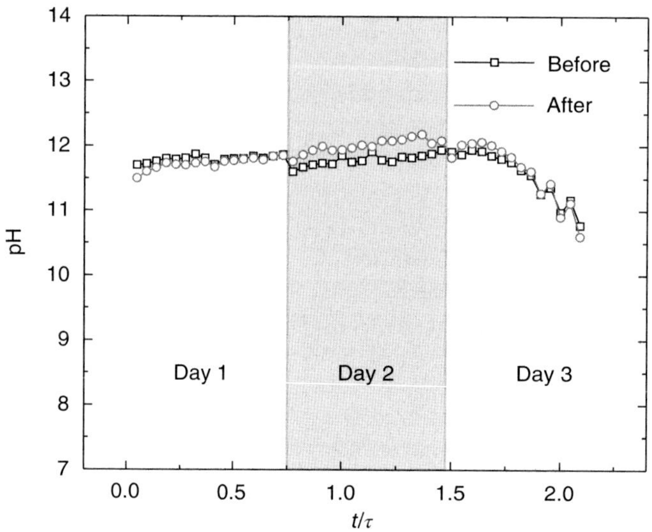

Figure 6.8 pH of supernatant liquid as a function of scaled elapsed time corresponding to the results presented in Figure 6.6. Source: Kong and Orazem 2014 [75]. Reproduced with permission of Elsevier.

at the bottom of the vials. The settled solids were suspended in the supernatant water and settled rapidly to the bottom of the vials.

The pH of the supernatant was measured and plotted in a similar manner. At $t = 0$, the pH of the cell contents was 7. As shown in Figure 6.8, the pH of the supernatant increased to 11.7 at the beginning of the experiment and remained stable around this value until the solids storage capability of the container was exceeded. There was no significant difference between the pH measured immediately after the sample was collected and after 24 hours. The alkalinity of the supernatant liquid was a result of the electrochemical reaction at the cathode, dissociation of water to create hydrogen gas and hydroxide ions, OH^-.

A minimum turbidity was achieved with an applied electric field of approximately 2 V cm^{-1}. The results support the conclusion that electrokinetic processes employing intermediate electric fields can produce supernatant water with turbidity < 10 NTU. The energy consumption in the semi-continuous operation was consistent with that for the batch operation. Experimental results showed that the combination of electrokinetic treatment followed by gravity free-settling can yield streams of low-turbidity water (pH 11.7) as solids accumulated within the cell. This free settling observed with the supernatant water could not be achieved by changing the pH and was therefore attributed to changes in the solids characteristics caused by the electrokinetic treatment [75].

6.3.3 Semi-continuous Operation to Recover Solids

The next step toward development of a continuous process was to design a method to remove dewatered solids while preventing resuspension. A schematic representation of the semi-continuous system for removal of solids is presented

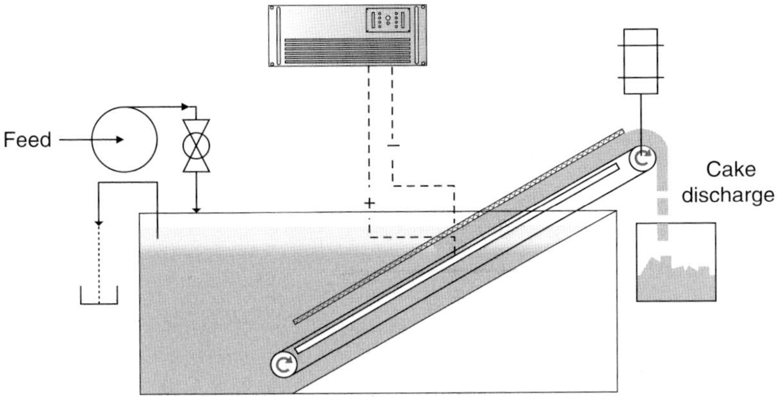

Figure 6.9 A schematic representation of the semi-continuous prototype for solids recovery. Source: Kong and Orazem 2014 [75]. Reproduced with permission of Elsevier.

in Figure 6.9, which comprised a belt-driven transfer system that lifted the consolidated cake out of the tank [76].

The belt system used to lift the clay from the tank must expose the feed clay to an electric field. The first design idea considered was to use a titanium wire mesh as a moving anode. The bench-top cell was used to test the stability of the titanium wire mesh anode. The titanium electrode corroded under service conditions and was therefore unsuited for use as an anode for electrokinetic dewatering. As a second approach, a non-conductive polyester mesh was used as a belt moving over a solid dimensionally stable anode. A belt with 50% open area was found to provide minimal resistance to flow of electrical current while preventing adhesion of the clay to the anode. As shown in Figure 6.9, the motion of the belt carried the solids from the tank.

In operation, the semi-continuous prototype was filled with a 10 wt% feed. Separation of supernatant and clay was visible after two hours of operation at 8 V cm^{-1}. After 3.3 hours of elapsed operational time, a 24.4 wt% clay cake was collected as shown in Figure 6.10. The results of the experiment show that the production of dewatered cake was achievable with the semi-continuous prototype for solids recovery.

The semi-continuous prototype shown in Figure 6.9 did not produce the clear supernatant that was observed in the previous design, shown in Figure 6.5. Photographs illustrating the operation of the semi-continuous prototype for solids removal are presented in Figure 6.11. The initial operation, before introduction of feed, presented in Figure 6.11a, showed that a layer of supernatant water was formed above the cathode, similar to the results obtained for the semi-continuous prototype used to recover clear supernatant water. Under pseudo-steady-state operation, Figure 6.11b, a turbid region with a 2–3 wt% solids content formed above the cathode due to introduction of the feed and mixing by gases produced by the electrochemical reactions. As shown in Figure 6.11c, a clear supernatant was formed above the cathode after the feed and electrical current was stopped.

The semi-continuous prototype for solids removal showed that solids could be removed continuously from the electrokinetic cell. The constitutive relationship

Figure 6.10 A thickened clay cake with an average solids content of 24.4 wt% was produced under the electric field of 8 V cm^{-1}. The electrode separation was 3.2 cm, and the residence time was 3.3 hours. Source: Taken from [76].

obtained from the batch experiments and given as Eq. (6.1) could be used to provide a relationship between final solids content, electric field, and elapsed time. The electric field required to achieve solids content greater than 20 wt% was on the order of 8 V cm^{-1}, and the energy and power requirements were too large to allow this system to be considered viable for solids–liquid separation.

6.3.4 Continuous Operation

Our fully continuous design represented a combination of the semi-continuous designs presented in Figures 6.5 and 6.9. As a clear supernatant could be best collected from a horizontal cathode and removal of solids was obtained by use of a inclined belt, the continuous prototype, shown in Figure 6.12, comprised a horizontal cake formation zone coupled to an inclined cake dewatering zone. A polyester belt was moved over the anode, transporting the solids from the horizontal zone to the inclined cake dewatering zone. In the cake dewatering zone, the cathode was lowered to make contact with the cake being carried by the belt. Supernatant water was collected at the beginning of the cake formation zone by use of a small pump.

An earlier design that did not include the electrodes on the inclined portion of the prototype yielded cake with solids content lower than 22 wt%, and the electric field of 8 V cm^{-1} required to achieve these values resulting in ohmic heating of the suspension to temperatures above 90 °C, nearing the boiling point of water. The addition of a second pair of electrodes on the inclined section greatly improved the efficiency of the process, yielding solid contents on the order 35 wt% with

(a) (b) (c)

Figure 6.11 Photographs illustrating the operation of the semi-continuous prototype for solids removal: (a) Initial operation before introduction of feed showed that a layer of supernatant water was formed; (b) pseudo-steady-state operation showed turbidity above the cathode due to introduction of the feed and mixing by gases produced by the electrochemical reactions; and (c) a clear supernatant was formed after the feed and electrical current was stopped. Source: Taken from [76].

an electric field of 4 V cm^{-1} or lower, and the suspension temperatures were unchanged from ambient values.

Operation of the fully continuous design was accompanied by challenges in balancing the feed rate, belt speed, and applied electric field. When operating at 4 V cm^{-1} and slow belt speeds, dewatered clay would accumulate at the transfer point from horizontal to inclined conveyor. When the belt speed was sufficient, a cake of 32 wt% was collected from the discharge. Under operation at 1 V cm^{-1} with a residence time of three hours, a cake of 35 wt% was collected at the discharge. These results show that the design could achieve large solids content with a smaller electric field, reducing the power required to produce a plastic cake [76].

A series of experiments with electrode spacing of 6.2 cm and electric field of 4 V cm^{-1} were conducted with residence time ranging from 0.59 to 2.4 h (belt speed from 0.5 to 2 m h^{-1}). The final solids content is shown in Figure 6.13 as a function of $E\tau$. For τ smaller than 1.2 h, the final solids content increased with the increase of $E\tau$, from 20 to 32 wt%. The final solids content reached the plateau with τ larger than 1.2 h. At the region where τ is larger than three hours, the final

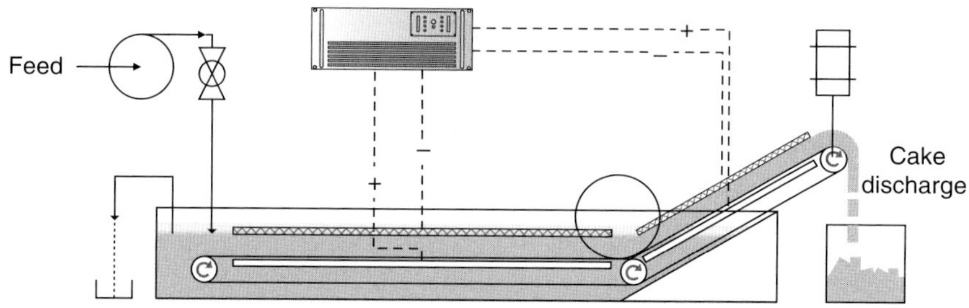

Figure 6.12 A schematic representation of the fully continuous prototype [76]. The horizontal electrode configuration (on the left) represents the cake formation zone, whereas the inclined section represents the cake dewatering zone.

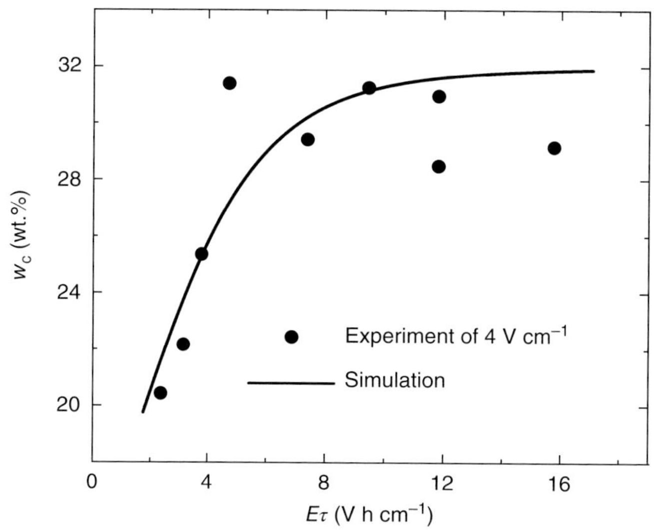

Figure 6.13 Final solids content as a function of $E\tau$ under an electric field of 4 V cm^{-1}. The scatter plot represents the experimental result, and the solid line represents the simulation result obtained from Eq. (6.1). Source: Taken from [76].

solids content appeared to decrease. The solid line in Figure 6.13 is the fitting result from the constitutive relationship given as Eq. (6.1).

With an electric field of 4 V cm^{-1} and a very slow belt speed, a build-up of solids appeared at the wheels used to redirect the motion of the belt. As the high solids content clay was blocked at the wheel region, the clay that traversed the wheel region and could be collected had a lower solids content. Thus, the average final solids content was underestimated. As the reduced solids content was caused by the mechanical design rather than the principle of electrokinetic dewatering, the performance can be improved by a future design of the prototype. Experiments using this fully continuous prototype with τ longer than three hours in 4 V cm^{-1} electric field were not suitable to evaluate the promise of electrokinetic dewatering.

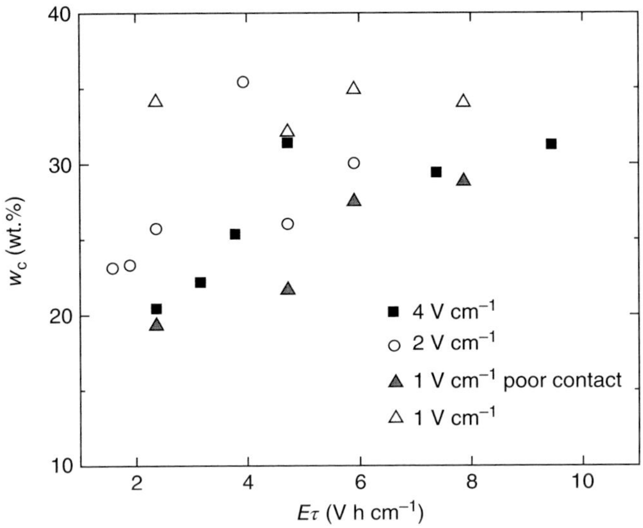

Figure 6.14 Final solids content as a function of $E\tau$ with electric field as a parameter. Source: Taken from [76].

In order to explore the influence of the applied electric field on the final solids content, experiments were conducted for electric fields of 1 and 2 V cm^{-1}. The reduced electric field was achieved by reducing the cell potential and keeping the same electrode spacing in the horizontal zone. The electrode spacing in the inclined zone was equal to the clay cake thickness, which is always controlled by the system itself. The final solids content is shown in Figure 6.14 as a function of $E\tau$ with electric fields of 1, 2, and 4 V cm^{-1}. In a few experiments at 1 V cm^{-1}, the cathode in the inclined zone did not properly contact the clay cake: these results are labeled as "1 V cm^{-1} poor contact" in Figure 6.14. In these experiments, the effective electrokinetic dewatering area was reduced, due to the poor contact, and the system produced clay with lower solids content.

The results from batch and semi-continuous prototypes showed that an increase of electric field increased the final solids content of thickened clay. However, an increase in solids content with increased electric field was not consistently observed for the fully continuous prototype. As shown in Figure 6.14, the experiments with an electric field of 1 V cm^{-1} produced the clay with a final solids content up to 35 wt%, and the experiments with electric field of 4 V cm^{-1} produced the clay with the final solids content only about 32 wt%, given a long enough time. This result suggests that the applied electric field of 1 V cm^{-1} is sufficient to achieve acceptable solids–water separation in the fully continuous prototype.

6.3.5 Energy and Power Requirements for All Prototypes Tested

The power consumption in the fully continuous experiment is defined by

$$P_{\text{req}} = V_{\text{cell}} I \tag{6.2}$$

which applies for the batch, semi-continuous, and continuous prototypes. In the field of clay dewatering, the energy consumption could be evaluated by the energy consumption per unit mass of removed water

$$E_{req,w} = \frac{P_{req}}{Q_w} \tag{6.3}$$

where $E_{req,w}$ is the energy consumption per unit mass of solids-free water removed, P_{req} is the power consumption, and Q_w is the effluent mass flow rate of supernatant water. In the present work, mass flow rates are represented by Q; whereas, volumetric flow rates are represented by Q_v. The energy consumption per unit mass of produced dry solids is given by

$$E_{req,s} = \frac{P_{req}}{Q_s} \tag{6.4}$$

where $E_{req,s}$ is the energy consumption per unit mass of dry clay produced and Q_s is the mass production rate of dry clay.

In a fully developed steady-state operation, the volumetric feed rate $Q_{v,in}$ of clay slurry can be related to the rate of production of solids by

$$Q_{v,in} = \frac{w_c Q_c}{w_{in} \rho_{in}} \tag{6.5}$$

where w_c is the final solids content of the thickened clay, Q_c is the mass production rate of the thickened clay, w_{in} is the solids content of the influent clay slurry, and ρ_{in} is the density of the influent clay slurry. The effluent mass flow rate of supernatant water is calculated from the materials balance, i.e.

$$Q_w = Q_c \left(\frac{w_c}{w_{in}} - 1 \right) \tag{6.6}$$

and the production rate of dry clay is calculated from

$$Q_s = w_c Q_c \tag{6.7}$$

Equations (6.2)–(6.7) were used to calculate energy and power requirements for each of the prototypes described in this section.

The energy consumption per kilogram of removed water is presented in Figure 6.15 as a function of applied electric field for experiments ranging from the batch to the fully continuous prototype. The squares represent the fully continuous experimental results under various operation conditions with electric field of 1, 2, and 4 V cm^{-1}. The results show that, in the fully continuous experiments, the increase of applied electric field increased the energy consumption, in agreement with all previous experiments. Under the same electric field, the fully continuous operation had the lowest energy cost when compared to other experiments.

The final solids content is shown in Figure 6.16 as a function of energy consumption with electric field as a parameter. The corresponding residence time is labeled for each point. The applied electric field varied from 1 to 4 V cm^{-1}, and τ varied from 0.59 to 3.9 h. As shown in Figure 6.16, the lower electric field is favorable for fully continuous operation. With the electric field of 1 V cm^{-1} and

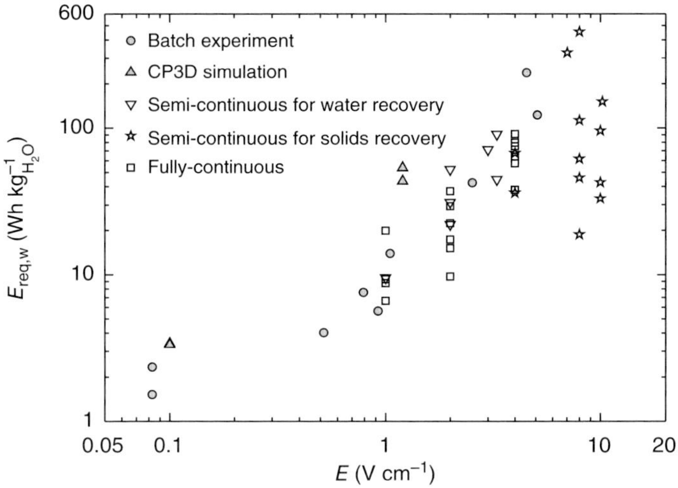

Figure 6.15 Energy consumption as a function of applied electric field. The batch experiments and CP3D simulation results were reported by McKinney and Orazem [71, 73], and the semi-continuous and continuous results were presented by Kong [76].

Table 6.2 Summary of prototype performance scaled to the output of the Four Corners phosphate mine.

Prototype	E (V cm^{-1})	w_c (wt%)	P_{req} (MW)	E_{req} (Wh kg^{-1}_{solids})
Bench-top cell, Section 6.3.1	1.2	25	44 000	248
Semi-continuous, Section 6.3.3	10	26	141	210
Continuous, Section 6.3.4	1	35	29	47

Source: Taken from [76].

residence time of 3.0 h, the solids content w_f reached 35 wt% with the energy cost $E_{req,w}$ smaller than 10 Wh $kg^{-1}_{H_2O}$. As the electric field increased to 4 V cm^{-1}, an energy of at least 63 Wh $kg^{-1}_{H_2O}$ was required to achieve the final solids content of 32 wt%. The experiments with 2 V cm^{-1} had a performance between the experiments at 1 and 4 V cm^{-1}. The experiments with a poor contact between the cathode and cake in the inclined dewatering zone had reduced final solids content, but the energy consumption was not reduced.

The power consumption for each generation of the electrokinetic dewatering process was quantified using the experimental data collected during operation. A summary of the estimated power requirement for each process is shown in Table 6.2.

The bench-top cell represents a batch process; thus, the power and energy requirements were estimated under the assumption that a 1 square mile CSA was processed in place [73]. The projected power demand for the semi-continuous and continuous processes were calculated for the estimated output of phosphatic clay from the Mosaic Four Corners mine. The calculations for the continuous

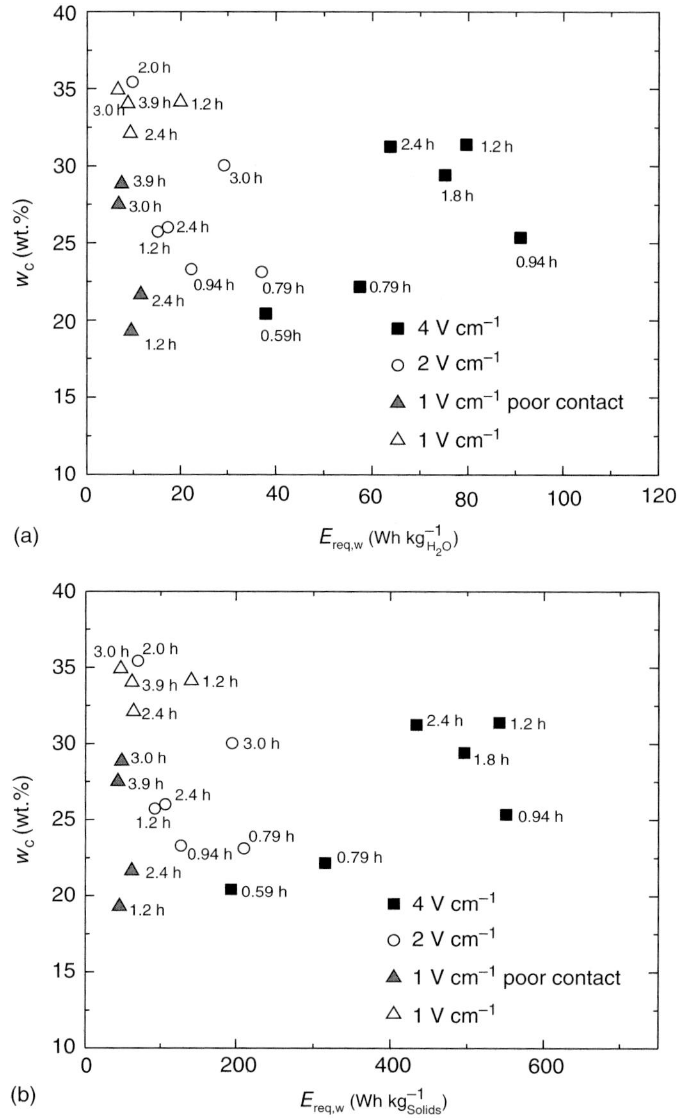

Figure 6.16 Relationship between final solids content and energy consumption for the fully continuous prototype: (a) The energy consumption was normalized by mass of water removed, and (b) the energy consumption was normalized by mass of dry clay produced. Source: Taken from [76].

process are detailed in Section 6.4.3. The power requirement for each process is represented by P_{req}. To account for the throughput of each dewatering process, the energy was scaled to the solid cake output and is represented by E_{req}. The power requirement of 29 MW estimated for scaling the continuous prototype to the output of the Four Corners Mine can be compared to the roughly 5 MW required to operate a drag line.

6.4 Economic Assessment for On-site Implementation

The fully continuous prototype developed for electrokinetic dewatering produced low-turbidity water and thickened clay. Hydrogen and oxygen gas generated on the electrode surface from the electrochemical reactions represent a by-product of this process. The hydrogen and oxygen, continuously generated during the operation, are released immediately into the atmosphere and hence represent a potential flammability hazard.

The concentration of hydrogen was calculated for the lab prototype and estimated for a full-scale on-site installation, and both were found to be below the lower flammable limit. The power and energy consumption for processing the output of the Mosaic Four Corners beneficiation plant was predicted using data generated from the fully continuous prototype. The assumptions and procedures used for estimating the hydrogen concentration and economic assessment for mine site treatment are discussed below.

6.4.1 Hydrogen Emission

Material balances were used to estimate the rate of hydrogen production for the prototype and for the full-scale implementation. Standard ventilation models were used to estimate the resulting hydrogen concentration in the cell area. The mathematical development is presented in this section, and calculated values are presented in Section 6.4.3.

During electrokinetic dewatering, current is carried by ions in the solution and by electrons in the metal electrode. At the interface between the bulk solution and electrode, the electrons are exchanged via the electrochemical reaction. In the electrokinetic dewatering of phosphatic clay suspensions, the major electrochemical reactions at the anode are

$$2H_2O \rightarrow O_2 + 4H^+ + 4e^- \tag{6.8}$$

and, for a non-noble electrode M,

$$M \rightarrow M^{n+} + ne^- \tag{6.9}$$

At the cathode, the reaction is

$$2H_2O + 2e^- \rightarrow H_2 + 2OH^- \tag{6.10}$$

As shown in Eq. (6.8), oxygen evolution occurs at the anode, and the pH of the solution is decreased by the generation of H^+.

The anode material must be corrosion resistant because the anode can corrode as shown in Eq. (6.9). The anode, in all the prototypes discussed in the present work, was a dimensionally stable anode, which contains a layer of IrO_2 deposited on a titanium substrate. This anode is very stable during the oxygen evolution reaction, and, thus, the corrosion reaction represented by Eq. (6.9) was not considered. As shown in Eq. (6.10), the hydrogen evolution reaction occurs at the cathode, and the pH of the solution is increased from the generation of hydroxide ions. The hydrogen and oxygen generated at the electrode surface would accumulate if the system is operated in an enclosed space and could result in an explosion in extreme circumstances. Therefore, the electrokinetic dewatering process must operate with ample ventilation, requiring a fume hood in the lab model and appropriate ventilation during plant operations. The quantitative estimation of hydrogen concentration for the lab prototype and on-site operations are illustrated separately, under the assumption of an ideal gas. For the lab-scale prototype, the hydrogen evolution rate is represented by

$$Q_{H_2,1} = \frac{I_1 M_{H_2}}{n_e F} \tag{6.11}$$

where $Q_{H_2,1}$ is hydrogen evolution rate and I_1 is the total current.

The fume hood ventilation rate is represented by

$$Q_{v,1} = A_1 \overline{u}_1 \tag{6.12}$$

where A_1 is duct cross-sectional area and $\overline{u}_1$ is the average air velocity in the fume hood. The concentration of hydrogen gas in fume hood is obtained from the balance between hydrogen evolution and ventilation, represented by Crowl and Louvar [77]

$$C_{H_2,1} = \frac{Q_{H_2,1} RT}{k_1 Q_{v,1} P M_{H_2}} \times 10^6 \tag{6.13}$$

where $C_{H_2,1}$ is the hydrogen concentration in the fume hood and k_1 is the non-ideal mixing factor in the fume hood (ranges between 0 and 1). The term k_1 was assumed to be 0.5 in these calculations.

For on-site implementation, the hydrogen evolution rate is proportionally scaled by the feed flow rate, represented by

$$Q_{H_2,2} = \frac{Q_{H_2,1} Q_{in,2}}{Q_{in,1}} \tag{6.14}$$

where the subscript 1 denotes the lab-scale prototype and 2 denotes the on-site implementation. The ventilation rate and hydrogen concentration are defined similarly to the lab-scale prototype, represented by

$$Q_{v,2} = A_2 \overline{u}_2 \tag{6.15}$$

and

$$C_{H_2,2} = \frac{Q_{H_2,2} RT}{k_2 Q_{v,2} P M_{H_2}} \times 10^6 \tag{6.16}$$

where the subscript 2 denotes the on-site prototype operation, and $\overline{u}_2$ is the wind velocity. Calculated values are presented in Section 6.4.3.

6.4.2 Capital and Operation Costs

The economic assessment for electrokinetic dewatering of phosphatic clay suspensions for plant operation requires consideration of capital and operation costs. The power and energy consumption for Mosaic's Four Corners plant was predicted using data generated from the fully continuous prototype. The assumptions and procedures used for economic assessment for mine site treatment are discussed below. An estimation of capital cost is discussed, and a graphic chart for total cost estimation under desired equipment budget and operation conditions is provided. The calculated values for specific prototypes are presented in Section 6.4.3.

6.4.2.1 Power and Energy consumption for On-site Operations

For on-site implementation, calculations were performed under the assumption that the electrode distance and electric field are similar to that used for the lab prototype and that the increased treatment capacity is achieved by a proportional increase in the electrode surface area. Under these assumptions, the total current for on-site operations is proportionally increased by an increase in the feed flow rate, represented by

$$I_2 = \frac{I_1 Q_{\text{in},2}}{Q_{\text{in},1}} \tag{6.17}$$

The total power consumption is given by

$$P_{\text{req}} = I_2 V_{\text{cell}} \tag{6.18}$$

and the total energy consumption is given by

$$E_{\text{req,w}} = \frac{P_{\text{req}}}{Q_{\text{w},2}} \tag{6.19}$$

or

$$E_{\text{req,s}} = \frac{P_{\text{req}}}{Q_{\text{s},2}} \tag{6.20}$$

where $Q_{\text{w},2}$ and $Q_{\text{s},2}$ could be obtained from Eqs. (6.6) and (6.7), respectively. The total operating cost for on-site operations is calculated under assumption of an industrial electric power rate of \$0.062/kWh to be

$$\text{operation cost} = E_{\text{req}} \times \$0.062/\text{kWh} \tag{6.21}$$

Equation (6.21) provides the operating cost for electrical power. The annualized capital expenditure is not included in these calculations.

6.4.2.2 Operation cost

The operation cost was estimated to be that of energy consumption under an industrial electricity rate of \$0.062/kWh. The data used for energy consumption were obtained from the experimental results of the fully continuous prototype. The results for each experiment were influenced by operating conditions such as the electric field, electrode spacing, and belt speed. The operation cost and the corresponding dry clay production rate is shown in Figure 6.17. The increase of dry clay production rate reduced the operation cost, and the increase of electric field increased the operation cost. The corresponding final solids content w_c for each experiment is illustrated in Figure 6.18, as a function of dry clay production

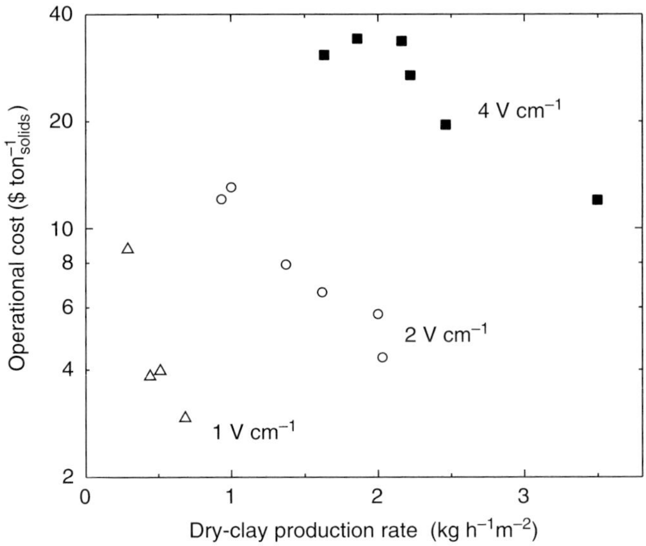

Figure 6.17 Operation cost as a function of dry clay production rate with applied electric field as a parameter.

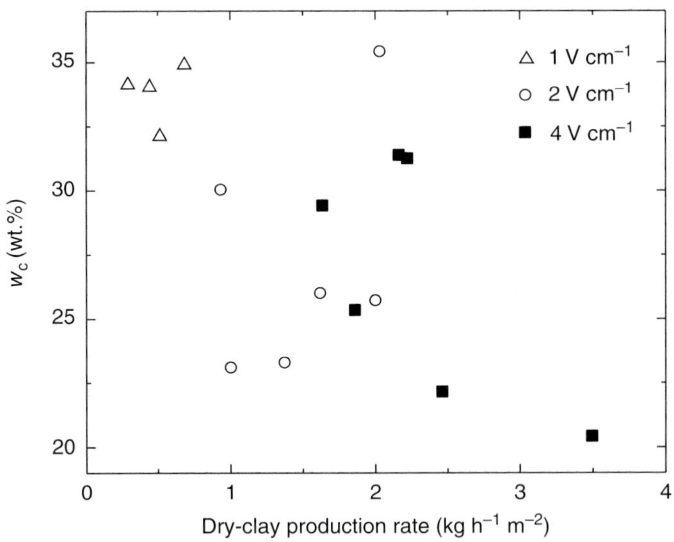

Figure 6.18 Final solids content obtained as a function of dry clay production rate with applied electric field as a parameter.

rate and with applied electric field as a parameter. In the fully continuous prototype, a high final solids content of clay could be obtained under a low electric field, but the corresponding dry clay production rate is lower as compared to the experiments using a higher electric field.

6.4.2.3 Capital Cost

The capital cost was evaluated in terms of dollars per ton of dry clay produced. The annualized capital cost A was obtained from [78]

$$A = P\frac{i(1+i)^n}{(1+i)^n - 1} \quad (6.22)$$

where P represents the present value of capital investment (equipment construction cost), i represents the interest rate, assumed to be 10%, and n represents the number of years for equipment life, assumed here to be 20 years.

Under the assumption that the equipment is operated 24 hours a day and 365 days a year, the equivalent capital cost to produce 1 ton of dry clay could be obtained as

$$C = \frac{A \times 1000\frac{\text{kg}}{\text{ton}}}{Q_s \times 24\frac{\text{hour}}{\text{day}} \times 365\frac{\text{day}}{\text{year}}} \quad (6.23)$$

where Q_s represents the dry clay production rate of the fully continuous prototype. The capital cost C is shown in Figure 6.19 as a function of dry clay production rate Q_s with the equipment construction cost P as a parameter. For a given equipment construction cost per unit area, an increase of dry clay production rate could dramatically reduce the capital cost. Thus, the reduction of capital cost could be obtained from the reduced equipment construction cost as well as an improved clay production rate.

The total cost for fully continuous operation is the combination of capital cost and operation cost, both of which depend on the dry clay production rate. As shown in Figure 6.19, an improvement of performance that increases the dry clay production rate in unit surface area could reduce the capital cost dramatically.

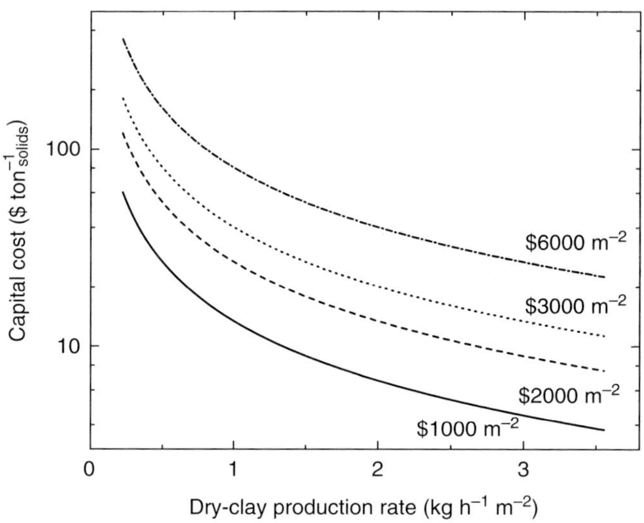

Figure 6.19 Capital cost as a function of dry clay production rate with estimated equipment construction cost as a parameter.

The relationship between operation cost and dry clay production rate is complex, and both are influenced by operation parameters such as electric field, electrode spacing, and belt speed. Figure 6.17 may serve as a reference that could be used to estimate the operation cost for a known operating condition. A future experimental optimization is expected to reduce the operation cost and increase the dry production rate. The final solids content, provided in Figure 6.18 for all experiments, provides additional information that narrows the desired range of dry clay production rate.

6.4.3 Results

A representative experiment under fully continuous operation is used as an example for the estimation of hydrogen concentration and energy consumption for on-site operations. This experiment was conducted using an applied electric field of 1 V cm^{-1} with 6.2 cm electrode spacing and three hours residence time. From the experimental results, the production rate of thickened clay was 0.884 kg h^{-1}, and this corresponded to a dry clay production rate of 0.687 kg h^{-1} m^2. The final solids content was 35 wt%, the cell potential was 6.2 V, and the average total current was 2.35 A.

The on-site projection was based on the Mosaic Four Corners Mine, in which 130 000 GPM clay slurry is produced at a solids content of 2 wt%. The solids content was assumed to be increased by flocculation or free-settling to 10 wt%; thus, influent flow rate was converted to 26 000 GPM with a solids content of 10 wt%. The equivalent flow of dry solids is about 600 ton/h. On site, ventilation was expected to be produced from wind, and the dewatering areas were anticipated to be covered by a roof structure 3 m high. Following EPA recommendations for a worst-case analysis, a wind velocity of 1.5 m s^{-1} was used for the calculations [77].

The hydrogen concentration in the fume hood was calculated following Eq. (6.13), and the hydrogen concentration in the plant was obtained from Eq. (6.14). The hydrogen generated by the electrochemical reaction produced a concentration of 3.2 ppm (0.00032 vol.%) for lab operations and 900 ppm (0.09 vol.%) for on-site operations. As the lower flammability limit of hydrogen gas is 4 vol.% in air [77], the hydrogen concentration generated by the electrokinetic dewatering process is much lower than that required for combustion. Thus, the process should not represent an explosion hazard. Hydrogen is, however, reported to be an indirect greenhouse gas with the potential to increase global warming [79]; thus, it should be collected and combusted to form water.

The estimated total power requirement for the mine site is 29 MW, and the energy requirement is 47 Wh kg_{solids}^{-1}. Based on an electricity rate of \$ 0.062 kWh^{-1}, the operation cost is \$ 2.9/10^3 kg_{solids}. The capital cost could be found from Figure 6.19 or calculated by Eq. (6.23). Assuming a construction cost of \$6000 m^{-2}, the capital cost is \$ 117/10^3 kg_{solids}. If the construction cost is reduced to be \$1000 m^{-2}, the capital cost is \$ 19/10^3 kg_{solids}. These estimates may be improved substantially by improvements in separation efficiency as discussed in Section 6.5.

6.5 Our Next Prototype: Dual-zone Continuous Operation

Improved process economics may be achieved by reducing the capital cost, which scales with surface area. The IrO_2-coated titanium cathode can, for example, be replaced by carbon steel cathodes. Improvement can also be achieved by increasing the throughput by improving process efficiency. The prototype described in Section 6.3.4 represented a major improvement over the batch prototype, but suffered from several deficiencies. The water separated from the clay in the cake dewatering zone rehydrated the clay that was separated in the cake formation zone. The resistance to movement caused by the floating cathode and the wheels reduced the throughput of solids.

These problems are being addressed by separating the cake formation and cake dewatering processes into two unit operations. The first unit, shown in Figure 6.20, employs horizontal electrodes with a belt over the anode that transports the solids. This system provides collection of clear supernatant water and accumulation of thickened sludge. Preliminary experiments show that a sludge with solids content of roughly 20 wt% can be collected and removed from the process by use of a screw conveyor. At a solids content of 20 wt%, the sludge is an appropriate feed for a cake dewatering process, shown in Figure 6.21, in which the field may be reversed such that the anode is placed on top and the cathode on the bottom. With this configuration, the water drains readily from the cake, producing cake with higher solids content. The motion of belts covering the anode and cathode eliminates the drag that reduced throughput in the prototype described in Section 6.3.4. The separation of the process into two units has the additional advantage that the two zones may be independently optimized.

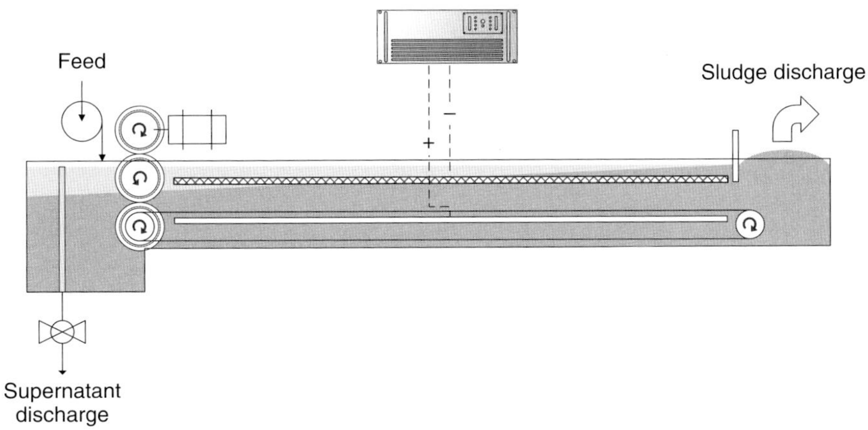

Figure 6.20 A schematic representation of the thickening zone of the dual-zone fully continuous prototype.

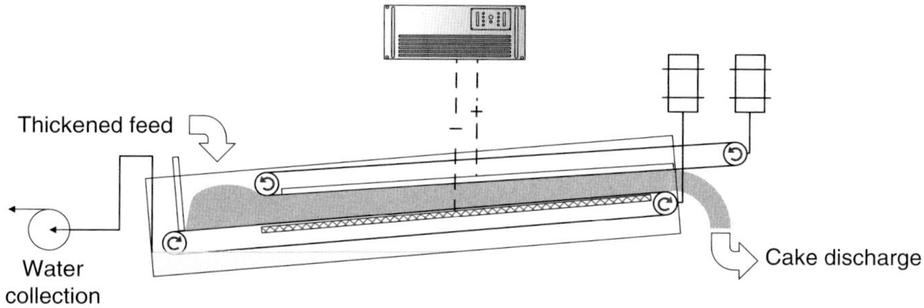

Figure 6.21 A schematic representation of the sludge dewatering zone of the dual-zone fully continuous prototype.

6.6 Conclusions

Many mining operations produce a dilute suspension of solids that, due to the properties of the suspended solids, take from 25 to 50 years to consolidate by hindered settling. The objective of this work was to develop an electrokinetic separation process that could be applied to phosphatic clay suspensions generated by the mining and beneficiation of phosphate ores. The use of an electric field to effect separation is attractive because the stability of the suspensions is caused, in part, by the charged character of the suspended particles. Previously reported work demonstrated the feasibility of batch-wise electrokinetic separation of phosphatic clay suspensions, but the process is widely regarded to be uneconomical. The present work was predicated on the assumption that an economically feasible separation requires development of a fully continuous electrokinetic separator.

A systematic progression of prototypes was used to explore the concepts needed to develop a fully continuous separator. The evolution of prototype design was guided by a constitutive relationship that was discovered for the batch system and found to apply, with different values for adjustable parameters, to each of the prototypes. To explore the production of clarified water, a semi-continuous prototype was designed that retained thickened solids and produced a stream of low-turbidity water. This process demonstrated that, when combined with gravity settling, clarified water could be produced with turbidity levels lower than 10 NTU. The residence time required for gravity settling was less than 24 hours. The subsequent semi-continuous prototype was designed to remove solids by use of a fabric belt that moved across the anode. Design parameters for the belt were identified. The key parameter was found to be the open area of the belt, which, to minimize potential losses, should be at least 50%. The belt design proved to be efficient, and a solids content of roughly 25 wt% was produced. The belt design was implemented in the first-generation fully continuous prototype, in which solids were collected on the belt in a horizontal configuration and then removed from the cell by redirecting the belt to an incline. The performance of the first-generation prototype was poor. A solids content of only 20 wt% could be achieved. The second-generation fully continuous prototype comprised a cake formation zone and a cake dewatering zone. A solids content as high as 35 wt% was achieved.

The improvement in economic feasibility is evident in a comparison of benchmark values to those obtained by application of previous bench-scale experiments to estimation of power and energy requirements for batch-wise dewatering of a clay-settling area. McKinney and Orazem [73] estimated the power requirement for dewatering a clay-settling pond at the Four Corners Mine in Bartow, Florida, from a solids content of 10-25 wt% to be 44 000 MW. The power requirement for continuous dewatering of the output of the Four Corners Mine to a higher solids content of 35 wt% was estimated to be as low as 29 MW. The improvement of efficiency is seen in the reduction of the energy consumption from 41.3 Wh $kg_{H_2O}^{-1}$ of reclaimed water for the batch dewatering of a clay settling pond to 6.6 Wh $kg_{H_2O}^{-1}$ for the continuous process. These figures may be expressed in terms of dry solids produced as 248 Wh kg_{solids}^{-1} and 47 Wh kg_{solids}^{-1}, respectively. It is important to recognize that the improvement in energy efficiency was obtained while increasing the final solids content from 25 to 35 wt%.

The efficiency of the present configuration can be improved further by separating the cake formation and cake dewatering zones. While the process was developed for processing phosphatic clay suspensions, it is anticipated that the electrokinetic dewatering may be applied as well for other mining operations such as associated with the oil sand tailings in Canada.

Acknowledgments

This work was supported by Mosaic Fertilizer LLC, Paul Kucera, program monitor. The authors express their gratitude for the support of Dr. Yichang Charles Guan, Mosaic Fertilizer LLC, and Profs. Guerry H. McClellan and David Bloomquist, University of Florida.

References

1 Reuss, F.F. (1809). Notice sur un nouvel effet de l'électricité galvanique. *Mém. Soc. Impér. Nat. Moscou* 2: 327–337.
2 Volta, A. (1800). On the electricity excited by the mere contact of conducting substances of different kinds. In a letter from Mr. Alexander Volta, F. R. S. professor of natural philosophy in the University of Pavia, to the Rt. Hon. Sir Joseph Banks, Bart. K. B. P. R. S. *Philos. Trans. R. Soc. London* 90: 403–431.
3 U.S. Geological Survey (2017). Phosphate rock. In: *Mineral Commodity Summaries*, 124–125. U.S. Geological Survey.
4 Gilbert, N. (2009). Environment: the disappearing nutrient. *Nat. News* 461 (7265): 716–718.
5 U.S. Geological Survey (1996). Phosphate rock. In: *Mineral Commodity Summaries*, 122–123. U.S. Geological Survey.

6 U.S. Geological Survey (1997). Phosphate rock. In: *Mineral Commodity Summaries*, 124–125. U.S. Geological Survey.
7 U.S. Geological Survey (1998). Phosphate rock. In: *Mineral Commodity Summaries*, 124–125. U.S. Geological Survey.
8 U.S. Geological Survey (1999). Phosphate rock. In: *Mineral Commodity Summaries*, 126–127. U.S. Geological Survey.
9 U. S. Geological Survey (2000). Phosphate rock. In: *Mineral Commodity Summaries*, 124–125. U.S. Geological Survey.
10 U.S. Geological Survey (2001). Phosphate rock. In: *Mineral Commodity Summaries*, 120–121. U.S. Geological Survey.
11 U. S. Geological Survey (2002). Phosphate rock. In: *Mineral Commodity Summaries*, 122–123. U.S. Geological Survey.
12 U.S. Geological Survey (2003). Phosphate rock. In: *Mineral Commodity Summaries*, 124–125. U.S. Geological Survey.
13 U.S. Geological Survey (2004). Phosphate rock. In: *Mineral Commodity Summaries*, 122–123. U.S. Geological Survey.
14 U.S. Geological Survey (2005). Phosphate rock. In: *Mineral Commodity Summaries*, 122–123. U.S. Geological Survey.
15 U.S. Geological Survey (2006). Phosphate rock. In: *Mineral Commodity Summaries*, 124–125. U.S. Geological Survey.
16 U.S. Geological Survey (2007). Phosphate rock. In: *Mineral Commodity Summaries*, 120–121. U.S. Geological Survey.
17 U.S. Geological Survey (2008). Phosphate rock. In: *Mineral Commodity Summaries*, 124–125. U.S. Geological Survey.
18 U.S. Geological Survey (2009). Phosphate rock. In: *Mineral Commodity Summaries*, 120–121. U.S. Geological Survey.
19 U.S. Geological Survey (2010). Phosphate rock. In: *Mineral Commodity Summaries*, 118–119. U.S. Geological Survey.
20 U.S. Geological Survey (2011). Phosphate rock. In: *Mineral Commodity Summaries*, 118–119. U.S. Geological Survey.
21 U.S. Geological Survey (2012). Phosphate rock. In: *Mineral Commodity Summaries*, 118–119. U.S. Geological Survey.
22 U.S. Geological Survey (2013). Phosphate rock. In: *Mineral Commodity Summaries*, 118–119. U.S. Geological Survey.
23 U.S. Geological Survey (2014). Phosphate rock. In: *Mineral Commodity Summaries*, 118–119. U.S. Geological Survey.
24 U.S. Geological Survey (2015). Phosphate rock. In: *Mineral Commodity Summaries*, 118–119. U.S. Geological Survey.
25 U.S. Geological Survey (2016). Phosphate rock. In: *Mineral Commodity Summaries*, 124–125. U.S. Geological Survey.
26 U.S. Geological Survey (2011). Phosphate rock. In: *Minerals Yearbook*, 56-1–56-11. U.S. Geological Survey.
27 Brown, M.T. (2005). Landscape restoration following phosphate mining: 30 years of co-evolution of science, industry and regulation. *Ecol. Eng.* 24 (4): 309–329.
28 Barnett, C. (2008). Mine field. Florida Trend: The Magazine of Florida Business, 84–90.

29 Van Kauwenbergh, S.J., Cathcart, J.B., and McClellan, G.H. (1990). *Mineralogy and Alteration of the Phosphate Deposits of Florida*, U.S. Geological Survey Bulletin 1914, U.S. Geological Survey.
30 Wilson, M. and Hanlon, E.A. (2012). Florida Reclaimed Phosphate Mine Soils: Characteristics, Potential Uses, and Management Considerations. Tech. Rep. SL370. Soil and Water Science Department, Florida Cooperative Extension Service, Institute of Food and Agricultural Sciences, University of Florida.
31 Moudgil, B.M. (1992). Enhanced Recovery of Coarse Particles During Phosphate Flotation. Tech. Rep. Florida Institute of Phosphate Research.
32 Tyler, P.M. and Waggaman, W.M.H. (1954). Phosphatic slime. *Ind. Eng. Chem.* 46 (5): 1049–1056.
33 Vesilind, P.A. (1994). The role of water in sludge dewatering. *Water Environ. Res.* 66 (1): 4–11.
34 Santamarina, J.C., Klein, K.A., Wang, Y.H., and Prencke, E. (2002). Specific surface: determination and relevance. *Can. Geotech. J.* 39 (1): 233–241.
35 Peng, C., Lai, H., Orazem, M.E., and Moghaddam, S. (2018). Microstructure of clay fabric in electrokinetic dewatering of phosphatic clay suspensions. *Appl. Clay Sci.* 158: 94–101.
36 Tripathy, T. and De, B.R. (2006). Flocculation: a new way to treat the waste water. *J. Phys. Sci.* 10: 93–127.
37 Abd-El Rahman, M.K. (2000). Dewatering of phosphatic clay waste by flocculation. *Chem. Eng. Technol.* 23 (5): 457–461.
38 Lu, L., Pan, Z., Hao, N., and Peng, W. (2014). A novel acrylamide-free flocculant and its application for sludge dewatering. *Water Res.* 57: 304–312.
39 Pearse, M.J., Weir, S., Adkins, S.J., and Moody, G.M. (2001). Advances in mineral flocculation. *Miner. Eng.* 14 (11): 1505–1511.
40 Vaxelaire, J. and Cézac, P. (2004). Moisture distribution in activated sludges: a review. *Water Res.* 38 (9): 2215–2230.
41 EPA (2000). Biosolids Technology Fact Sheet Belt Filter Press. Tech. Rep. EPA 832-F-00-057. United States Environmental Protection Agency.
42 Novak, J.T. (2006). Dewatering of sewage sludge. *Drying Technol.* 24: 1257–1262.
43 EPA (2000). Biosolids Technology Fact Sheet Centrifuge Thickening and Dewatering. Tech. Rep. EPA 832-F-00-053. United States Environmental Protection Agency.
44 Snyman, H.G., Forssman, P., Kafaar, A., and Smollen, M. (2000). The feasibility of electro-osmotic belt filter dewatering technology at pilot scale. *Water Sci. Technol.* 41: 137–144.
45 Peeters, B. (2010). Mechanical dewatering and thermal drying of sludge in a single apparatus. *Drying Technol.* 28 (4): 454–459.
46 Mahmoud, A., Olivier, J., Vaxelaire, J., and Hoadley, A.F.A. (2011). Electro-dewatering of wastewater sludge: influence of the operating conditions and their interactions effects. *Water Res.* 45 (9): 2795–2810.
47 Horwitz, W. (1939). The theory of electrokinetic phenomena. *J. Chem. Edu.* 16 (11): 519–535.
48 Speil, S. and Thompson, M.R. (1942). Electrophoretic dewatering of clay. *J. Electrochem. Soc.* 81: 119–145.

49 Kim, D.H., Ryu, B.G., Park, S.W. et al. (2009). Electrokinetic remediation of Zn and Ni-contaminated soil. *J. Hazard. Mater.* 165 (1): 501–505.
50 Buchireddy, P.R., Bricka, R.M., and Gent, D.B. (2009). Electrokinetic remediation of wood preservative contaminated soil containing copper, chromium, and arsenic. *J. Hazard. Mater.* 162 (1): 490–497.
51 Baek, K., Kim, D.H., Park, S.W. et al. (2009). Electrolyte conditioning-enhanced electrokinetic remediation of arsenic-contaminated mine tailing. *J. Hazard. Mater.* 161 (1): 457–462.
52 Mao, X., Wang, J., Ciblak, A. et al. (2012). Electrokinetic-enhanced bioaugmentation for remediation of chlorinated solvents contaminated clay. *J. Hazard. Mater.* 213: 311–317.
53 Guo, Y. and Shang, J.Q. (2014). A study on electrokinetic dewatering of oil sands tailings. *Environ. Geotech.* 1: 121–134.
54 Bourgès-Gastaud, S., Stoltz, G., Dolez, P. et al. (2014). Laboratory device to characterize electrokinetic geocomposites for fluid fine tailings dewatering. *Can. Geotech. J.* 52 (4): 505–514.
55 Lamont-Black, J., Jones, C.J.F.P., Glendinning, S. et al. (2007). Laboratory evaluation of the potential for electrokinetic belt filter press dewatering of kimberlite slimes. In: Paste, 147–152.
56 Bayat, O., Kilic, O., Bayat, B. et al. (2006). Electrokinetic dewatering of Turkish glass sand plant tailings. *Water Res.* 40 (1): 61–66.
57 Fourie, A.B., Johns, D.G., and Jones, C.J.F.P. (2007). Dewatering of mine tailings using electrokinetic geosynthetics. *Can. Geotech. J.* 44: 160–172.
58 Shang, J.Q. and Mohamedelhassan, E. (2001). Electrokinetic dewatering of eneabba west mine tailings: a laboratory experimental study. In: *Soft Ground Technology*, Chapter 26, 346–357. American Society of Civil Engineers.
59 Johnson, T.J. and Davis, E.J. (2000). Experimental data and theoretical predictions for the rate of electrophoretic clarification of colloidal suspensions. *Environ. Sci. Technol.* 34 (9): 1806–1812.
60 Johnson, T.J. and Davis, E.J. (1999). Electrokinetic clarification of colloidal suspensions. *Environ. Sci. Technol.* 33 (8): 1250–1255.
61 Johnson, T.J. and Davis, E.J. (1999). An analysis of electrophoresis of concentrated suspensions of colloidal particles. *J. Colloid Interface Sci.* 215 (2): 397–408.
62 Shang, J.Q. (1997). Electrokinetic sedimentation: a theoretical and experimental study. *Can. Geotech. J.* 34 (2): 305–314.
63 Shang, J.Q. (1997). Electokinetic dewatering of clay slurries as engineered soil covers. *Can. Geotech. J.* 34 (1): 78–86.
64 Shang, J.Q. and Lo, K.Y. (1997). Electrokinetic dewatering of a phosphate clay. *J. Hazard. Mater.* 55: 117–133.
65 Glendinning, S., Lamont-Black, J., and Jones, C.J.F.P. (2007). Treatment of sewage sludge using electrokinetic geosynthetics. *J. Hazard. Mater.* 139 (3): 491–499.
66 Ho, M.Y. and Chen, G. (2001). Enhanced electro-osmotic dewatering of fine particle suspension using a rotating anode. *Ind. Eng. Chem. Res.* 40 (8): 1859–1863.

67 Hwang, S. and Min, K.S. (2003). Improved sludge dewatering by addition of electro-osmosis to belt filter press. *J. Environ. Eng. Sci.* 2 (2): 149–153.

68 Raats, M.H.M., van Diemen, A.J.G., Lavén, J., and Stein, H.N. (2002). Full scale electrokinetic dewatering of waste sludge. *Colloids Surf., A* 210 (2–3): 231–241.

69 Jones, C.J.F.P., Lamont-Black, J., and Glendinning, S. (2011). Electrokinetic geosynthetics in hydraulic applications. *Geotext. Geomembr.* 29 (4): 381–390.

70 Fourie, A. (2006). Harnessing the power: opportunities for electrokinetic dewatering of mine tailings. *Geotech. News* 24 (2): 27–32.

71 McKinney, J.P. and Orazem, M.E. (2011). A constitutive relationship for electrokinetic dewatering of phosphatic clay slurries. *Miner. Metall. Process.* 28: 49–54.

72 McKinney, J.P. (2010). Design of electrolytic dewatering systems for phosphatic clay suspensions. PhD thesis. University of Florida.

73 McKinney, J.P. and Orazem, M.E. (2011). Electrokinetic dewatering of phosphatic clay settling areas: numerical simulation and economic assessment. *Miner. Metall. Process.* 28: 71–76.

74 EKS (2015). Electro-kinetic reclamation (EKR) testing and development phase II. Final report. EKS Electrokinetic Solutions.

75 Kong, R. and Orazem, M.E. (2014). Semi-continuous electrokinetic dewatering of phosphatic clay suspensions. *Electrochim. Acta* 140: 438–446.

76 Kong, R. (2015). Continuous electrokinetic dewatering of phosphatic clay suspensions. PhD thesis. University of Florida.

77 Crowl, D.A. and Louvar, J.F. (2011). *Chemical Process Safety: Fundamentals with Applications*, 3e. Upper Saddle River, NJ: Prentice Hall.

78 Peters, M.S. and Timmerhaus, K.D. (1968). *Plant Design and Economics for Chemical Engineers*, 2e. New York: McGraw-Hill.

79 Derwent, R., Simmonds, P., O'Doherty, S. et al. (2006). Global environmental impacts of the hydrogen economy. *Int. J. Nucl. Hydrogen Prod. Appl.* 1: 57–67.

7

Breaking the Chemical Paradigm in Electrochemical Engineering: Case Studies and Lessons Learned from Plating to Polishing

E. Jennings Taylor, Maria E. Inman, Holly M. Garich, Heather A. McCrabb, Stephen T. Snyder, and Timothy D. Hall

Faraday Technology, Inc., 315 Huls Drive, Englewood, OH 45315, USA

7.1 Introduction

In his seminal work titled *The Structure of Scientific Revolutions* [1], history of science philosopher Thomas Kuhn discusses scientific progress and introduces the concept of "paradigms." Since Kuhn's work, the use of the word paradigm has dramatically increased without a clear understanding of its meaning [2]. Specifically, a paradigm is not the same as a hypothesis or the scientific method. Simply stated, scientific paradigms are universally recognized concepts providing a framework of assumptions, methods, and ways of thinking by a community of practitioners. As such, scientific paradigms are open-ended, providing the tools for a community of scientists, engineers, and technologists to solve new problems within their particular field of endeavor. The resulting successful application of the scientific paradigm toward solving emerging scientific, engineering, and technological problems further strengthens the paradigm within the community of practitioners and increases the number of practitioners advocating the paradigm.

However, with the establishment of a scientific paradigm, the community of practitioners often discounts observations that do not conform to the accepted paradigm. As a consequence, peer-reviewed manuscripts and grant proposals building on non-conforming observations often go unpublished and unfunded, respectively. Kuhn summarizes this evolving state of affairs by noting that scientific, engineering, and technological advancement based on established paradigms [3],

> …does not aim at novelties of fact or theory and when successful, finds none.

In essence, the entrenched scientific community becomes so reliant on the established paradigm that observations that do not conform to the paradigm are ignored or discounted.

In summary, paradigms provide the basis for scientific discovery and engineering and technological advancements while simultaneously suppressing

Electrochemical Engineering: The Path from Discovery to Product,
First Edition. Edited by Richard C. Alkire, Philip N. Bartlett, and Marc T. Koper.

non-conforming observations that could lead to new or modified paradigms. In this contribution, events at Faraday Technology, Inc., spanning a period of approximately 22 years (1991–2013) are described that led to the development of an industrially compatible electropolishing process for niobium superconducting radio frequency (SRF) cavities using pulse reverse voltage waveforms in conjunction with low concentration aqueous acid electrolytes devoid of hydrofluoric acid. This development is a paradigm shift from the widely accepted viscous salt film electropolishing model of Jacquet [4].

One of the authors' objectives is to report the specific technical discoveries and observations that led to an environmentally benign and industrially compatible pulse reverse voltage electropolishing process for SRF cavities. While acknowledging that scientific, engineering, and technological advances occur by building on well-established paradigms, another of the authors' objectives is to provide useful insights related to the pursuit of non-conforming observations that can lead to potentially paradigm-shifting technologies. The authors conclude that scientific discovery of paradigm-shifting innovations require a careful balance between current fundamental understandings combined with the pursuit of observations that do not necessarily fit the existing paradigm.

7.1.1 Perspective

The authors' perspective is based on over 20 years of experience at a small research, development, and engineering company that was founded in 1991 to conceive, develop, and commercialize novel electrodeposition/plating technologies based on pulse current/pulse reverse current (PC/PRC) electrolytic principles. Historically, progress in the plating industry has focused on a reliance on "electrochemical art" of chemical additives, which is perpetuated by chemical suppliers acting as "gate keepers" to the metal finishing industry. Specifically, the current reliance on electrochemical art may be traced to approximately 150 years ago when the understanding of electrochemical principles was nascent. Plating history is full of stories documenting the serendipitous "discovery" of chemical additions to plating baths leading to the desired coating properties. These discoveries ultimately served as models for development of new plating bath chemistries, which led to the proprietary chemical additives of today's chemical formulators. While considerable progress was made using this chemical paradigm, possible solutions that did not involve additives were deemphasized. By shifting the paradigm from the "electrochemical art" of chemical additives to the electrochemical science of nucleation, growth, kinetics, and mass transport phenomena, the authors speculated that pulse reverse waveforms could be utilized to develop robust and environmentally benign metal finishing processes, while retaining the advantages of electrodeposition based on chemical additions such as high deposition speed and low capital investment.

The company's mission is to conceive and commercialize electrochemical processes based on the PC/PRC technology platform. To support these activities, funding is obtained through the highly competitive US Small Business Innovative Research (SBIR) and Small Business Technology Transfer (STTR) programs. These programs consist of a Phase I technology feasibility demonstration activity of approximately \$70 000–150 000 for 6–12 months period of performance. Approximately 10% of the Phase I proposals are awarded. After the successful conclusion of an awarded Phase I program, the small business may then apply for Phase II technology validation funding of \$300 000–1 000 000 for a duration of two years. The Phase I to awarded Phase II conversion rate is approximately 30%, effectively making the "odds" of receiving Phase II funding 1 in 30. Because of the emphasis on innovation in seeking highly competitive SBIR/STTR funding, small business researchers must often look outside traditional scientific paradigms both in the proposal phase and while gathering data that demonstrates the proof of concept of their proposed technology. However, these observations generally must be tempered with acknowledgement of the traditional scientific paradigms, especially given that proposal reviewers may base their evaluation on their knowledge and acceptance of those traditional paradigms.

Once the Phase II development activity has been completed, additional funding to commercialize the technology for private sector use or to transition the technology for government use comes from other sources of private sector or non-SBIR government funding, respectively. While the SBIR/STTR programs represent a highly competitive source of technology demonstration and validation funding, importantly, these programs allow the small business to retain assignment of the inventions "conceived or first actually reduced to practice" with said funds [5].

The company considers itself to be an "innovation factory" based on its pulse/pulse reverse electrolytic technology platform, and its product is intellectual property in the form of know-how, trademarks, and patents. Strategic partners fund the pre-production α-scale validation of the subject technology with the option to acquire the competitive advantage associated with the intellectual property relevant to the technology. Generally speaking, the strategic partner does not require a detailed mechanistic understanding of the technology. However, the strategic partner does require sufficient understanding to enable implementation and robust process control of the technology. This experience is similar to that reported for the development of the nickel–iron plating process for thin film head production, which was in production years before [6],

> ...a satisfactory mechanism of the phenomena had been achieved.

The competitive advantage associated with the technology is generally provided to the strategic commercial partner in the form of "field-of-use licenses," "early adapter licenses," or "supply chain licenses," although patents are sometimes sold [7]. In the case of governmental strategic partners, the technology is freely available without the need for acquiring the rights to the patents and/or know-how [5]. In summary, the company's business model utilizes "seed" funding to advance the pulse/pulse reverse electrolytic platform technology via the highly competitive SBIR/STTR environment, identifying university collaborators as appropriate, engaging commercial and governmental strategic partners, and establishing intellectual property in the form of patents and know-how.

In order to provide a complete picture of this approach, this chapter will begin the discussion by exploring some initial successes in pulse reverse plating of copper. As will be discussed below, a paradigm-shift was identified in the process for deposition of copper, specifically with respect to the conventional reliance on additive chemistries to achieve the desired deposit characteristics. This recognition that the use of pulse reverse electric fields may obviate the need for chemical additives in cathodic (electrodeposition and plating) technologies was later applied to a wide range of anodic (electrochemical deburring, electrochemical machining, and electropolishing) technologies to enable performance-based processes using environmentally benign and worker safe electrolytes. Although not discussed further, pulse reverse electrolytic principles have also been applied to facilitate electrophoretic deposition from electrolytes containing higher water contents than permitted in direct current (DC) electrophoretic deposition [8, 9].

Consequently, while the company's initial founding vision was:

> ...to change the focus of **electrodeposition processes** from the use of electrolytes with multicomponent chemical additions to the use of simpler electrolytes enabled by pulse current/pulse reverse current electric fields...

This vision has evolved to include PC/PRC anodic dissolution or surface-finishing processes such as deburring, electrochemical machining, electropolishing, and electrochemical through-mask etching. This broadening of the vision was based on observations and speculations that the lessons learned from cathodic PC/PRC electrodeposition processes could be adapted to anodic PC/PRC surface-finishing processes. Consequently, the current company vision is:

> ...to change the focus of **electrochemical manufacturing/engineering processes** from the use of electrolytes with multicomponent chemical additions to the use of simpler electrolytes enabled by pulse current/pulse reverse current electric fields...

This vision has led to approximately 30 US patents as well as approximately 15 foreign patents covering various embodiments of pulse/pulse reverse electrochemical processes as well as electrochemical apparatus. In addition, some of the technical successes have been recognized with awards including the 2007 Blum Scientific Achievement award, a 2011 R&D 100 award, and a 2013 Presidential Green Chemistry Challenge award.

7.2 A Brief Overview of Pulse Reverse Current Plating

The concept of PRC plating is not new and was first reported at least in the early part of the nineteenth century [10]. The guiding principles of PC/PRC plating, presented in 1986 in the classic compendium *Theory and Practice of Pulse Plating* edited by Jean-Claude Puippe and Frank Leaman, are still relevant today [11]. In spite of numerous theoretical and experimental studies characterizing pulse current plating reported in the 1970s (these will be briefly summarized below), most of the studies directed toward practical applications of PC/PRC plating used the existing plating baths containing chemical additions optimized for DC plating. Regarding the practical applications of PC/PRC plating, two questions guided the pursuit of novel PC/PRC electrodeposition technologies:

1. Why should one expect that plating bath chemical additives optimized for DC plating would be optimum for PC/PRC plating?
2. Compared to DC plating, can PC/PRC enable plating from a simpler plating baths with low or no chemical additives and thereby lead to a more robust plating process?

A brief review of pulse current plating is provided to establish the basis for evolving plating and metal-finishing activities, ultimately leading to the development of the niobium electropolishing process in low-viscosity solutions. This discussion, while perhaps technically dry, sets the stage by outlining the guiding principles that enabled us to design waveforms for the specific applications described later. A more recent review of pulse current plating is also available [12].

Figure 7.1 presents a generic PRC waveform for a net cathodic or plating process. The generic description illustrates a cathodic (forward) pulse followed by an off time, followed by an anodic (reverse) pulse and followed by a second off time. The cathodic peak current density ($i_{cathodic}$), cathodic on time ($t_{cathodic}$), cathodic off time ($t_{off,cathodic}$), anodic peak current density (i_{anodic}), anodic on time (t_{anodic}), and anodic off time ($t_{off,anodic}$) are individual variables for process control. There are numerous embodiments of PC/PRC waveforms [13], a common one being waveforms consisting of only cathodic pulses. Additionally, other common embodiments that include the anodic pulse may eliminate one or both off times.

The sum of the cathodic on time, anodic on time, and off time(s) is the period (T) of the pulse and the inverse of the period is the frequency (f). Specifically,

$$T = (t_{cathodic}) + (t_{off,cathodic}) + (t_{anodic}) + (t_{off,anodic}) \tag{7.1}$$

$$f = (1/T) \tag{7.2}$$

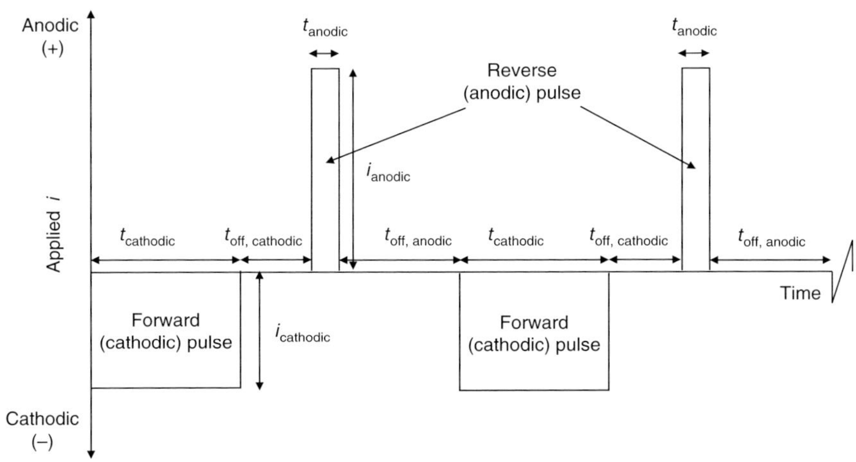

Figure 7.1 Schematic illustration of a generalized pulse reverse current waveform.

The cathodic duty cycle (γ_{cathodic}) is the ratio of the cathodic on time to the pulse period, and the ratio of the anodic on time to the pulse period is the anodic duty cycle (γ_{anodic}). The frequency and duty cycles are additional variables for process control. The average current density (i_{average}) or electrodeposition rate is given by

$$i_{\text{average}} = (i_{\text{cathodic}})(\gamma_{\text{cathodic}}) - (i_{\text{anodic}})(\gamma_{\text{anodic}}) \tag{7.3}$$

It should be noted that even though PC/PRC waveforms contain off times and anodic periods, the plating rate is typically the same or higher than in DC plating. As discussed below, this is attributed to the fact that the "instantaneous" peak currents attained during the pulse on time can be much higher than that attained during DC plating.

7.2.1 Mass Transport Effects in Pulse Current Plating

Mass transport in pulse current plating is a combination of steady-state and non-steady-state diffusion processes [14]. The mass-transfer-limited current density (i_l) is related to the reactant concentration gradient ($C_b - C_s$) and to the diffusion layer thickness (δ) by

$$i_l = -nFD\,(\partial C/\partial x)_{x=0} = -nFD[(C_b - C_s)/\delta] \tag{7.4}$$

where n, F, and D are the number of equivalents exchanged, Faraday's constant, and diffusivity of the reacting species, respectively. In DC electrolysis, δ is a time-invariant quantity for a given electrode geometry and hydrodynamic condition. In pulsed electrolysis, however, δ varies from 0 at the beginning of the pulse to its steady-state value when the Nernst diffusion layer, δ_N, is fully established [15–17]. The corresponding mass-transport-limiting current density is equal to an infinite value at $t = 0$ and decreases to a steady-state value of the DC-limiting current density. The advantage of pulse electrolysis is that the current is interrupted before δ reaches steady state, allowing the reacting ions to diffuse back to the surface and partially or completely replenish the surface concentration to its original value before the next current interruption. Therefore, the concentration of reacting species in the vicinity of the electrode changes with the pulse frequency. During pulse electrolysis, Ibl [18, 19] proposed a "duplex diffusion layer" consisting of a pulsating layer, δ_p, and a stationary layer, δ_s for a deposition process (Figure 7.2a). Figure 7.2b shows the equivalent diffusion layers for an anodic process, e.g. metal removal.

By assuming a linear concentration gradient across the pulsating diffusion layer and conducting a mass balance, Ibl derived the pulsating diffusion layer thickness (δ_p) as

$$\delta_p = (2Dt_{\text{on}})^{1/2} \tag{7.5}$$

where t_{on} is the pulse on time. When the pulse on time is equal to the transition time, the concentration of reacting species at the interface drops to zero at the end of the pulse. An approximate expression for the transition time, τ, is

$$\tau = ((nF)^2 \cdot C_b{}^2 D)/2i_c{}^2 \tag{7.6}$$

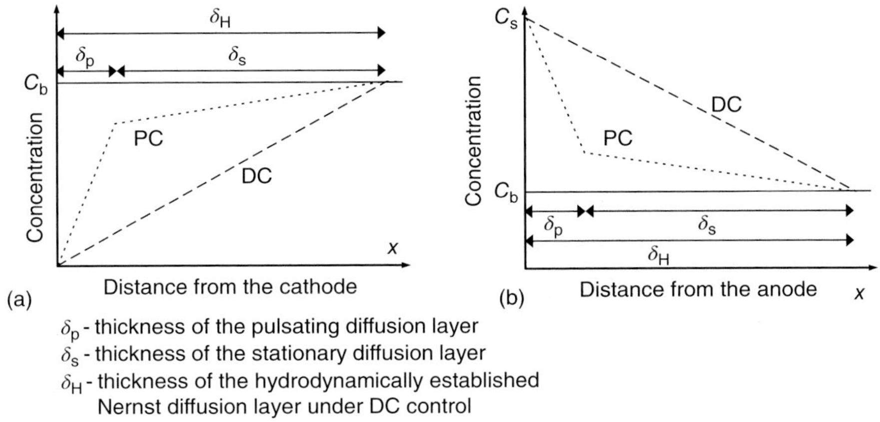

Figure 7.2 Schematic representation delineating the pulsating vs stationary diffusion layer for (a) a cathodic (metal deposition) process and (b) an anodic (metal removal) process.

More exact solutions are given by integrating Fick's diffusion equation

$$\delta_p = 2((Dt_{on})/\pi)^{1/2} \tag{7.7}$$

$$\tau = \pi((nF)^2\, C_b{}^2 D)/4i_c{}^2 \tag{7.8}$$

Figure 7.3 shows the change in pulsating diffusion layer thickness as a function of the pulse on time. For on times less than the transition time, τ, the concentration of the reacting metal species at the interface remains above zero. For on times equal to or greater than τ, the concentration of the reacting metal species at the interface is zero, and the process is mass transport limited. For a DC process, a hydrodynamic Nernst diffusion layer is established.

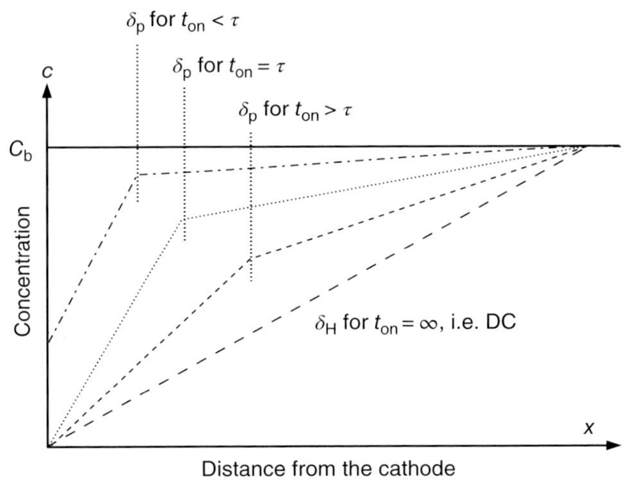

Figure 7.3 Schematic representation of the change in pulsating diffusion layer thickness as a function of the relationship between t_{on} and the transition time, τ.

Yin [20] derived a similar relationship for the pulsating diffusion layer in the case of PRC plating. To emphasize that the thickness of the pulsating diffusion layer is related to the pulse current on time, this is referred to as the "electrodynamic" diffusion layer. The key points used in the development of a pulsed process for electrodeposition are (i) the electrodynamic diffusion layer thickness is proportional to pulse on time and (ii) transition time is inversely proportional to the pulse current.

7.2.2 Current Distribution Effects in Pulse Current Plating

Current distribution and hence metal distribution are also a function of the pulse current plating parameters [21, 22]. Current distribution is an important parameter in coating electrodeposition. Primary current is governed solely by the geometric effects of the plating apparatus. Secondary current distribution is governed by kinetic effects and activation overpotentials are considered. Tertiary current distribution is governed by mass transport effects and both activation and concentration overpotentials are considered [23]. The addition of secondary or tertiary current distribution effects tend to make the current distribution more uniform, as compared to primary current distribution alone.

For the purposes of this discussion, it is helpful to examine the balance of primary and secondary current distribution in terms of the Wagner number. The Wagner number (W_a) is the ratio of the activation polarization (kinetic effects) to the ohmic polarization (geometric effects):

$$W_a = (d\eta/di) \cdot (\kappa/l) \tag{7.9}$$

where "$d\eta/di$" is the Tafel slope for the reaction of interest, "κ" is the solution conductivity, and "l" is the characteristic length over which the electrochemical reaction is assumed to take place. For $W_a \ll 1$, primary current distribution prevails and the current distribution is less uniform. For $W_a \gg 1$, secondary current distribution prevails and the current distribution is more uniform. According to the Wagner number relationship, the tools with which to understand and perhaps manipulate the balance of primary, secondary, and tertiary current distribution are

- Tafel slope
- Solution conductivity
- Characteristic length

The characteristic length, l, may change over time as a surface is roughened or smoothed. The solution conductivity is a function of electrolyte chemistry and temperature. Manipulation of the Tafel slope may be done using pulsed electric fields.

The work of Prof. Der-Tau Chin's group at Clarkson University related to selective plating of solder and gold bumps helps to understand the effect of pulsed currents on the Tafel slope [24]. Solder and gold bumps are used as lead frame contacts in the electronics industry. The bumps need to be a certain height, l (the characteristic length in the Wagner equation), in order to effectively function as a lead frame contact. The electronics industry desired to achieve the required

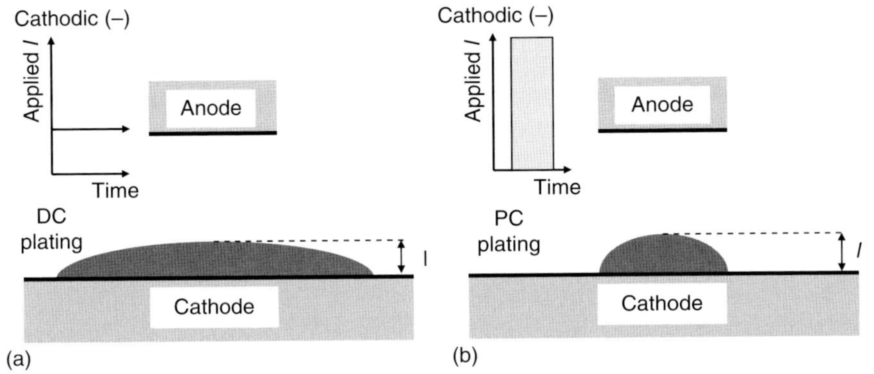

Figure 7.4 Schematic illustration of "bump" plating under (a) DC and (b) PC conditions at the same average current density.

height with minimum spreading of the bump in order to conserve lead frame "real estate" and material cost. Modeling and experimental studies were conducted of bump plating under DC and PC conditions at the same average current densities [24]. Using the geometrical setup in Figure 7.4, they reported that by decreasing the duty cycle during pulse current plating, the spread of the bump was reduced for a constant height. The results were rationalized in that during pulse current plating, the "time-averaged" Tafel slope was reduced compared to DC. Consequently, W_a trended toward being much less than 1, and therefore, the current distribution was less uniform.

A further consideration with regard to current distribution and the balance of primary, secondary, and tertiary effects is the relationship between the characteristic length, l, and the diffusion layer. In a macroprofile (Figure 7.5a), the roughness of the surface is large compared with the thickness of the hydrodynamic diffusion layer, δ_H, and when a pulsed electric field is applied, the diffusion layer is compressed to form a pulsating or electrodynamic diffusion layer, δ_p. The pulsating diffusion layer, δ_p, tends to follow the surface contour and becomes more compressed and thinner as the pulse on time becomes shorter. In a microprofile (Figure 7.5b), the roughness of the surface is small compared with the thickness of the hydrodynamic diffusion layer, δ_H. In this case, for a long pulse on

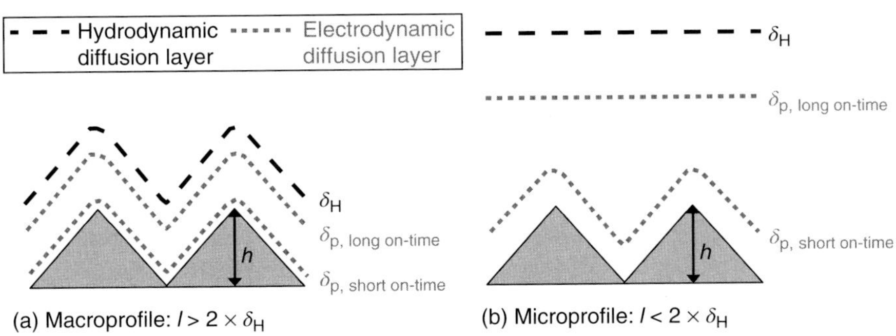

Figure 7.5 Schematic illustration of (a) macroprofile and (b) microprofile.

time, the pulsating diffusion layer is compressed but is still much larger than the characteristic length, and the microprofile is maintained. For very short pulse on times, the pulsating diffusion layer is compressed to the point at which it follows the surface contour, and the system effectively mimics a macroprofile.

A final consideration is that if the applied waveform is designed such that the pulse on time is much longer than the transition time, tertiary current distribution will play an important role in the process. With the addition of tertiary control, the concepts of macroprofile and microprofile and their influence on current distribution become important.

Under DC conditions and mass transport control, a macroprofile results in the most uniform current distribution and a nearly conformal deposit. As illustrated in Figure 7.5a, the application of pulse currents generates a smaller macroprofile. Based on experimental observations, for a macroprofile boundary layer condition, relatively long pulse on times can yield a slightly non-uniform current distribution compared to DC conditions, and relatively short pulse on times can yield a significantly more non-uniform current distribution than DC conditions. Assuming the same average current, for shorter pulse on times, the relative influence on current distribution shifts from tertiary current distribution control to secondary as well as primary current distribution control. Consequently, as concentration polarization effects are removed, the current distribution becomes less uniform. These observations are consistent with Prof. Chin's work [24] under a macroprofile where they observed a more non-uniform deposit with shorter pulse on times.

Under DC conditions and mass transport control, a microprofile results in the most non-uniform current distribution and a non-conformal deposit. As illustrated in Figure 7.5b, the application of pulse currents with a small enough on time can convert a microprofile to a macroprofile, establishing a small δ_p Eq. (7.5). For a microprofile diffusion layer condition, assuming tertiary current distribution control is maintained, short pulse on times sufficient to convert the microprofile to a macroprofile results in a significantly more uniform current distribution. Conversely, long pulse on times sufficient to maintain the microprofile results in a slightly more uniform current distribution compared to DC, assuming that tertiary control is maintained by selecting on times and peak currents that ensure $t_{on} \gg \tau$ per Eqs. (7.6) and (7.8).

Consequently, by applying the appropriately designed current pulse in terms of peak current and on time, the current distribution can be effectively focused or de-focused to create more uniform or more non-uniform deposition, respectively. This enhanced process control, enabled by the additional parameters associated with pulse processes relative to DC, is especially attractive for complex-shaped substrates requiring uniform coatings. As these concepts are based on the influence of pulse currents on current distribution, they are applicable to metal deposition as well as metal dissolution.

There are four pulse types, independent of anodic or cathodic orientation, that can be used to tailor waveforms for specific applications, and these are summarized in Figure 7.6. While there are additional permutations that may be

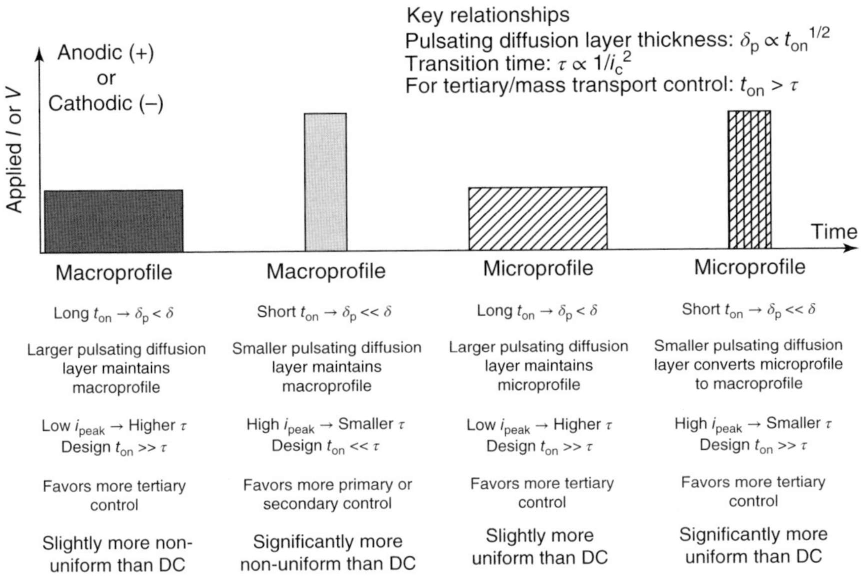

Figure 7.6 Guidelines for designing pulse parameters as a function of macroprofile/microprofile and need for more uniformity or more non-uniformity.

considered, these are the ones that have proven most useful in the development activities to date.

To summarize, while the Wagner number is not being used to directly design the pulse waveform parameters, the principles involved are being used to be cognizant of the effect of both the characteristic length and how the pulse parameters change the Tafel slope on the balance of primary and secondary as well as tertiary current distribution. That knowledge can be combined with an understanding of electrochemical processes operating under either macroprofile or microprofile boundary layer conditions to design the pulse parameters to achieve more uniformity or non-uniformity in each process, whether it be metal deposition or metal removal. These are guiding principles only, based on a review of the literature combined with experimental observations, and some iteration is generally required to determine appropriate peak currents and voltages and on times. It has proven challenging to date to assign exact numerical values for these parameters a priori, although numerical simulations may be used in the future to achieve predictive capabilities.

For electrochemical processes that are close to 100% faradaic efficiency for both metal deposition and metal dissolution, combining these cathodic and anodic pulses can create net waveforms that will achieve the overall desired process result. Combining multiple waveforms into sequences may also be required, as will be shown in the examples to follow. This is easily done with modern programmable rectifiers – a far simpler change than having to modify electrolytes or electrode geometries to achieve the same effect.

7.2.3 Grain Size Effects in Pulse Current Plating

Grain size in electrodeposited coatings, and hence mechanical properties such as ductility, is also a function of the pulse current plating parameters [25]. During the plating process, as ions enter the electrified interface between the solution and the cathode, a charge transfer reaction results in the formation of adatoms on the surface of the cathode. Electrocrystallization is the mechanism by which adatoms are incorporated into a crystal lattice during plating. Electrocrystallization can occur by either growth on previously deposited crystals or nucleation of new crystals. If electrocrystallization occurs by growth on previously deposited crystals, the resulting deposit will consist of large grains. In contrast, if electrocrystallization occurs by nucleation of new crystals, the resulting deposit will consist of small grains or even amorphous deposits. The nucleation rate (ν) is given by

$$v = k_1 \cdot \exp(-k_2/|\eta|) \tag{7.10}$$

where k_1 is proportionality constant, k_2 is related to the amount of energy needed for the two-dimensional nucleation, and η is the overpotential. The nucleation rate increases exponentially with increasing overpotential.

Pulse current plating generally utilizes peak current densities that are much higher than the current densities used in DC plating [25]. Therefore, the instantaneous current or voltage pulses, and hence the overpotentials during pulse current plating, may be higher than during DC plating. Consequently, pulse current plating can promote nucleation and a finer grained structure compared to DC plating. In fact, the "grain size" concept has been taken to the extreme by forming 4–5 nm catalyst particles for gas diffusion electrode applications [26–28]. Numerous studies have reported the effect of increased current density on grain size during pulse current plating [25]. These investigations included (i) increasing the pulse peak current density while maintaining a constant pulse on time and average current density, as illustrated in Figure 7.7a, and (ii) increasing the pulse

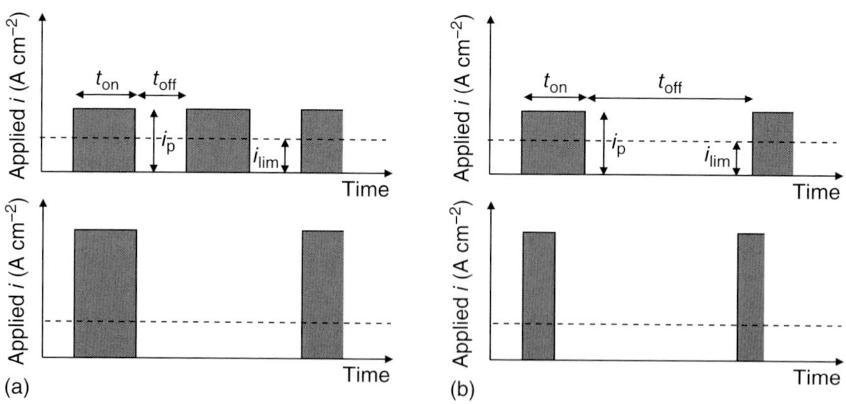

Figure 7.7 Illustration of two methodologies for studying the impact of increasing pulse peak current density on grain size while maintaining the same average current density, (a) cathodic charge per pulse and off time increase and (b) cathodic charge per pulse and off time are constant.

peak current density while maintaining a constant average current density with a shortened on time, as illustrated in Figure 7.7b. In the methodology illustrated in Figure 7.7a, the cathodic charge per pulse and off time as well as the pulse peak current density are increased. In the methodology illustrated in Figure 7.7b, the off time is increased slightly as the pulse peak current density is increased, while the cathodic charge per pulse is maintained. In both cases, an increase in pulse peak current density resulted in a deposit with smaller grain sizes, even though the total charge for each waveform remained the same. Finally, the pulse on time and pulse off time also affect the grain size of the deposit during pulse current plating. The impact of pulse on time and off time is more complicated and is generally dependent on the presence or absence of species in the plating bath that inhibit or promote grain growth.

7.2.4 Current Efficiency Effects in Pulse Current Plating

Current efficiency during electrodeposition is also a function of the pulse current plating parameters [29]. As is well known in the plating art, hydrogen evolution occurs during plating of many materials. The fraction of total current attributed to the desired metal deposition reaction is termed the current efficiency. Often, low current efficiency plating processes adversely impact deposit morphology such as increased porosity and increased crack formation. In addition, the hydrogen evolution reaction inherently increases the pH at the electrode surface even in an acidic plating bath. The current efficiency and hence the property of the deposits can be significantly impacted by pulse current plating parameters such as pulse peak current density, on time, off time, as well as PRC parameters.

7.2.5 Concluding Remarks for Pulse Current Plating

PC/PRC plating offers considerable parametric flexibility relative to DC plating. Just as there are infinite combinations of height, width, and length to obtain a given volume, in pulse and pulse reverse processes, there are unlimited combinations of peak current densities, duty cycles, and frequencies to obtain a given electrodeposition rate [30]. Pulse current plating parameters impact mass transport, current distribution and hence metal distribution, deposit grain size and hence mechanical properties, and current efficiency and hence deposit morphological characteristics. The key challenge is to identify the key PC/PRC waveform parameters to tune for a specific plating application without initiating an exhaustive experimental development program. The above review suggests that the key waveform parameters are the pulse on times and pulse peak currents based on their impact on the "electrodynamic" pulsating boundary layer Eqs. (7.5) and (7.7) and the transition time Eqs. (7.6) and (7.8), respectively. While waveform parameters such as duty cycle and frequency are often reported, these parameters are derived from the more physically meaningful pulse peak currents and pulse on times. Based on this "tangible" understanding as well as evolving observations, guiding principles have been reported for developing pulse and pulse reverse waveform parameters for specific electrodeposition applications [31].

7.3 Early Developments in Pulse Plating

In current state-of-the-art metal-plating operations, chemical additives are used to produce the desired properties and/or geometry of the metal coating. The additives influence a number of deposit properties, including (i) brightness, (ii) hardness, (iii) adhesion, (iv) corrosion resistance, (v) wear resistance, (vi) internal stress, and (vii) mechanical characteristics, i.e. strength and ductility. These chemical additives commonly include organic, metallic, ionic, and/or nonionic substances in concentrations ranging from several $mg\,l^{-1}$ to a few percent. The influence of additives is dramatic in that one additive molecule may affect the electrodeposition of several thousand metal ions [32]. During plating, the additives are consumed and/or incorporated into the deposit and must be quantified and replenished. The "discovery" of plating additives was patented in 1847 and even Edison experimented with plating additives [33]. Currently, a vibrant industrial sector of chemical formulators has evolved to supply the plating industry with proprietary additives for various applications.

In the mid-1990s, Faraday was approached by the research director of chemical formulator who supplied proprietary additives for copper plating baths used in the printed circuit board (PCB) industry. One of their competitors had implemented a PRC plating process for "plated through-holes" (PTHs) used as z-axis interconnects in the PCB industry. The PRC plating process resulted in more uniform plating of PTHs, as measured by the throwing power. Generally, the throwing power is the ratio of the plated copper thickness on the surface of the PCB to the plated copper thickness in the center of the PTH. The PCB industry specification for throwing power was ~80% and the use of PRC compared to DC copper plating permitted higher current densities and hence higher throughput while maintaining the required throwing power. The chemical formulator funded a "research for hire" activity to demonstrate the benefit of PRC copper plating of PTHs using their proprietary additives. The PTH diameter of interest at that time was 325 μm at an aspect ratio of ~4 : 1. The initial attempts using the PRC waveform parameters of their competitor resulted in a powdery burnt copper deposit that was unacceptable and consequently throwing power was irrelevant. The fact that slight changes in the proprietary additives of one chemical formulator would not be optimum using the same PRC parameters was not surprising due to the complex impact of PRC parameters on the adsorption/desorption of ionic organic additives during the cathodic and anodic pulse on times during plating. Using a small design of experiments approach, the PRC parameters that successfully plated bright copper deposits using the client's proprietary additives were established, with the target throwing power for the PTH at the same current density as reported by the competitor.

In addition, with the PRC waveform parameters identified for the chemical formulator client, the effect of the leveling additive concentration on the throwing power for the PTHs was investigated. Interestingly, as the leveling additive concentration was increased from the specified concentration, the throwing power decreased. Conversely, as the leveling additive concentration was decreased from the specified concentration, the throwing power increased. In fact, the throwing power was over 100% for a leveling additive concentration of near zero.

In summary, PRC copper plating yielded better throwing power without the leveling additive, than either DC or PRC with the leveling additive! As providing chemistry without additives was outside the business model of the chemical formulator client, the project was discontinued. In spite of the lack of interest in moving forward, this outcome was encouraging in that it showed the potential to realize the company vision of using PRC processing to enable simpler plating bath chemistries for an industrially important application.

7.3.1 Leveling Without Levelers Using Pulse Reverse Current Plating

Due to the need for increased packing density, the PCB industry has been continually driven to incorporate smaller *z*-axis interconnect features alongside the larger PTHs. These features include microvias of ~100 μm diameter and smaller. Researchers had attempted to plate copper into 100 μm microvias using pulse reverse waveform parameters that had been previously been employed to plate PTHs of >200 μm. These waveforms were applied in conventional plating baths containing the typical chemical additions; the existing paradigm for copper plating in the electronics industry focuses first on the impact of chemistry on the process, and latterly on the impact of the electric field parameters.

Based on the prior experience and observations with the chemical formulator, which suggested that a new paradigm of predominantly electric field process control may be emerging, it was felt that the challenge of plating both these new *z*-axis interconnect features as well as the larger PTHs could be addressed in simple copper plating baths without the addition of brighteners and levelers, using specifically tailored pulse reverse electric fields. The boundary layer in a typical PCB plating tank was estimated to be approximately 50–75 μm. Essentially, any feature with an asperity higher than approximately two times the boundary layer thickness would be plated under a macroprofile condition and any feature with an asperity less than approximately two times the boundary layer would be plated under a microprofile condition. Therefore, it was assumed that 200 μm PTHs would be plated under a macroprofile or conformal boundary layer condition and the 100 μm microvias would be plated under a microprofile or non-conformal condition. As such, based on the discussion above, it would be surprising if similar waveform parameters would be applicable to the different sized features under both macroprofile and microprofile boundary layer conditions.

In essence, plating a *z*-axis interconnect is similar to plating around a corner. Inhibition of plating at the high current density corner was the reason for the development of plating additives for DC plating. The additives block deposition on the high current density regions, essentially leveling the deposit. However, when using PRC plating, as copper is approximately 100% current efficient for both the cathodic deposition of copper and anodic dissolution of copper, the problem may be thought of as designing the cathodic pulse to minimize the copper over-plating at the corner, and as designing the anodic pulse to accentuate copper dissolution at the corner that had over-plated during the preceding cathodic pulse. In this manner, the net effect would be a conformal coating around the corner.

By combining this strategy with a conceptual understanding of the influence of a macroprofile (a PTH) or microprofile (a microvia) on the effect of an applied current pulse on current distribution, it seemed possible that conformal plating of PTHs and microvias could be done in a single process step using a sequence of two waveforms: a generalized "macroprofile waveform" (Figure 7.8a) and "microprofile waveform" (Figure 7.8b), respectively. While these concepts do not provide precise waveform parameters, they did provide general guidance for examination of the PRC waveform parameter space for plating the z-axis interconnects for electronic applications.

Per the discussion summarized in Figure 7.6, the macroprofile waveform for the PTHs was designed as a relatively long on time, low current cathodic pulse resulting in slightly more non-uniform deposition (compared to DC) around the corner by focusing on mass transport control, followed by a relatively short on time, high current anodic pulse to provide significantly more non-uniform removal of copper from the corner by deemphasizing mass transport control and preferentially focusing on primary and secondary current distribution. The net effect was good throwing power at high rates for PTHs.

Conversely, the microprofile waveform for the microvias was designed as a relatively short on time, high current cathodic pulse to achieve significantly more uniform deposition around the corner by converting the microprofile to a macroprofile by establishing a small δ_p via a small t_c (see Eq. (7.5)), and maintaining large component of tertiary current distribution control by establishing a relatively small τ, so that $(t_{cathodic}/\tau) \gg 1$. This was accomplished by using a large $I_{cathodic}$ (see Eq. (7.6)). This was followed by a relatively long on time, low current anodic pulse to provide non-uniform removal of copper from the corner by maintaining the microprofile and mass transport control. $I_{cathodic}$ was low, so to maintain $(t_{cathodic}/\tau) \gg 1$, $t_{cathodic}$ had to be long. The net effect was good throwing power at high rates for microvias.

From the previous work with the chemical formulator, the macroprofile waveform had been validated for plating 325 μm PTHs under macroprofile boundary layer conditions in the absence of leveling additives. To test the microprofile

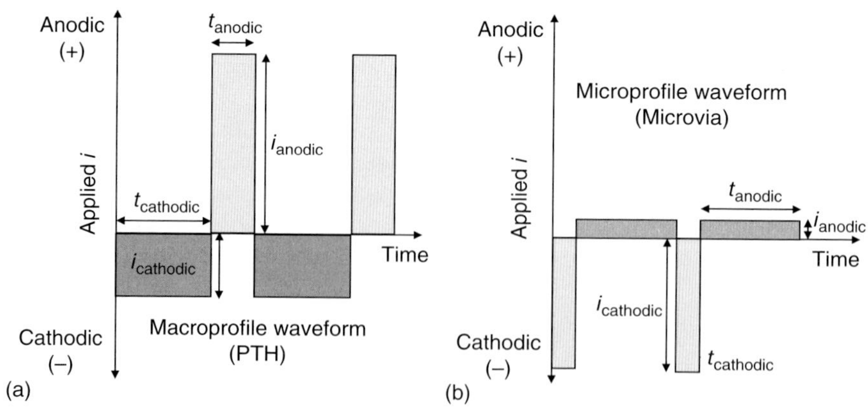

Figure 7.8 Illustration of (a) macroprofile waveform and (b) microprofile waveform.

waveform concept under microprofile boundary layer conditions, copper plating experiments were conducted in a bath without leveling additives on brass substrates into which 100 μm holes had been mechanically drilled [34–37]. In all cases, the average current was 35 mA cm^{-2}. In Figure 7.9a–f, the results of these experiments are summarized. In Figure 7.9a is a cross section for DC plating. As expected, a "key-hole" is formed due to preferential plating at the high current density corner. Figure 7.9b is a cross section for PC plating. Again, severe voiding is evident in the cross section. Figure 7.9c is a cross section for PRC plating with the macroprofile waveform. Under these conditions, the copper deposit essentially "bridges" across the opening of the feature and exhibits a worse condition than the DC plating. A PCB process engineer observed this data at a tradeshow and verified that this is what happens when their leveling additive is out of specification. Essentially, they were correcting a worse current distribution than DC with bath additives! Figure 7.9d is a cross section for PC plating at a higher frequency. The higher frequency appears to improve the plating of the feature and does employ a short cathodic on time similar to the microprofile waveform. However, the situation is still unacceptable. Figure 7.9e is a cross section for PRC plating with the microprofile waveform. The desired conformal coating is observed. Finally, continued plating with the microprofile waveform leads to filling of the feature with a level overplate as evident in Figure 7.9f. The flexibility inherent in the process enables cases where conformal coating of small features is desired as well as other cases featuring filling is desired.

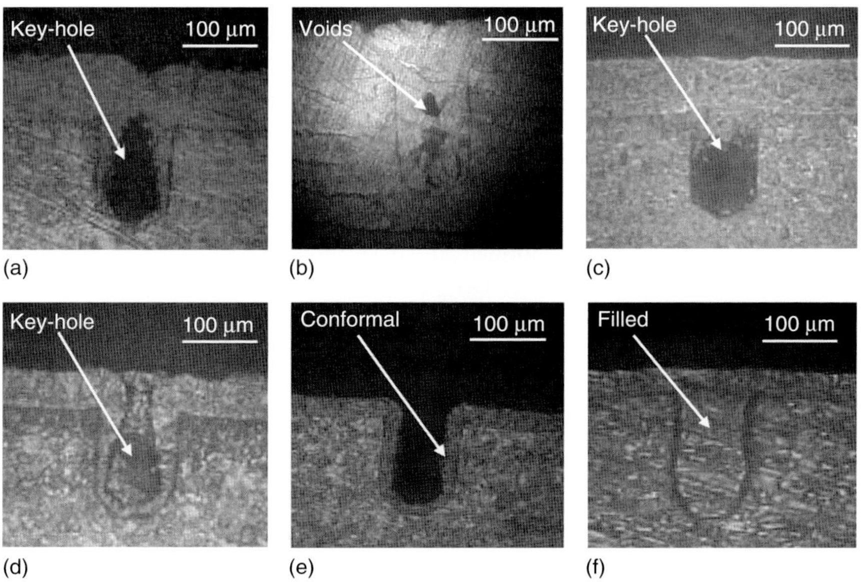

Figure 7.9 Plating of a 100 μm feature under microprofile boundary layer conditions from a copper bath without brightener (a) DC plating, (b) PC plating, (c) PRC plating with a macroprofile waveform, (d) PRC plating with the macroprofile waveform at higher frequency, (e) PRC plating with a microprofile waveform, and (f) PRC plating with the microprofile waveform for a longer period of time.

Using the macroprofile waveform approach, the ability to copper plate PTHs with high aspect ratios was demonstrated [38]. Using the microprofile waveform approach, the ability to copper fill features as small as 0.25 μm without additives was demonstrated [39]. In addition, the ability to plate multiple feature sizes representing PTHs and microvias was demonstrated, on the same board, in the same process step, using a sequence consisting of the macroprofile waveform followed by the microprofile waveform [40–43]. While these activities demonstrated the utility of PRC plating for controlling current distribution, at technical conferences, chemical formulators would still question the ability of PRC plating of copper to deliver good mechanical properties from plating bath chemistries that did not contain additives.

7.3.2 Ductility Without Brighteners Using Pulse Current Plating

Copper-plated *z*-axis interconnects for PCB applications need to exhibit sufficient ductility in order to sustain the thermal shock of soldering operations without cracking. In conventional DC plating, chemical formulators provide additives, such as "brighteners," to ensure that the plated copper exhibits sufficient ductility to avoid cracking. The ductility of the copper deposits is understood to result from small grained copper deposits. As noted above, high pulse current at short on times would be expected to favor nucleation vis-à-vis growth and consequently result in fine grained deposits.

A PCB manufacturer funded work to prepare standard tensile samples using PRC plating from a copper plating bath devoid of brighteners [44]. The mechanical properties of the PRC-plated tensile strips were compared to the mechanical properties of tensile strips plated under DC conditions in the conventional additive chemistry of the PCB manufacturer. Both tensile test strips were plated at an average current density of 25 mA cm^{-2}. The stress vs elongation data for the DC and PRC plated tensile strips are presented in Figure 7.10a,b, respectively. Both samples passed the Institute for Printed Circuits specification of an ultimate tensile strength of at least 36 000 psi and an elongation of at least 12%. Interestingly, the stress elongation curves for the PRC-plated samples exhibited a lower standard deviation. More studies would be required to establish that the enhanced reproducibility of the PRC samples is real. However, these observations demonstrated that PRC-plated copper could yield acceptable mechanical properties without the addition of brighteners to the plating bath.

In summary, the combination of cathodic and anodic pulses can be net cathodic or net anodic. The principles outlined above apply equally to both cases, in terms of the effect on the balance of primary, secondary, and tertiary current distribution. However, the effect of that current on the surface state and electrochemical reactions may range from very simple to very complex. It is perhaps serendipitous that this journey began in the copper plating system, as copper deposition is a simple, 100% current efficient, and reversible process. It was relatively easier to develop the preliminary guidelines that allowed the manipulation of the pulse reverse parameters to obtain the desired result in a system where there are little or no side reactions, and what is readily deposited may be readily removed. Other

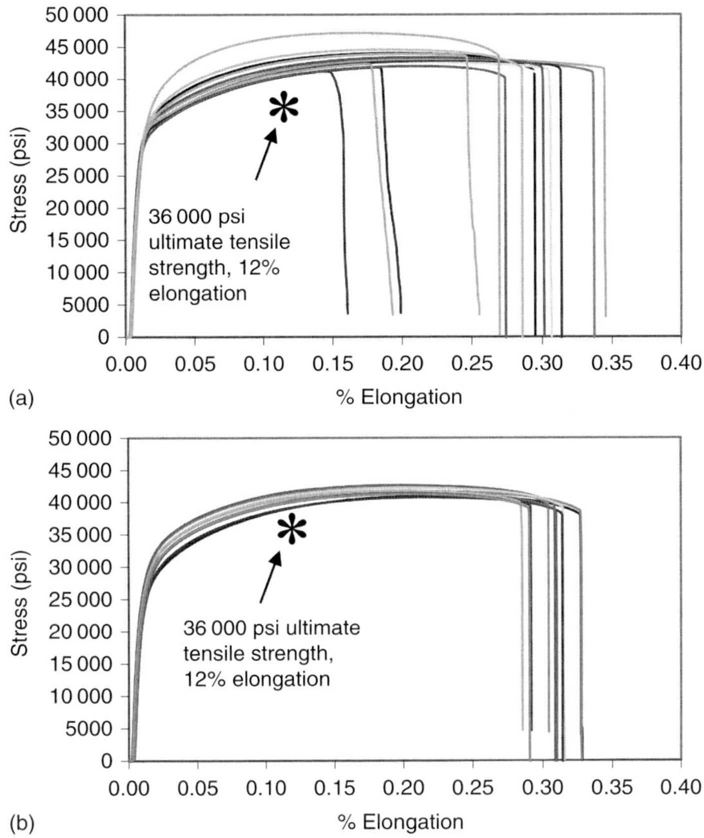

Figure 7.10 Stress–strain data for copper tensile specimens plated with (a) DC from a bath-containing brighteners and (b) PRC from a bath without brighteners.

systems, such as electropolishing of strongly passive materials, are more complex and the current response to a voltage pulse may be more difficult to interpret.

7.4 Transition of Pulse Current Plating Concepts to Surface Finishing

Within a few years of founding, Faraday was beginning to establish a reputation in the field of electrochemical research, development, and engineering. Specifically, an approach to electrochemical process engineering was being developed that challenged the conventional paradigm of DC processing with additive chemistry control. In addition to being approached by companies interested in solving plating problems, the company was also approached by companies with surface finishing challenges, such as electrochemical deburring and electrochemical polishing. Historically, like electroplating, conventional electrochemical surface finishing has also been based on the chemical control paradigm. As will be discussed below, a similar paradigm shift was observed in this technology area, by manipulation of electric field parameters to reduce or eliminate difficult-to-control and/or toxic chemistries, while maintaining or improving performance based metrics.

7.4.1 Pulse Voltage Deburring of Automotive Planetary Gears

Removing rough edges and burrs from manufactured parts is an important industrial challenge. Deburring is often accomplished with manual labor using rudimentary tools and implements, resulting in issues in terms of cost, quality, and worker repetitive motion injury. In the mid-1990s, the company was approached by a machine builder who had been contracted to install an electrochemical deburring process for planetary gears at Ford Motor Co.'s Power Train Division in Livonia, Michigan. Ford manufacturing engineers sought a reproducible cost-competitive process to replace their current manual deburring activities. The part of interest was a cast iron (SAE 1010 steel) planetary carrier with oil grooves that had rollover burrs from the milling process. Initially, Ford engineers worked on an electrochemical deburring process based on an electrolyte primarily consisting of ethylene glycol with additions of ammonium salts and a small amount of water [45]. The process had been successfully employed for a number of electrochemical surface-finishing problems, generally at limited volume batch-level processing. Ford's initial bench scale experiments indicated that the planetary gears could be deburred in ~45 seconds or less.

The electrochemical dissolution of relatively large features (arbitrarily defined as >1 μm) is termed anodic leveling or macrosmoothing while the removal of relatively small features (arbitrarily defined as <1 μm) is defined as anodic brightening or microsmoothing [46]. The deburring of Ford planetary gears falls into the category of macrosmoothing. For macrosmoothing using a DC process, the current is focused on the burr by altering the primary current distribution by the use of a highly resistive electrolyte, such as ethylene glycol. This adheres to the conventional electrochemical engineering paradigm of chemical process control under a DC field. As shown schematically in Figure 7.11, by using the highly

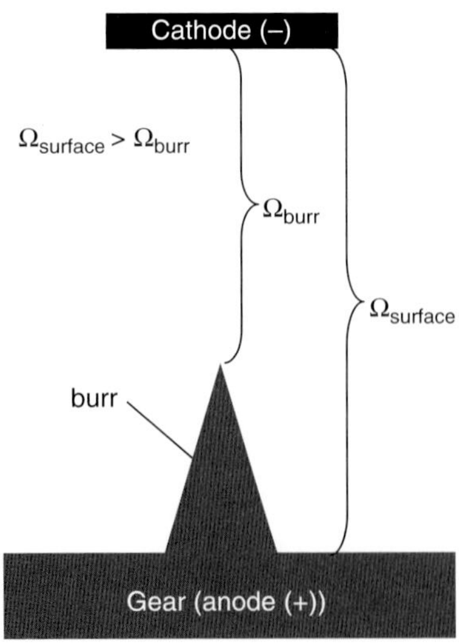

Figure 7.11 Schematic illustration of electrochemical deburring with a highly resistive electrolyte.

resistive electrolyte, the burr is preferentially dissolved relative to the surface of the gear via the generalized reaction:

$$M^0 \rightarrow M^{n+} + ne^- \tag{7.11}$$

The Ford application required a throughput of ~400 parts per hour. To accomplish this high-volume throughput, the machine tool maker constructed a 16-station automated programmable logic controller (PLC) machine. Eight stations were electrochemically deburring planetary gears while the other eight stations were unloading processed planetary gears and loading planetary gears to be processed. During initial production trials, the electrolyte heated up due to the large amount of current being passed thought the highly resistive ethylene glycol electrolyte and thermal runaway occurred within about 30 minutes. Even with active chilling of the electrolyte, several problems were noted, specifically, (i) a noticeable ammonia odor, (ii) limited tool (cathode) lifetime, and (iii) the electrolyte was difficult to maintain and expensive to replace.

Engineers from Ford and the machine tool maker funded work to investigate whether their electrochemical deburring process could be modified to meet the ~400 parts per hour designed throughput. From a current distribution perspective, electrochemical deburring is simply bump plating in reverse. Applying the principles determined in copper plating, it was speculated that by using pulse currents (voltages) instead of DC, the Ford planetary gears could be effectively deburred in a simpler non-resistive electrolyte. Note, while surface finishing is generally conducted under voltage control, the waveform is generally referred to as PC/PRC surface finishing similar to the use of DC surface finishing.

The same benchtop apparatus and cathode tooling as that used by the Ford engineers for the ethylene glycol-based process was used to study pulse current deburring. The planetary gear was fabricated from SAE 1010 steel and after some initial trials, a ~12% (w/w) NaCl aqueous solution was identified as the ideal electrolyte. Several experimental trials in the benchtop apparatus indicated that pulse current waveforms of ~25% duty cycle could effectively remove the burrs within the established part dwell time of <45 seconds [47, 48]. A picture of the planetary gear and cathode tool as well as an oil groove before and after deburring is shown in Figure 7.12. The process was transitioned to the fully automated production machine where a 5000-part production run-off with zero failures was

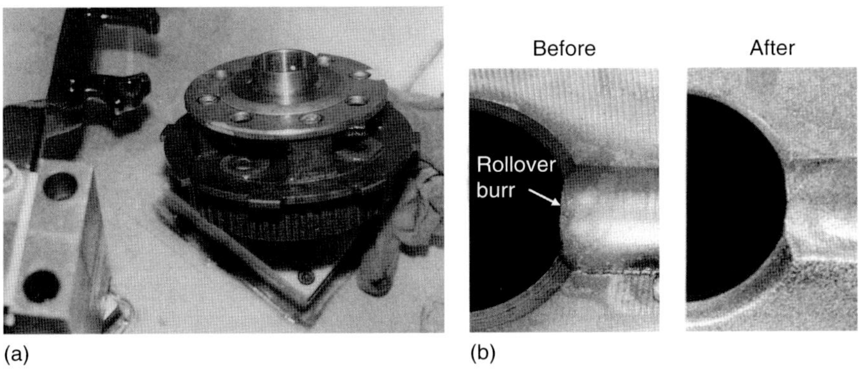

Figure 7.12 Picture of (a) planetary gear and cathode tool and (b) oil-groove before and after deburring.

completed. The process has been in full production at Ford Power Train Division since early 1996 [49]. Ford has reported that the "saltwater" electrolyte costs a fraction of the cost of the proprietary ethylene glycol electrolyte and electrolyte maintenance is simple. Specifically, water is added twice monthly to replace the water lost by evaporation and the electrolyte is replaced about every six months primarily to avoid adverse effects associated with oil drag-in. Thus, the paradigm of DC deburring in a highly resistive electrolyte was changed to PC deburring in a simple, conductive electrolyte.

7.4.2 Transition to Pulse Reverse Voltage Electropolishing of Passive Materials

The successful development and implementation of a pulse current (voltage) process for deburring of automotive planetary gears opened up another set of applications and the company vision was expanding from PC/PRC plating processes to PC/PRC surface-finishing processes, such as leveling (macrosmoothing) and brightening (microsmoothing) of metal surfaces by electrochemical anodic dissolution. In addition to deburring, electrochemical machining, electrochemical through-mask etching, and electropolishing are an industrially important surface-finishing applications [4]. The Jacquet paradigm established that electropolishing, as opposed to electrochemical etching (or pitting as originally described by Jacquet), is accomplished by polarizing the DC voltage in the mass transport region of the polarization curve illustrated in Figure 7.13. With time, a viscous salt film builds up in the valleys and the asperities are removed under mass transport control. The nature of the mass transport-controlled species is generally speculated to be either the dissolving metal ion or an "acceptor" species [50]. The mass transport-limited dissolving metal ion through a viscous salt film generally occurs in high water content electrolytes such as those used in electrochemical machining. The mass transport-limited acceptor species generally occurs in low water content electrolytes, such as viscous concentrated phosphoric and sulfuric acids used in electropolishing. The acceptor species can be water itself. A number of studies attempt to elucidate the mechanistic details of the mass transport-limited process, such as mass transport-limited metal ion diffusion through a viscous salt film [51], mass transport-limited acceptor

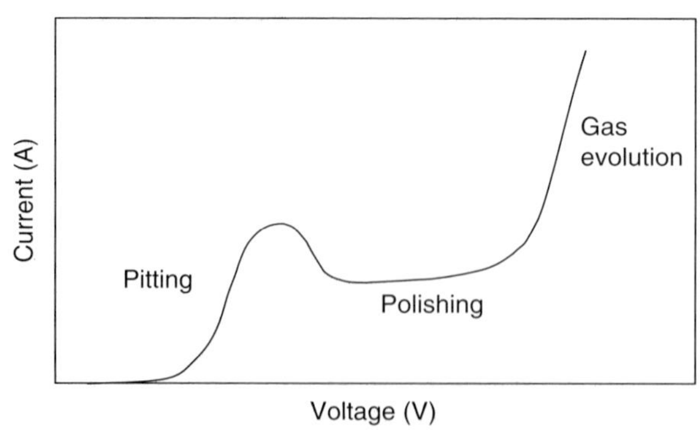

Figure 7.13 Schematic illustration of the region of the polarization curve purported by Jacquet to enable electropolishing [3].

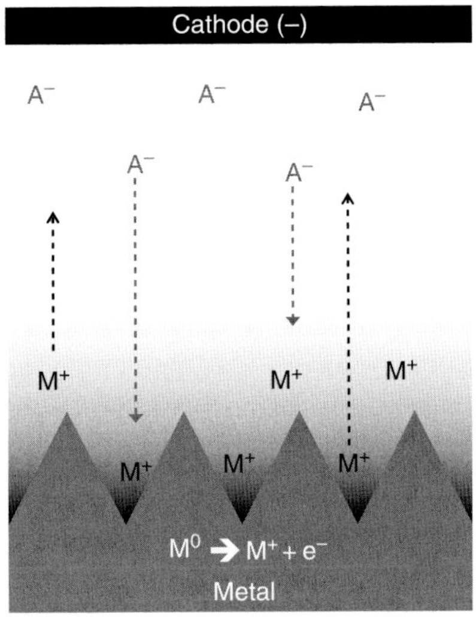

Figure 7.14 Schematic illustration of the Jacquet paradigm of mass transport.

species [52], and mass transport-limited water acceptor species [53, 54]. Regardless of the exact mechanism, it is widely accepted that mass transport plays an important role in surface finishing in general and electropolishing in particular as schematically illustrated in Figure 7.14.

With the success of PC deburring of automotive planetary gears, SBIR funding from the National Science Foundation and US Air Force SBIR programs was obtained to develop surface-finishing processes for materials of commercial and Department of Defense military interest. These materials included stainless steel, nickel alloys, and titanium alloys. During electrochemical anodic dissolution, these materials would develop a passive oxide layer via the competing anodic reaction, generally described as

$$xM^0 + yH_2O \rightarrow M_xO_y + 2yH^+ + 2ye^- \tag{7.12}$$

The conventional DC electropolishing paradigm to mitigating the adverse effect of the oxide layer formation during electropolishing involves either (i) concentrated acid solutions, such as mixtures of sulfuric and phosphoric acid, to chemically remove the oxide layer for mildly passive materials such as stainless steels [55–57], (ii) addition of fluoride salts or hydrofluoric acid to chemically remove the oxide layer for strongly passive materials such as niobium, nickel-based, and titanium-based alloys [58], or (iii) use of low water content or non-aqueous ionic liquid solutions to avoid or minimize oxide formation for strongly passive materials [59, 60].

Initial work focused on electropolishing of INCONEL®,[1] alloy 718 coupons in simple aqueous solutions, such as the salt water electrolyte developed for the automotive gear, using anodic only pulse currents. However, the use of a PC waveform resulted in a rougher surface [61], attributed to localized breakdown of the oxide film, analogous to pitting corrosion.

1 INCONEL is a registered trademark of the special metals group of companies.

Prior observations with PRC plating were applied to address the issue of oxide layer formation during anodic dissolution in aqueous electrolytes without the use of concentrated acids and/or addition of fluoride species. At about this stage of finishing process development, funding was awarded from the Environmental Protection Agency SBIR program to replace chromium plating from hexavalent chrome plating baths with an environmentally and worker friendly trivalent chrome plating bath. While decorative chromium coatings were plated from trivalent chrome plating baths, functional chromium coatings were only plated from hexavalent plating baths. As reported in the literature, chromium coatings from a trivalent chrome plating bath were limited to thicknesses that were not sufficient for functional applications requiring properties such as high hardness and wear resistance. The faradaic efficiency for chromium plating from trivalent chrome plating baths was quite low, with ~15% of the current attributed to the desired chromium plating Reaction (7.13) and ~85% of the current attributed to the hydrogen evolution Reaction (7.14):

$$Cr^{3+} + 3e^- \rightarrow Cr^0 \tag{7.13}$$

$$2H^+ + 2e^- \rightarrow H_2 \uparrow \tag{7.14}$$

Generally, after plating of less than ~2.5 μm, the chromium plating ceased. The generation of hydrogen due to the low faradaic efficiency could have resulted in an alkaline pH at the interface leading to a species that adversely affected the continued plating of the chromium [62]. Consequently, there was speculation that insertion of an anodic reverse current pulse during pulse current plating could "re-acidify" the interface and allow plating to continue. Using PRC plating, the ability to plate thick chromium coatings from a trivalent chrome plating bath was successfully demonstrated [63].

In the case of surface finishing of passive or oxide forming materials, there is an analogous adverse Reaction (7.12) of oxide formation. Therefore, it was speculated that the adverse effects of the oxide formation reaction during surface finishing could similarly be eliminated by inserting a cathodic reverse current pulse during pulse current surface finishing to remove the oxide layer formed during the anodic forward pulse. The parallels between eliminating the adverse effects of hydrogen evolution during plating and eliminating the adverse effects of oxide layer formation during surface finishing are illustrated in Figure 7.15a,b, respectively. While the exact mechanisms of these two approaches have not yet be fully elucidated, this approach was used by analogy to successfully demonstrate PRC electropolishing of stainless steels, nickel alloys, and titanium alloys in aqueous sodium chloride/sodium nitrate salt electrolyte [64, 65].

7.4.3 Sequenced Pulse Reverse Voltage Electropolishing of Semiconductor Valves

Soon after presenting and publishing the pulse reverse voltage electropolishing results for passive materials, the company was approached by a manufacturing engineer from Swagelok Corporation. Swagelok was using conventional direct voltage electropolishing as the final surface-finishing step for their stainless

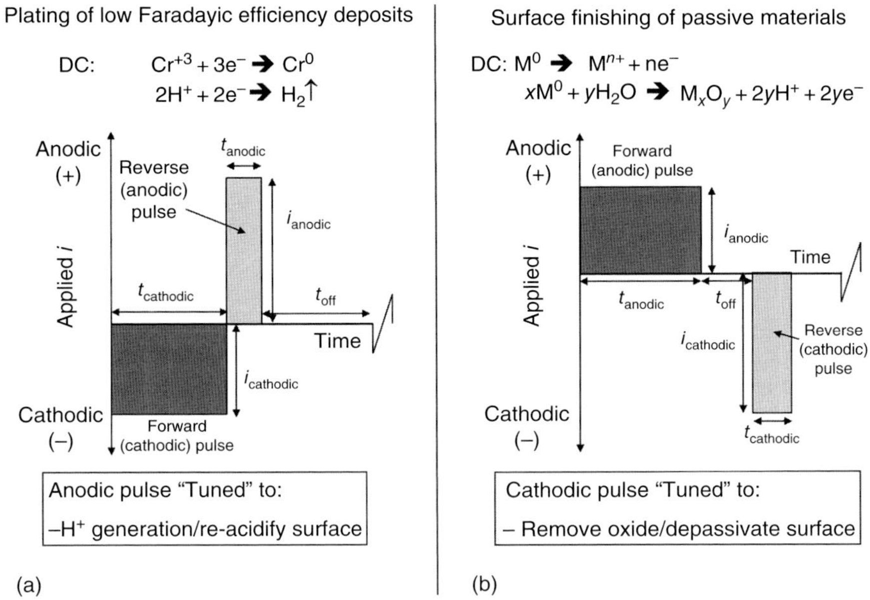

Figure 7.15 Schematic illustration of PRC approaches to mitigate (a) adverse effect of hydrogen evolution during plating and (b) adverse effect of oxide layer formation during surface finishing.

(300 series) steel valves, fittings, and tubular products. These were specialty products for the semiconductor and biomedical industries, and the internal surfaces of these valves required a mirror-like finish. Due to the slow material removal rate of conventional electropolishing (viscous electrolytes operating under mass transport control), Swagelok used a two-step surface finishing process: (i) abrasive flow machining (AFM) for deburring and bulk material removal of tool lines, followed by (ii) conventional electropolishing to achieve the final mirror-like surface finish. The electropolishing process used a chilled electrolyte solution consisting of concentrated sulfuric/phosphoric acid as well as proprietary additives that may have included fluoride species. The electropolishing process required ∼160 seconds to achieve the final mirror-like finish. The AFM media was expensive and the combined AFM/electropolishing process was difficult to control.

Under a stage-gated research-for-hire activity funded by Swagelok, electropolishing of stainless steel coupons of their specific 300 series valves in an aqueous solution of ∼12% $NaCl/NaNO_3$ salt electrolyte without addition of fluorides was demonstrated. The next stage-gate involved transitioning the process to actual production hardware and actual products. The focus was to replace both the AFM and electropolishing process steps with a single surface-finishing process based on pulse reverse voltage in a simpler electrolyte. Swagelok moved one of their electropolishing manufacturing cells with a representative valve to the company's α-scale prototyping area. Swagelok's concentrated acid electropolishing electrolyte was switched out with the simpler aqueous salt solution and replaced their DC rectifier with a PRC rectifier. Initial experimental electropolishing trials indicated that the pulse reverse voltage waveforms could remove the burrs and reduce the *Ra* from ~1 to ~0.2 μm within about 30 seconds. Continued electropolishing up to about 2.5 minutes resulted in no improvement in the *Ra* and the surface exhibited an "orange-peel" structure indicative of electrochemical etching in lieu of electropolishing.

To better understand this transition from surface smoothing to etching, previous theoretical and experimental studies by Landolt and collaborators were considered that examined anodic dissolution of well-defined triangular surface profiles under DC conditions [66, 67]. The geometrical setup considered both experimentally and theoretically is depicted in Figure 7.16 where "L" is the gap between the surface and the electrode, "β" is the angle of the model surface asperity, and "l" is the amplitude of the surface asperity.

The experimental and modeling studies maintained $L \gg l$, and l was considered the characteristic length in the Wagner number (see Eq. (7.9)). The parameters "β" and "l" represent the surface roughness and as electropolishing progresses "β" to increase and "l" were expected to decrease. As "l" decreases the Wagner number increases, the current distribution becomes more uniform. Landolt and collaborators experimentally verified that as "l" decreased and "β" increased, the effectiveness of the electropolishing process decreased.

As noted above, the same behavior was observed when electropolishing the Swagelok valves. Initially, the surface roughness was reduced, i.e. "l" decreased, but after 30 seconds of processing, the surface roughness stopped improving. As "l" decreased, the Wagner number increased, resulting in a shift toward more secondary current distribution, and therefore a more uniform material removal rate at the peaks and valleys of the asperities. To continue to improve the surface finish, the balance of primary, secondary, and tertiary current distribution needed

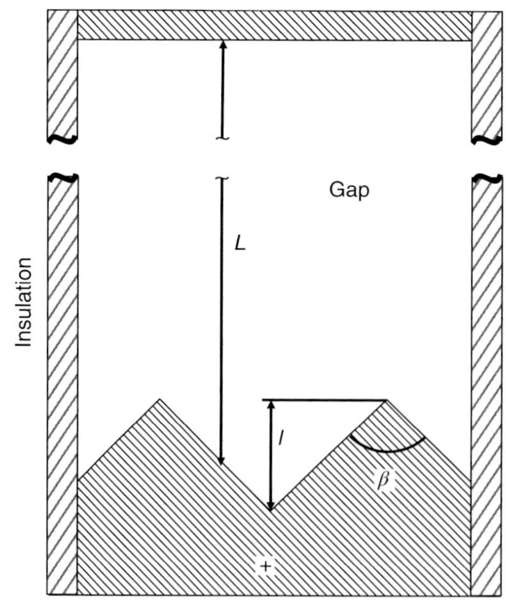

Figure 7.16 Model geometry used for theoretical and experimental studies of DC electropolishing.

adjustment back toward primary control. Using DC processing, this would only have been possible by changing the tool geometry and/or solution conductivity. However, in the spot plating of gold example described earlier, Chin et al. [24] showed that pulsed processing could decrease the Tafel slope and therefore the Wagner number toward zero, making it less uniform. In other words, pulsed processing gives us another tool to manipulate the current distribution to achieve more or less uniformity as required, with consideration of the impact of macroprofile or microprofile.

Although the prior work was under conventional DC "leveling" conditions, it was assumed that similar phenomena were impacting the ability to electropolish the Swagelok valves. It was noted that the manufacturing apparatus did not submerge the complete valve into the electrolyte but rather "jetted" the polishing solution into the internal diameter of the valve body. Initially, the boundary layer resulting from the high flow velocity represented a macroprofile boundary layer during the initial stages of electropolishing. After ~30 seconds of electropolishing, the speculation was that the surface roughness was sufficiently smoothed, whereby the boundary layer became a microprofile. At this point, the "macroprofile waveform" induced uniform electropolishing of the peaks and valleys of the asperities, rather than peaks only.

The waveform was changed per the design guidelines described in Figure 7.6 so as to induce significantly more non-uniform polishing under a microprofile condition to continue the reduction in surface roughness. Consequently, after 30 seconds of electropolishing, the pulse reverse voltage waveform was sequenced from short-anodic/long-cathodic pulses to long-anodic/short-cathodic pulses for an additional 15 seconds. As illustrated in Figure 7.17 [68], using the sequenced waveform parameters, the surface *Ra* was further reduced from 0.2 to 0.026 μm with a mirror-like finish and complete absence of the "orange-peel" structure (Figure 7.18). Consequently, the pulse

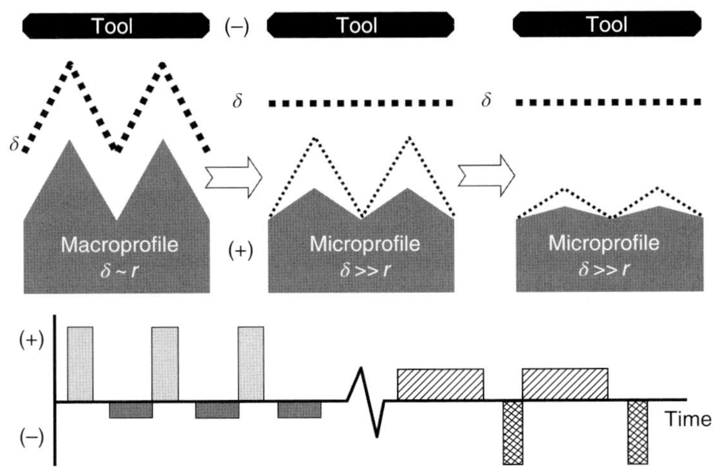

Figure 7.17 Schematic representation of the sequenced waveform used to electropolish the stainless valves.

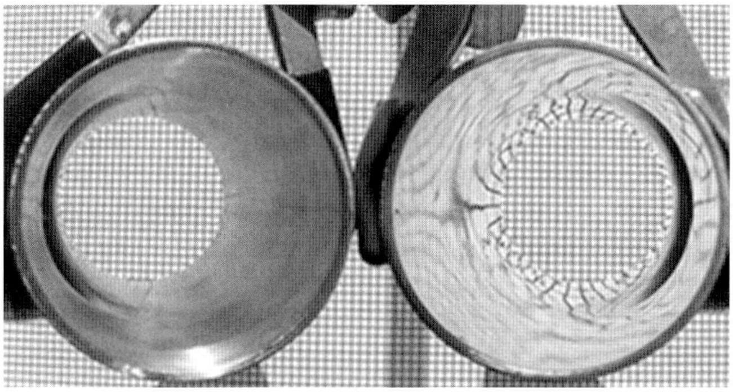

Figure 7.18 Sample part before (left) and after (right) electropolishing in an aqueous salt solution.

reverse voltage approach replaced both the AFM and the conventional DC electropolishing steps and

(1) enabled a simpler aqueous electrolyte without the need for concentrated, viscous, or chilled acids with the addition of fluorides,
(2) required ~45 seconds compared to the ~160 seconds required for conventional electropolishing step alone,
(3) lead to a more robust process, and
(4) replaced an expensive consumable.

In summary, the similar reasoning that lead to the macroprofile and microprofile waveforms for plating various size electronic z-axis interconnects lead to an analogous approach to surface finishing. The pulse reverse voltage electropolishing process using the low viscosity aqueous salt electrolyte has been in production by Swagelok Corporation since 2000.

7.4.4 Pulse Reverse Voltage Electropolishing of Strongly Passive Materials

The successful demonstration of pulse reverse voltage surface finishing of stainless steel-, nickel-, and titanium-based alloys as well as the implementation of PRC surface finishing of Swagelok valves opened the door for numerous applications for surface finishing of these and other materials. In the mid-2000s, an SBIR project from the Department of Energy (DOE) was awarded to apply the PRC approach to electropolishing of niobium. Niobium is a superconducting metal and is used to fabricate SRF cavities for particle accelerators, such as the International Linear Collider, as well as other high energy physics projects. Maximizing the particle accelerating electric fields in these SRF cavities requires eliminating or minimizing surface defects on the interior surface of the cavities.

For more than 30 years, electrochemical surface finishing or electropolishing has been the fundamental surface preparation process to achieve high gradients and quality factors for current and proposed SRF-based accelerators using elliptical cavities. The benefits of electropolishing niobium cavities are well known to the SRF community [69, 70]. Electropolishing of SRF cavities is based on the well-established "viscous salt film" model by Jacquet [4] and utilizes a viscous electrolyte consisting of a mixture of sulfuric acid (95–98%) and hydrofluoric acid (49%) in an approximately 9 : 1 volume ratio [71].

Jacquet's initial research was directed toward the electropolishing of copper and his "viscous salt film model" fits the definition of a paradigm. The Jacquet paradigm was open-ended and provided a framework for subsequent researchers to develop electropolishing processes for other metals and alloys. These subsequent studies enhanced the Jacquet paradigm by clarifying the nature of the mass transport limited species for specific material and electrolyte systems. Consistent with the Jacquet paradigm, the concentrated sulfuric acid establishes a thick boundary layer that leads to surface smoothing when a direct anodic current is applied such that the niobium dissolution operates under mass transport control. The concentrated hydrofluoric acid function is to form a soluble species with the dissolved niobium and to remove the oxide film formed during niobium dissolution [58, 72]. A mechanistic study reported by a research team including the DOE Thomas Jefferson National Accelerator Facility definitively demonstrated that state-of-the-art electropolishing of niobium occurs by diffusion-limited transport of fluoride ions to a compact salt film [71]. This finding is consistent with other studies generally indicating that an acceptor ion is the mass transport-limited species for low water content electrolytes. Due to the slow material removal rate of electropolishing, SRF cavity surface finishing utilizes a two-step surface-finishing process similar to Swagelok. The first step is "buffered chemical polishing" (1 : 1 : 2 HNO_3(69%):HF(49%):H_3PO_4(85%)) to remove ~100 μm of material to a *Ra* of ~1 μm. This is followed by an electropolishing step that removes ~30 μm of material to a *Ra* of <0.2 μm, although the final surface finish is typically mirror-like.

Significant negative aspects are associated with niobium SRF cavity electropolishing as currently practiced [73]. Specifically, electropolishing with a mixture of concentrated sulfuric and hydrofluoric acids carries significant personnel safety risk, as well as negative environmental impact. Consequently, SRF cavity processors must abide by stringent regulations from both the Occupational Safety and Health Administration (OSHA) and the Environmental Protection Agency (EPA). In addition, the use of the concentrated hydrofluoric and sulfuric acid mixture significantly impacts both the capital and operating costs of cavity processing and is a barrier to establishing an industrially compatible cavity surface finishing process. Consequently, there has been a long-felt need for an environmentally friendly electropolishing process for niobium SRF cavities that is also compatible with industrial practice. The DOE SBIR topic call was directed toward an "eco-friendly" niobium SRF cavity electropolishing process.

While the PRC approach had previously been used to replace concentrated acid electropolishing electrolytes, including those with fluoride salt additions, this approach had not been used to replace electropolishing processes where concentrated hydrofluoric acid was directly added as a significant component of the electropolishing electrolyte. These materials generally exhibit <−600 kJ mol^{-1} free energy of oxide formation on the Ellingham diagram and in addition to niobium, include titanium, tantalum, tungsten, and molybdenum. Within the DOE SBIR Phase I feasibility activity, the ability to electropolish niobium coupons to a

Figure 7.19 Image of niobium coupon electropolished in 5% sulfuric acid by weight using pulse reverse voltage.

mirror finish (Figure 7.19) had been successfully demonstrated in aqueous solutions of sulfuric acid of 5–30% (w/w), without hydrofluoric acid! Atomic force microscope scans by Dr. Charles Reece of DOE Thomas Jefferson National Accelerator Facility of 2500, 100 and 4 μm^2 areas indicated *Ra* values ranging from 3.5 to 0.36 nm [74]. The surfaces were comparable to those obtained using the conventional process based on hydrofluoric acid.

While the earlier work was highly successful, it was limited to flat niobium coupon surfaces of no greater than $1'' \times 1''$ or an area of $<6.25\ cm^2$. In contrast, a single-cell SRF cavity has an internal surface area of $\sim 1230\ cm^2$ consisting of tubular and elliptical regions. Therefore, the logical next step was to submit a proposal directed toward gaining an understanding of how the pulse reverse voltage niobium electropolishing would scale to the larger surface areas and more complicated geometries of single-cell cavities. A DOE Phase II proposal was submitted that would transition the coupon pulse reverse voltage parameters to internal tubes and curved half-cell hemispherical surfaces representative of the SRF single-cell cavities.

About the same time, a request for proposal under the America Reinvestment in Research Act (ARRA) directed towards demonstration of an "eco-friendly" process for surface finishing of niobium SRF cavities was released from DOE Fermi National Accelerator Laboratory (FNAL). Discussions were held with the FNAL technical point of contact, Mr. Allan Rowe, who was very interested in the approach and encouraged submission of a proposal. The parameters of the ARRA solicitation did not permit extensive research and development activities, instead being directed toward validation of an eco-friendly process on single-cell cavities using the existing electropolishing fixturing and apparatus. Ideally, the timing of the DOE Phase II SBIR and the ARRA programs would coincide and the SBIR program results would synergistically permit the research and development activities that could be applied to the ARRA single-cell SRF cavity validation.

The nature of the current review system is such that it must use experts in the scientific community who are familiar with the relevant paradigms in order to ensure scientific quality. When these experts are faced with data or observations that do not agree with the current paradigms, they may receive the new data with some skepticism. This seemed to be the case with the Phase II SBIR proposal, which was declined, based on the assumption that the demonstration of electropolishing of niobium coupons did not imply future success with full-size cavities. As conventional electropolishing had originally been demonstrated on coupons and transitioned to SRF cavities and the new process was still, in fact, employing electropolishing, this seemed surprising. However, this skepticism was not limited to the review panels. A niobium coupon electropolished in aqueous sulfuric acid using pulse reverse voltage was shown to a well-respected electrochemical colleague who proclaimed.

...you can't do that...

While initially perplexing, upon reflection, it seemed that, in effect, this colleague could not "observe" the niobium electropolishing result because it did not fit the well-established Jacquet electropolishing paradigm.

7.4.5 Pulse Reverse Voltage Electropolishing of Niobium Superconducting Radio Frequency Cavities

While the DOE Phase II SBIR was not funded, the ARRA proposal did receive an award, and the effort was able to continue. As noted above, the ARRA program did not permit "extensive" research and development activities and required that an electropolishing apparatus be constructed similar to the one used in current SRF cavity electropolishing and placed in a cleanroom environment [75]. The idea was that their hazardous electrolyte and DC power supply would simply be replaced with the eco-friendly electrolyte and pulse reverse rectifier, analogous to the Swagelok activity. During the initial part of the ARRA program, work continued on pulse reverse voltage waveform development on flat niobium coupons up to 3″ × 3″ (~60 cm^2) in scale. While this was a long way from the internal surface area of a single-cell SRF cavity (1800 cm^2) and even further removed from the internal surface area of the nine-cell SRF cavities used in actual particle accelerators, these experiments were used to transition the pulse reverse voltage electropolishing process from "coupons to cavities."

The DOE Phase I SBIR approach used high-voltage and high-frequency waveform parameters. However, due to the high surface areas inherent in the single-cell and nine-cell SRF cavities, pulse on times and peak voltages were limited at the scale for processing single-cell and nine-cell SRF cavities by the technical constraints of commercially available pulse reverse rectifiers.

During the investigation of pulse reverse waveforms for electropolishing of 1″ × 1″ niobium coupons, effective polishing occurred when a current transition was observed in the anodic current response as shown in Figure 7.20 [76]. The speculation was that during the first part of the anodic voltage pulse, a niobium oxide film is formed on the niobium surface. After the "complete" formation of the niobium oxide film, oxygen evolution occurs during the remainder of the anodic voltage pulse. The shift from oxide formation to oxygen evolution could account for the observed anodic current transition. Further speculation suggested that the niobium oxide film is subsequently removed during the

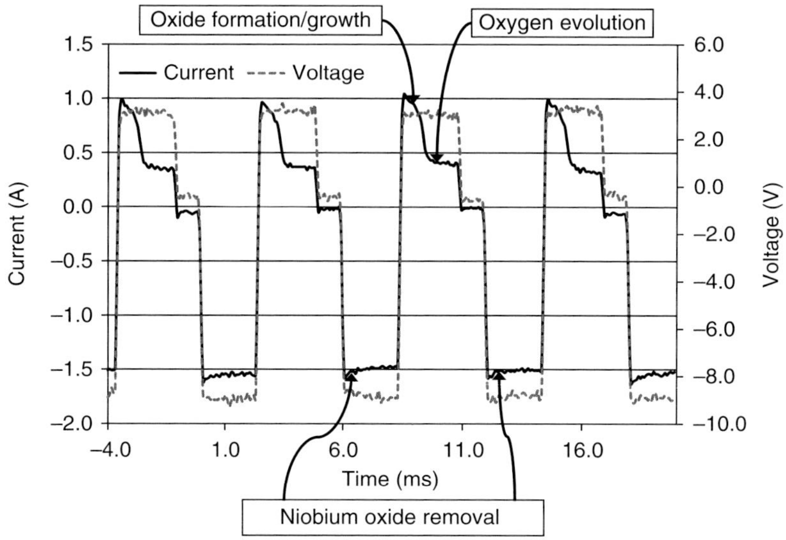

Figure 7.20 Anodic current transition observed for effective electropolishing of niobium. Applied waveform: anodic pulse of 3 V for 2.5 ms, off time of 1 ms, cathodic pulse of 9 V for 2.5 ms.

cathodic voltage pulse. This mechanism was termed "cathodic electropolishing." A similar anodic current transition has been observed for other strongly passive materials such as tantalum and nickel–titanium shape memory alloys. Synergistically, in parallel to the niobium electropolishing activities, funding was awarded from the National Institutes of Health (NIH) SBIR program directed toward electropolishing of nickel–titanium materials for medical shape set devices, such as stents. This work has led to successful development and licensing of a PRC process for electropolishing of nickel–titanium wire.

Note, the applied waveform in Figure 7.20 appears to be net cathodic. However, it can be assumed that both the anodic and cathodic pulses are not 100% current efficient and water electrolysis reactions are taking place. While additional studies and optimization are required to elucidate the contributions of the water electrolysis during the anodic and cathodic pulses, material is removed from the surface, even though the charge associated with the anodic and cathodic portions of the waveform may suggest otherwise. Finally, consistent with niobium oxide removal during the cathodic pulse, an accumulation of niobium oxide powder on the cathode was observed, presumably deposited electrophoretically from the niobium oxide removed from the niobium coupon during the cathodic pulse. While much more work needs to be conducted to elucidate the exact nature of the anodic current transition, the anodic current transition appears to correlate strongly with effective electropolishing.

During the establishment of clean room and cavity-processing apparatus, as well as the extensive pulse reverse voltage waveform study, word was received from Mr. Rowe (ARRA TPOC) that there was concern that the proposed approach would not lead to a practical solution due to the heat buildup during processing. This seemed unlikely as the anodic voltage pulses were typically 2–4 V and the cathodic voltage pulses were ~10 V in a relatively conductive electrolyte. This is in contrast to the DC process using ~18 V in a relatively

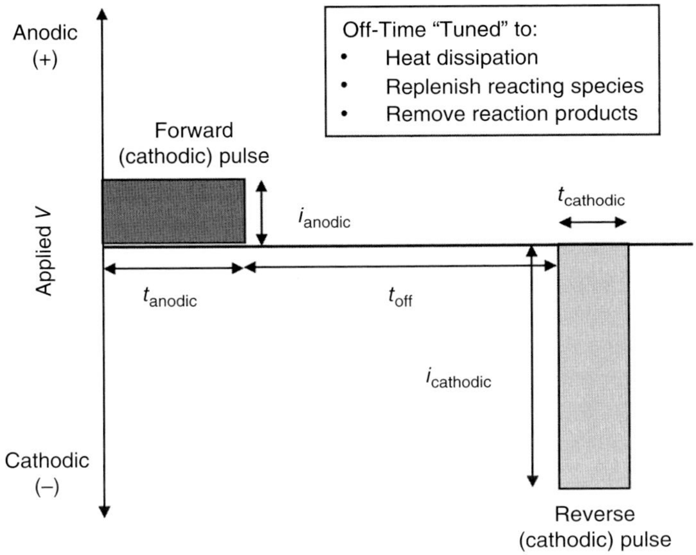

t_{off} (ms)	Total time (min)	T_i (°C)	T_f (°C)
0.3	40	15	33
0.6	120	14	35
1.0	120	14	28
5.0	560	16	19

Figure 7.21 Elimination/minimization of heat buildup during pulse reverse voltage electropolishing by inserting longer pulse off times.

resistive electrolyte. However, in order to demonstrate that PRC waveforms could be designed to minimize heat buildup during processing, experiments were conducted using a high power waveform (30 V anodic and 35 V cathodic) in the bench-scale coupon cell where significant heating was observed during a short processing time. As shown in Figure 7.21, increasing the off time between pulses from 0.3 to 5 ms demonstrated that the heat could be dissipated, even after 560 minutes of operation. For the lower power waveforms that are currently being used, heat buildup can be effectively controlled. This demonstrates another strong advantage of the pulse/pulse reverse voltage approach to electropolishing.

With the development of pulse reverse voltage waveform parameters for electropolishing niobium coupons and within the constraints of commercially available rectifiers, attention was focused on transitioning the electropolishing process to single-cell SRF cavities. In addition to the difference in surface area between the 1″ × 1″ and 3″ × 3″ coupons and single-cell cavities, the geometry of the cavity was different. The geometric relationship between the cathode tool and the cavity was characterized by two regions: two cylindrical "beam tube" regions and one "equator" region, as shown in Figure 7.22.

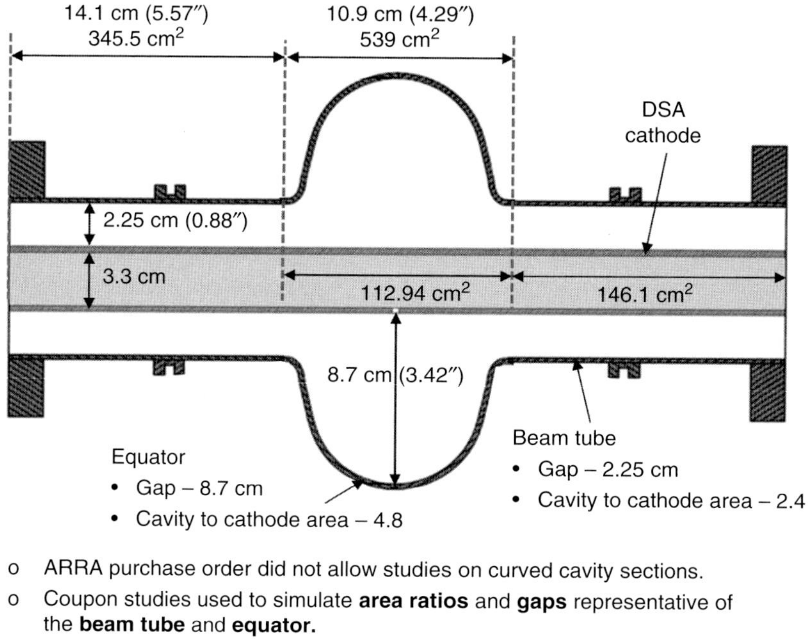

Figure 7.22 Niobium SRF cavity geometry gaps and area ratios for the beam tube and equator regions.

In the beam tube regions, the cathode to cavity gap was 2.25 cm and the ratio of cavity area to cathode area was 2.4. In the equator region, the cathode to cavity gap was 8.7 cm and the ratio of cavity area to cathode area was 4.8. To "simulate" some of these geometrical aspects of the SRF cavity beam tube and equator regions, a limited number of experiments were conducted with flat coupons at gaps and anode to cathode area ratios representative of these regions. The resulting data suggested that in order to achieve effective polishing:

(1) As the anode–cathode gap increases, the anodic transition time increases, and
(2) As the anode–cathode surface area ratio increases, the anodic transition time increases.

With these observations, future efforts were focused on electropolishing of single-cell niobium SRF cavities. Subsequent review of this and additional data and new observations have suggested refinements to this model. However, at this early stage of development, these observations provided the basis for moving forward.

7.4.6 Transition Pulse Reverse Voltage Electropolishing to Niobium Superconducting Radio Frequency Cavities

Due to the expense, only four single-cell niobium SRF cavities could be provided by FNAL under the ARRA program. Specifically, a "developmental" single-cell cavity would be used to transition the pulse reverse voltage process to the niobium cavity by appropriate adjustment of the waveform parameters. The remaining three "performance" single-cell cavities were used to validate equivalent performance to single-cell cavities processed using the standard sulfuric/hydrofluoric acid process [77]. In addition, the ARRA purchase order stipulated that the single-cell cavities would be processed in the "standard" cavity processing apparatus based on the design specifications of Argonne National Laboratory/FNAL [74]. Therefore, the developmental cavity was processed in a horizontal orientation (Figure 7.23) with ~60% of the volume filled with 5–10% (w/w) aqueous sulfuric acid and rotated at 1 revolution per minute per standard practice.

After approximately seven unsuccessful trials in the developmental cavity, based on visual observations, the original Siemens literature upon which current cavity processing is derived was revisited [78]. The current cavity electropolishing protocols (horizontal orientation, partial electrolyte fill, and rotation) were developed for the highly viscous sulfuric/hydrofluoric acid solution where according to the original Siemens patent,

the problem of

> ...electrolytic polishing hollow niobium bodies of a complicated geometrical structure...where development of gases...rise from the cathode...forming gas pockets...resulting in portions of the inside surface not polished...

is solved by

> ...horizontally orienting the hollow niobium body...partially filling said hollow body with polishing solution and slowly rotating said hollow body...

As the viscosity of the low concentration sulfuric acid solution (~1 cp) is much less than that of the conventional sulfuric/hydrofluoric acid solution (~25 cp),

Figure 7.23 SRF cavity processing apparatus with rotating horizontal cavity orientation and partially filled with electrolyte.

it was speculated that the rotation rate used for the high viscosity standard solution may not be optimal for the low viscosity solution. Furthermore, horizontal operation with partial cavity fill and rotation may not be required for the pulse reverse voltage niobium SRF cavity electropolishing process. Additionally, vertical cavity processing is simpler and due to capital cost considerations would be more desirable from an industrial perspective. Consequently, the cavity-processing apparatus was positioned in the electrolyte "dump" mode and additional electropolishing experiments were conducted with the cavity in a vertical orientation completely filled with electrolyte and without rotation (Figure 7.24).

Additional pulse reverse voltage electropolishing trials were conducted on the developmental cavity in the vertically oriented dump mode without rotation and completely filled with electrolyte. In these additional developmental cavity trials, the waveform timing was adjusted to account for the increased anode to cathode gap and area ratios until the anodic current transition was observed as an indication of electropolishing. In addition, as shown in Figure 7.25, regions of the cavity were progressively isolated to expose

(1) one beam tube region,
(2) the equator region,
(3) one beam tube and the equator regions,
(4) the complete cavity, i.e. both beam tube regions and the equator

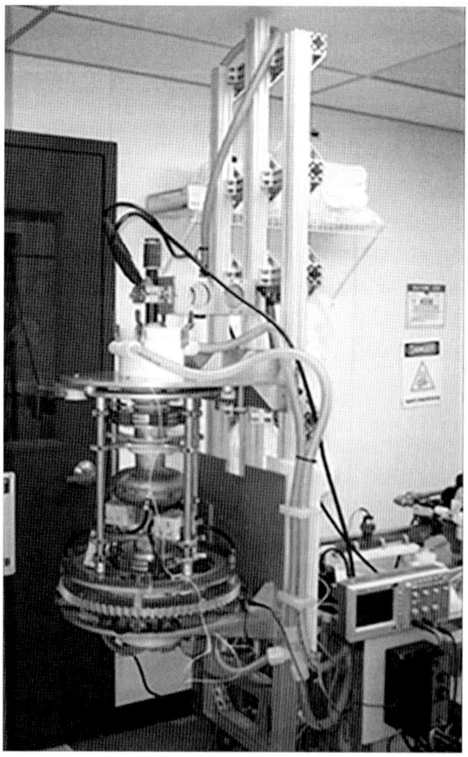

Figure 7.24 SRF cavity processing apparatus operation in the "dump" mode oriented vertically without rotation and completely filled with electrolyte.

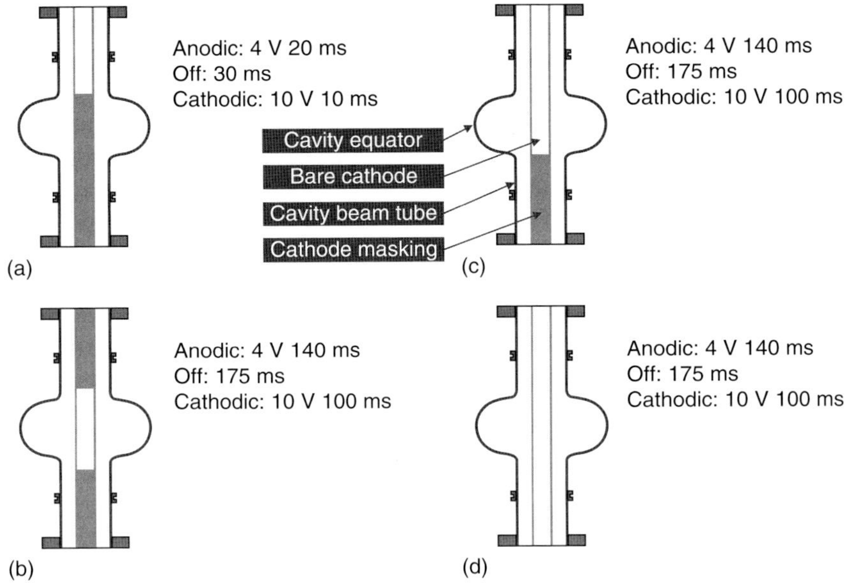

Figure 7.25 SRF developmental cavity processing sequence. Shaded areas on the cathode indicate masking. (a) 1 beam tube, (b) equator, (c) 1 beam tube + equator, and (d) 2 beam tubes + equator (unmasked).

Consistent with the experimental observations of the effect of the electrode gaps and area ratios in the coupon studies, the anodic pulse period was extended in order to observe the anodic current transition in both the beam tube and equator regions. As shown in Figure 7.26, in the single beam tube test (no. 1), a single transition at a relatively short transition time was observed, from oxide formation in the beam tube to water electrolysis. In the equator-only test (no. 2), a single transition at a relatively long transition time was observed, from oxide formation in

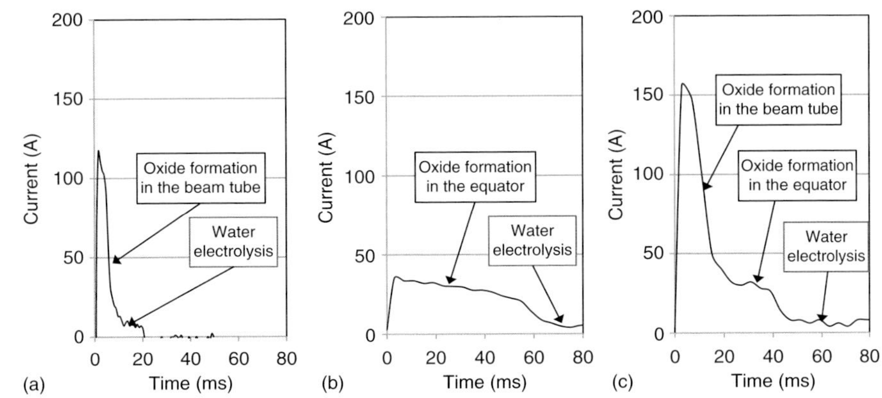

Figure 7.26 Anodic current transitions for the SRF developmental cavity per the experiments shown in Figure 7.25: (a) beam tube only, (b) equator only, and (c) 2 beam tubes + equator. Note, for clarity, the anodic and cathodic voltage pulses and cathodic current responses are not shown.

the equator to water electrolysis. In the full cavity test (no. 4), two anodic transitions were observed: the first at a relatively short transition time from oxide formation in the beam tube to oxide formation in the equator, followed by a second transition from oxide formation in the equator to water electrolysis. These were termed the Beam Tube Transition and the Equator Transition.

When the anodic current responses were overlaid (Figure 7.27) for each of the four trials in the developmental cavity (Figure 7.25), the observation was made that as the total area exposed increased, the Beam Tube Transition time increased, while the Equator Transition time decreased. Based on the difference in anode to cathode gap and area ratio at the beam tube vs the equator, this suggests that the beam tube dominates the geometric current distribution over the equator. This would result in some non-uniformity in removal from the beam tube to the equator, and in fact that has been observed in the pulse reverse cavity electropolishing experiments as well as conventional practice [79].

After electropolishing the complete single-cell cavity, visual inspection indicated acceptable electropolishing and this was confirmed at FNAL [80]. At least from a visual inspection perspective, the coupon study observations proved critical in developing the appropriate pulse period. Subsequently, additional analysis of the coupon experimental data would provide further observations of the

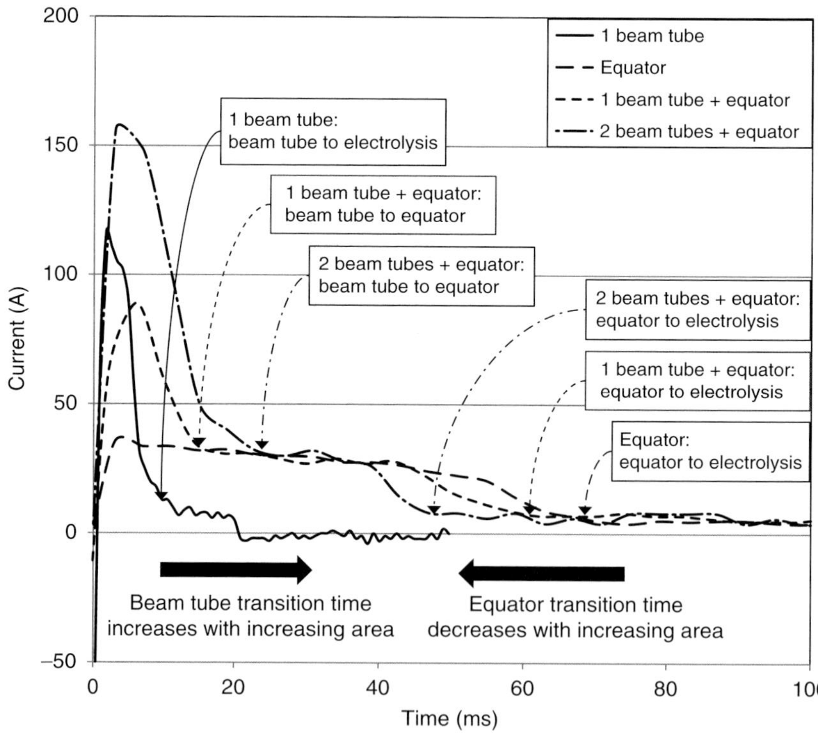

Figure 7.27 Anodic current transition times for SRF developmental cavity processing sequence described in Figure 7.25. Note, for clarity, the anodic and cathodic voltage pulses and cathodic current responses are not shown.

impact of anodic and cathodic pulse charge on niobium electropolishing. With the acceptable visible electropolishing confirmed, "performance" cavities were sent by FNAL for electropolishing. All subsequent performance cavity processing trials were conducted with the cavity in a vertical orientation, filled with electrolyte and without rotation.

The first performance cavity had previously been processed using the standard two-step surface-finishing protocol of buffered chemical polishing followed by electropolishing. The cavity superconducting and accelerating gradient characteristics were validated at FNAL before sending the cavity to Faraday, where "light" electropolishing was conducted to remove ~30 μm of material. The cavity was sent back to FNAL and the cavity superconducting and accelerating gradient characteristics were verified to be within the acceptable range of cavities subjected to the standard surface-finishing protocols [80]. This experiment demonstrated that the pulse reverse voltage electropolishing process did not adversely impact niobium SRF cavity performance.

The second performance cavity had not been processed using the standard surface finishing protocol and did not exhibit acceptable superconducting and accelerating gradient characteristics. "Heavy" electropolishing was conducted whereby ~100 μm of material was removed. The superconducting and accelerating gradient characteristics were evaluated at FNAL and the cavity superconducting and accelerating gradient characteristics were verified to be within the acceptable range of cavities subjected to the standard surface finishing protocols [80]. In addition, the cavity exhibited "Q-disease," which is indicative of hydrogen uptake and generally occurs during bulk (>~50 μm) material removal. In conventional cavity processing, Q-disease is eliminated with a high temperature (~800 °C) bake after bulk material removal [80]. "Light" material removal with the conventional electropolishing process does not result in Q-disease. While at FNAL, this cavity was subjected to the standard high temperature bake and the Q-disease was eliminated. This cavity was returned for "light" electropolishing with ~20 μm material removed. Subsequent testing at FNAL demonstrated that "light" material removal with the pulse reverse voltage electropolishing process similarly does not result in Q-disease [80].

The third and last performance cavity had previously been processed using the standard two-step surface-finishing protocol and was identified as a "high-performing" cavity by FNAL. "Light" electropolishing was conducted whereby ~25 μm of material was removed. After evaluation of the superconducting and accelerating gradient characteristics, FNAL reported that this cavity exhibited the highest performance of any cavity evaluated at FNAL [80]. While the source of this exceptional performance is not clear, this observation along with the observations on the previous performance cavities validate the applicability of pulse reverse voltage surface finishing for electropolishing niobium SRF cavities in a simple eco-friendly aqueous electrolyte [81]. Due to the lower capital and operating costs, the pulse reverse voltage electropolishing process is estimated to cost approximately 20% of the cost of the conventional process [82].

In summary, an electrochemical polishing process for niobium SRF single-cell cavities based on PRC principles has been demonstrated. The use of PRC enables a low-viscosity/low acid concentration aqueous electrolyte, which does not require hydrofluoric acid. This development activity occurred over a period of approximately five years, and the PRC-based process has been demonstrated to yield at least equivalent SRF cavity performance to the conventional process based on DC electrochemical polishing using high-viscosity/high acid concentration 9 : 1 sulfuric/hydrofluoric acid electrolyte. While some in the community accepted the potential of PRC electrochemical polishing of flat niobium coupons as fundamentally paradigm-changing, it took the demonstration of single-cell niobium cavity polishing and performance demonstration for the PRC electrochemical polishing process to be more widely embraced. With this successful demonstration, confidence in the ability of the PRC electrochemical polishing process to replace the current process is growing and work is ongoing independently or collaboratively with several DOE as well as international high energy physics facilities to further advance the PRC electrochemical polishing technology. A next step is to better understand the effects of anode to cathode gap and area ratios; another is to manipulate the geometry of the electrochemical polishing cell by shielding to get more uniform geometric current distribution throughout the cavity. It must be noted that while a pulse reverse waveform can influence the balance of primary, secondary, and tertiary current distribution at the microscale, it cannot overcome the effects of inefficient cell design. Ongoing work with curved sections, half cavity cells, and investigation of the effects of flow on the boundary layer throughout the cavity, as well as further manipulation of the pulse reverse waveform, will all lead to a better understanding of the suggested cathodic electropolishing mechanism and the optimization of the PRC electrochemical polishing process. These activities have lead to a demonstration of the PRC electrochemical polishing process on nine-cell niobium SRF cavities, which are a key component of accelerator technology.

7.5 Concluding Thoughts

Observations and milestones over a period of 22 years describe the development of PC/PRC processes that have led to simpler plating, deburring, and metal-finishing chemistries and the development of a new "eco-friendly" pulse reverse voltage electropolishing process for niobium SRF cavities. These developments required a paradigm shift in electrochemical plating and surface-finishing processes, specifically with respect to replacing the conventional reliance on additive and sometimes toxic chemistries for process control with the use of simple chemistries and pulse reverse electric fields. Guiding principles based on seminal works from the early development of pulse and pulse reverse plating has resulted in approximately 45 patents covering various embodiments pulse and pulse reverse plating and surface finishing.

The first success in identifying and taking advantage of this paradigm shift and these guiding principles was in pulse reverse plating of copper, where plating performance and deposit properties were improved with the application of PRC and a reduction in chemical additives. The recognition that the pulse reverse electric field may obviate the need for chemical additives in plating processes was then applied to a wide range of anodic processes. Initially, deburring of non-passive materials using forward-only anodic pulses was demonstrated. This approach was derived based on observations of localized current distribution during pulse plating of lead frame contacts for electronics applications. Subsequently, electropolishing of passive materials was demonstrated by incorporating a cathodic reverse pulse to remove the adverse effects of passive film formation. This approach was derived based on observations of elimination of adverse hydrogen evolution effects during pulse reverse plating of low faradaic efficiency materials such as trivalent chromium. From this accumulated knowledge, the pulse reverse voltage approach was successfully applied to strongly passive materials such as niobium. Finally, the pulse reverse voltage electropolishing process was transitioned to niobium SRF cavities based on the key observation that electropolishing correlated with an anodic current transition during the anodic voltage pulse.

Conventional direct voltage electropolishing of niobium SRF cavities utilizes a concentrated mixture of sulfuric and hydrofluoric acids and is understood to mechanistically function as described by the Jacquet "viscous salt film" model. Pulse reverse voltage electropolishing uses low-concentration aqueous sulfuric acid electrolyte devoid of hydrofluoric acid and is not easily rationalized with the viscous salt film model. As such, a paradigm shift or modification to the Jacquet model is emerging. While additional work is required to elucidate the mechanistic aspects of the pulse reverse process, this process is tentatively referred to as "cathodic electropolishing."

Upon its founding in 1991, Faraday was focused solely on the development of cathodic/electrodeposition processes using PC/PRC principles. Over time, lessons learned in cathodic processes were successfully applied to developing pulse and pulse reverse anodic processes and thus broadened the company's mission. These lessons learned will continue to be applied, as new electrochemical engineering technologies are investigated and developed in many other fields. The experience gained over 20 years of research and development suggests that scientists, engineers, and technologists should conduct their research and development activities with a careful balance between current fundamental understandings combined with a willingness to pursue evolving observations that do not necessarily fit existing paradigms. Finally, this discussion concludes with a quote by Eliezer Gileadi, Professor of Chemistry at Tel-Aviv University, a long-time friend and colleague, which is displayed in Faraday's lobby and serves as a guiding principle [83].

> ...in order to create, one must first question that which exists...

Acknowledgments

The authors gratefully acknowledge the US Small Business Innovation Research and Small Business Technology Transfer (SBIR/STTR) programs of the Department of Energy (DOE), National Science Foundation (NSF), Environmental

Protection Agency (EPA), Department of Defense (DoD), and National Institutes of Health (NIH) for providing the seed funding, which contributed to the advancements of pulse and pulse reverse electrochemical technologies. The authors also acknowledge the technology development funding of Ford Motor Company, Swagelok Corporation, as well as numerous confidential commercial clients and licensees. Specifically regarding the niobium electropolishing activities, the authors gratefully acknowledge the financial support and guidance of (i) DOE Program Manager Dr. Manouchehr Farkhondeh (SBIR Phase I Grant No. DE-SC0004588), (ii) DOE Program Manager Dr. L. K. Len (SBIR Phase I Grant No. DE-FG02-08ER85053), (iii) DOE Program Manager Mr. Allan Rowe (America Reinvestment in Research Act (ARRA) under DOE Fermi National Accelerator Laboratory (FNAL) P.O. No. 594128), (iv) DOE Program Manager Dr. Manouchehr Farkhondeh (SBIR Phase I/II Grant No. DE-SC0011235), and (v) DOE Program Manager Dr. Kenneth R. Marken, Jr. (SBIR Phase I/II Gant No. DE-SC0011342). In addition, the authors would like to acknowledge the technical guidance regarding SRF cavity processing, Dr. Charles Reece of the DOE Thomas Jefferson National Accelerator Facility, Mr. Allan Rowe of the DOE Fermi National Accelerator Laboratory, and Dr. John Mammosser and Dr. Jeff Saunders of the DOE Oak Ridge National Laboratory. The authors also acknowledge Prof. Patrick J. Moran of the U.S. Naval Academy, Prof. Der-Tau Chin of Clarkson University, Prof. C. C. Liu and Prof. Robert Savinell of Case Western Reserve University, Prof. Alan C. West of Columbia University and Prof. John Fortman of Wright State University, as well as numerous other university collaborators. Finally, the authors also acknowledge prior Faraday employees for their contributions to our pulse/pulse reverse electrochemical activities.

References

1 Kuhn, T. (1966). *The Structure of Scientific Revolutions*, 3e. University of Chicago Press.
2 Göktürk, E. (2011). *What Is "Paradigm"?* http://folk.uio.no/erek/essays/paradigm.pdf. Norway: Department of Informatics, University of Oslo accessed October 2016.
3 Kuhn, T. (1966). *The Structure of Scientific Revolutions*, 3e, 52. University of Chicago Press.
4 Jacquet, P.A. (1936). On the anodic behavior of copper in aqueous solutions of orthophosphoric acid. *Trans. Electrochem. Soc.* 69 (1): 629–655.
5 35 U.S.C. 201 (1980) Definitions, Bayh-Dole Public Law 96-517, December 12, 1980.
6 Romankiw, L.T. and Krongelb, S. (2015). The path from invention to product, Chapter 2. In: *Electrochemical Engineering across Scales* (ed. R.C. Alkire, P.N. Bartlette and J. Lipkowski). Weinheim, Germany: Wiley-VCH.
7 Taylor, E.J. and Miller, P. (2001). Innovation case studies at an R&D company…alignment of technology, intellectual property and financial matters. *Les Nouvelles* XXXVI (2).

8 McCrabb, H., Kell, J., and Kumar, B. (2012). Pulse electric fields for EPD of thermal barrier coatings. *Key Eng. Mater.* 507: 21–25.

9 Kell, J., McCrabb, H., and Kumar, B. (2012), Electrophoretic deposition of thermal barrier coatings by the Faradaic process, Material Science and Technology 2008 Conference, Pittsburgh, PA, 25–28 September 2005.

10 Gillis, J. (1918) Method of recovering metals from alloys, US Patent No. 1,260,661 filed 4 September 1917 and issued 26 March 1918.

11 Puippe, J.C. and Leaman, F. (1986). *Theory and Practice of Pulse Plating*. Orlando, FL: AESF.

12 Hansel, W.E.G. and Roy, S. (2012). *Pulse Plating*. Bad Saulgau, Germany: Leuze Verlag KG.

13 Puippe, J.C. (1986). Qualitative approach to pulse plating. In: *Theory and Practice of Pulse Plating* (ed. J.C. Puippe and F. Leaman). Orlando, FL: AESF.

14 Landolt, D. (1986). Mass transport in pulse plating. In: *Theory and Practice of Pulse Plating* (ed. J.C. Puippe and F. Leaman). Orlando, FL: AESF.

15 Cheh, H.Y. (1971). Electrodeposition of gold by pulsed current. *J. Electrochem. Soc.* 118: 551–557.

16 Cheh, H.Y. (1971). Limiting rate of deposition by P-R plating. *J. Electrochem. Soc.* 118: 1132–1134.

17 Viswanathan, K., Farrell-Epstein, M., and Cheh, H.Y. (1978). The application of pulsed current electrolysis to a rotating-disk electrode system: I. Mass transfer. *J. Electrochem Soc.* 125: 1772–1776.

18 Ibl, N., Puippe, J.C., and Angerer, H. (1978). Electrocrystallization in pulse electrolysis. *Surf. Technol.* 6: 287.

19 Ibl, N. (1980). Some theoretical aspects of pulse electrolysis. *Surf. Technol.* 10: 81.

20 Yin, K. (1996). Duplex diffusion layer model for pulse with reverse plating. *Surf. Coat. Technol.* 88: 162–164.

21 Dossenbach, O. (1986). Current distribution in pulse plating. In: *Theory and Practice of Pulse Plating* (ed. J.C. Puippe and F. Leaman). Orlando, FL: AESF.

22 Chin, D.T. and Balamurugan, D. (1992). An experimental study of metal distribution in pulse plating. *Electrochim. Acta* 37 (11): 1927–1934.

23 Newman, J. and Thomas-Alyea, K.E. (2004). *Electrochemical Systems*, 3e. New York: Wiley.

24 Chin, D.-T., Vilambi, N.R.K., and Sunkara, M.K. (1989). Current distribution in selective pulse plating. *Plating & Surface Finishing* 76 (74).

25 Puippe, J.C. (1986). Influence of pulse plating on crystallization. In: *Theory and Practice of Pulse Plating* (ed. J.C. Puippe and F. Leaman). Orlando, FL: AESF.

26 Taylor, E.J., Anderson, E.B., and Vilambi, N.R.K. (1992). Preparation of high-platinum-utilization gas diffusion electrodes for proton-exchange-membrane fuel cells. *J. Electrochem. Soc.* 139 (5): L45–L46.

27 Vilambi, N.R.K., Anderson, E.B. and Taylor, E.J. (1992), High utilization supported catalytic metal-containing gas-diffusion electrode, process for making it, and cells utilizing it, US Patent No. 5,084,144, issued 28 January.

28 Taylor, E.J. and Inman, M. (2000), Electrodeposition of catalytic metals using pulsed electric fields, US Patent No. 6,080,504, issued 27 June.

29 Puippe, J.C. (1986). Pulsed electrodeposition of rhenium. In: *Theory and Practice of Pulse Plating* (ed. J.C. Puippe and F. Leaman). Orlando, FL: AESF.

30 Puippe, J.C. (1986), Pulse Plating Versus DC Plating, 3rd International Pulse Plating Symposium, Washington, DC, October 28–29, AESF, Orlando, FL.

31 Taylor, E.J. (2008). Adventures in pulse/pulse reverse electrolytic processes: explorations and applications in surface finishing. *J. Appl. Sur. Fin.* 3 (4): 178–189.

32 Schwartz, M. (1982). Deposition from aqueous solutions: an overview. In: *Deposition Technologies for Films and Coatings* (ed. R.F. Bunshah). Park Ridge, NJ: Noyes Publications.

33 Dini, J.W. (1993). *Electrodeposition: The Materials Science of Coatings and Substrates*. Park Ridge, NJ: Noyes Publications.

34 Taylor, E.J., Sun, J.J. and Zhou, C., (2001) Electrodeposition of metals in small recesses using modulated electric fields, US Patent No. 6,303,014 filed 20 April 2000, issued 16 October 2001.

35 Taylor, E.J., Sun, J.J. and Zhou, C., (2001) Electrodeposition of metals in small recesses for manufacture of high density interconnects using reverse pulse plating, US Patent No. 6,210,555 filed 29 January 1999, issued 3 April 2001.

36 Taylor, E.J., Sun, J.J. and Zhou, C., (2003) Electrodeposition of metals in small recesses using modulated electric fields, US Patent No. 6,524,461 filed 3 April 2001, issued 25 February 2003.

37 Sun, J.J., Taylor, E.J., Leedy, K.D. et al. (1999). Modulated reverse electric field copper metallization for high density interconnects and very large scale integration application. In: *Proceedings Electrochemical Processing in ULSI Fabrication and Semiconductor/Metal Deposition II* (ed. P.C. Andricacos, P.C. Searson, C. Reidsema-Simpson, et al.) PV 99-9. Pennington, NJ: The Electrochemical Society.

38 Taylor, E.J., and Sun, J.J. (2004) Electrodeposition of metals in high-aspect ratio cavities using modulated reverse electric fields, US Patent No. 6,827,833 filed 15 October 2001, issued 7 December 2004.

39 Sun, J.J., Taylor, E.J., Inman, M.E. et al. (1999). Application of an electrochemical copper metallization-planarization process to sub-0.25μ features. In: *Proceedings Interconnect and Contact Metallization for ULSI* (ed. Y. Arita, G.S. Mathad and H.R. Rathore) PV 99-31. Pennington, NJ: The Electrochemical Society.

40 Sun, J.J., Taylor, E.J., and Inman, M.E. (1999). A sequential copper plating process for multiple feature sizes for HDI applications. In: *Proceedings Electrochemical Technology Applications in Electronics III* (ed. C. Madore, T. Osaka, L.T. Romankiw and Y. Yamazaki) PV 99-34. Pennington, NJ: The Electrochemical Society.

41 Taylor, E.J., Sun, J.J. and Inman, M.E. (2001) Sequential electrodeposition of metals using modulated electric fields for manufacture of circuit boards having features of different sizes, US Patent No. 6,309,528 filed 18 October 1999, issued 30 October 2001.

42 Taylor, E.J., Sun, J.J. and Inman, M.E. (2003) Sequential electrodeposition of metals using modulated electric fields for manufacture of circuit boards

having features of different sizes, US Patent No. 6,652,727 filed 26 April 2002, issued 25 November 2003.
43 Taylor, E.J., Sun, J.J. and Inman, M.E. (2005) Sequential electrodeposition of metals using modulated electric fields for manufacture of circuit boards having features of different sizes, US Patent No. 6,863,793 filed 3 April 2001, issued 8 March 2005.
44 Garich, H.M., Gebhart, L.E., Snyder, S.T., and Taylor, E.J. (2009). Novel plating cell geometry for uniform metal deposition. *ECS Trans.* 19 (26): 11–18.
45 Blomsterberg, K-I. (1983) Method of anodically deburring articles of steel or aluminum alloys in an electrolytic bath, and a bath for carrying out the method, US Patent No. 4,411,751 filed 14 September 1982, issued 25 October 1983.
46 Landolt, D. (1978). Fundamental aspects of electropolishing. *Electrochim. Acta* 32 (1): 1–11.
47 Sun, J.J., Gebhart, L.E., Renz, R.P., Taylor, E.J., Inman, M.E. (2000) The application of CM-ECM technology to metal surface finishing in Proceedings North American Manufacturing Research Institution of the Society of Manufacturing Engineers, MR00-209 NAMRC XXVIII May 24–26, Lexington, KY.
48 Taylor, E.J. and Inman, M.E. (2014). Electrochemical surface finishing. *Interface* 23 (3): 57.
49 Stacherski, K. (1996) Electrochemical Deburring in Cutting Tool News, Ford PowerTrain 2(1) Winter 1996 (available www.FaradayTechnology.com accessed May 26 2016)
50 Landolt, D. (1996). Research award of the electrodeposition division award address: mass transport processes in high rate metal dissolution and deposition. In: *Proceedings High Rate Metal Dissolution Processes* (ed. M. Datta, B.R. MacDougal and J.M. Fenton) PV 95-19. Pennington, NJ: The Electrochemical Society.
51 Elmore, W.C. (1939). Electrolytic polishing. *J. Appl. Phys.* 10: 724.
52 Edwards, J. (1953). The mechanism of electropolishing of copper in phosphoric acid solutions: II. The mechanism of smoothing. *J. Electrochem. Soc.* 100 (8): 508–512.
53 Glarum, S.H. and Marshal, J.H. (1985). The anodic dissolution of copper into phosphoric acid: I. Voltammetric and oscillatory behavior. *J. Electrochem. Soc.* 132 (12): 2872–2878.
54 Glarum, S.H. and Marshal, J.H. (1985). The anodic dissolution of copper into phosphoric acid: II. Impedance behavior. *J. Electrochem. Soc.* 132 (12): 2878–2885.
55 Gillespie, L.K. (1999). *Deburring and Edge Finishing Handbook*, 317. Dearborn, MI: Society of Manufacturing Engineers.
56 Ponto, L., Datta, M., and Landolt, D. (1987). Electropolishing of iron–chromium alloys in phosphoric acid–sulphuric acid electrolytes. *Surf. Coat. Technol.* 30 (3): 265–276.
57 Singh, V.B. and Arvind, U. (1995). Electrodissolution of AISI 304 stainless steel in concentrated acids leading to electropolishing. *Indian J. Chem. Technol.* 2: 211–216.

58 MacDougall, B. (1995). The importance of surface oxide films in corrosion, semiconductor and environmental research. In: *Proceedings of the Symposium on High Rate Metal Dissolution Processes*, vol. 95-19 (ed. M. Datta, B. MacDougall and J. Fenton), 16–31. Pennington, NJ: The Electrochemical Society.

59 Mathieu, J.B. and Landolt, D. (1978). Electropolishing of titanium in perchloric acid–acetic acid solutions. *J. Electrochem. Soc.* 125 (7): 1044.

60 Wixtrom, A.I., Buhler, J.E., Reece, C.E., and Abdel-Fattah, T.M. (2013). Electrochemical polishing applications and EIS of a Vitamin B4–based ionic liquid. *J. Electrochem. Soc.* 160 (3): E22–E26.

61 Zhou, C.D., Taylor, E.J., Sun, J.J. et al. (1997). Electrochemical machining of hard passive alloys with pulse reverse current. *Trans. NAMRI/SME* XXV: 147–152.

62 Renz, R.P., Fortman, J.J., Taylor, E.J., and Chalmers, P.D. (2006). Functional trivalent chromium plating: an alternative to hexavalent chromium plating. *J. Appl. Surf. Finish.* 1 (2): 155–162.

63 Kagajwala, B., Hall, T.D., Inman, M.E., Taylor, E.J., Griffin, B., Cushnie, G., Taylor, R., Jaworowski, M., Bonivel, J. (2013) Functional trivalent chromium electroplating of internal diameters, in Products Finishing posted 1-2-2013.

64 Zhou, C., Taylor, E.J., Sun, J.J., Gebhart, L.E. and Renz, R.P. (2002) Electrochemical machining method using modulated reverse electric fields. US Patent No. 6,402,931, filed 18 May 1998 and issued 11 June 2002.

65 Sun, J.J., Taylor, E.J., and Srinivasan, R. (2001). MREF-ECM process for hard passive materials surface finishing. *J. Mater. Process. Technol.* 108: 356–368.

66 Sautebin, R., Froidevaux, H. and Landolt, D. (1980) Theoretical and experimental modeling of surface leveling in ECM under primary current distributions.

67 Sautebin, R. and Landolt, D. (1982) Anodic leveling secondary and tertiary current distribution conditions.

68 Taylor, E.J. (2003) Sequential electromachining and electropolishing and the like using modulated electric fields. US Patent 6,558,231, filed 17 October 2000 and issued 6 May 2003.

69 Eozénou, F., Boudigou, Y., Carbonnier, P. et al. (2014). Development of vertical electropolishing process applied on 1300 and 704 MHz superconducting niobium resonators. *Phys. Rev. Spec. Top. Accel. Beams* 17: 083501.

70 K. Saito, H. Inoue, E. Kako, T. Fujino, S. Noguchi, M. Ono, and T. Shishido (1997), Superiority of electropolishing over chemical polishing on high gradients, in Proceedings of the 8th Workshop on RF Superconductivity (SRF 1997), AbanoTerme (Padova), Italy, 1997, V. Palmieri (INFN, Padova,1998), pp. 795–813.

71 Tian, H., Corcoran, S., Reece, C., and Kelly, M. (2008). The mechanism of electropolishing of niobium in hydrofluoric-sulfuric acid electrolyte. *J. Electrochem. Soc.* 155: D563–D568.

72 Kneisel, P. (1999). High gradient superconducting niobium cavities: a review of the present status. *IEEE Trans. Appl. Supercond.* 9 (2): 1023–1029.

73 Mammosser, J. (2015) Chemical safety for SRF work, US Particle Accelerator School, presented at Thomas Jefferson National Accelerator Facility, January.

74 Inman, M., Hall, T., Taylor, E.J., Reece, C.R. Trofimova, O. (2011) Niobium electropolishing in an aqueous, non-viscous HF-free electrolyte: a new polishing mechanism in Proceeding SRF2011, TUPO012 377-381 Chicago, IL.

75 Inman, M., Garich, H., Snyder, S., Taylor, E.J., Cooley, L., Cooper, C.A. and Rowe, A. (2011) Development and scale-up of an HF free electropolishing process in single-cell niobium SRF cavities in Proceeding SRF2011, TUPO017 397-400 Chicago, IL.

76 Inman, M.E., Taylor, E.J., and Hall, T.D. (2013). Electropolishing of passive materials in HF-free low viscosity aqueous electrolytes. *J. Electrochem. Soc.* 160 (9): E94–E98.

77 Taylor, E.J., Hall, T.D., Inman, M.E. and Snyder, S.T. (2013) Electropolishing of niobium SRF cavities in low viscosity aqueous electrolytes without hydrofluoric acid. Proceedings SRF2013, TUP054 Paris, France.

78 Roth, A. Schmidt, O. (1977) Method for the electrolytic polishing of the inside surface of hollow niobium bodies, US Patent No. 4,014,765, issued 29 March 1977.

79 Eozénou, F., Boudigou, Y., Carbonnier, P. et al. (2014). Development of vertical electropolishing process applied on 1300 and 704 MHz superconducting niobium resonators. *Phys. Rev. Spec. Top. Accel. Beams* 17: 083501-1–083501-9.

80 Rowe, A.M., Grassellino, A., Hall, T.D., Inman, M.E., Snyder, S.T. and Taylor, E.J. (2013) Bipolar EP: electropolishing without fluorine in a water based electrolyte. Proceedings SRF2013, TUIOC02 Paris, France.

81 Taylor, E.J., Inman, M.E. and Hall, T.D. (2015) Electrochemical system and method for electropolishing of superconductive radio frequency cavities. US Patent No. 9,006,147, filed 11 July 2012 and issued 14 April 2015.

82 Taylor, E.J., Inman, M.E., Hall, T.D., Snyder, S.T., Rowe, A.M. and Holmes, D. (2015) Economics of electropolishing of niobium SRF cavities in eco-friendly aqueous electrolytes without hydrofluoric acid. Proceedings SRF2015, MOPB092 Whistler, Canada.

83 Gileadi, E. (2004) Personal communication, 205th Meeting of The Electrochemical Society, San Antonio, TX.

8

The Interaction Between a Proton and the Atomic Network in Amorphous Silica Glass Made a Highly Sensitive Trace Moisture Sensor

Yusuke Tsukahara, Nobuo Takeda, Kazushi Yamanaka, and Shingo Akao

Ball Wave Inc., T-Biz 503, 6-6-40, Aza Aoba, Aramaki, Aoba, Sendai, Miyagi 9890-8579, Japan

8.1 Unexpected Long Propagation of Surface Acoustic Waves Around a Sphere

Lord Rayleigh described in his celebrated book on the sound wave [1] that a solid half space would support the propagation of vibrations confined in the vicinity of the surface. Simply said, an elastic wave propagates along the solid surface. It became known as a Rayleigh wave, an example of which you can definitely feel when an earthquake hits. Later on, many kinds of surface waves were found, theoretically and experimentally, and those are collectively named surface acoustic waves (SAW), or SAWs. In addition to the fracture of crust in the case of earthquakes, there are many other ways to generate SAWs. When the surface of a metal slab or a silicon wafer was illuminated with a high-intensity laser light, SAWs were generated. The surface absorbed the energy of laser illumination and was heated up instantly causing the thermal expansion of the surface that generated the SAWs. If the laser light is focused in a small spot on the surface, it makes a point source of SAWs that propagate in all directions. If the laser light is focused in a line shape, SAWs are generated and propagate perpendicular to the line in both directions. An invention, scanning interference fringes (SIF) method, was made to generate the SAW in only one direction effectively by using the interference of two arms of a split laser beam [2]. When one of the present authors, KY, was asked to detect the damage or a crack on the surface of a steel ball for ball bearings of a hard disc drive, he tried to detect a change in the propagation of SAWs by the damage and applied the SIF method because he was one of its inventors. He was surprised to see that the SAW propagated for an unexpectedly long period of time, which meant that the SAW went around the ball along its equator so many times [3]. It proved that the method would be useful in the very precise measurement of mechanical properties of solid balls.

In the experiment, steel bearing balls with 8 mm in diameter were used. One ball was used as received, while Ag films were deposited on the other two balls by vacuum evaporation to mimic surface damages. The maximum thicknesses of the

Electrochemical Engineering: The Path from Discovery to Product,
First Edition. Edited by Richard C. Alkire, Philip N. Bartlett, and Marc T. Koper.

two Ag films were ~50 and 150 nm, respectively, although the thicknesses were not uniform over the curved surfaces of the balls. Figure 8.1 shows the experimental setup. A yttrium aluminum garnet (YAG) laser beam was split into two arms, 3 mm in diameter, and one arm of the beam was frequency shifted by 30 MHz by using a Bragg cell. Then, the two arms of the laser beam were directed almost perpendicular to the surface of a steel ball. Due to the interference of the two coherent laser beams with different optical frequencies, scanning interference fringes were produced. By carefully adjusting the mechanical stage, the average spacing of the fringes was made equal to the wavelength and the scanning velocity of the fringes to the phase velocity of the SAW. This made the scanning of the interference fringes in a perfect match with the propagation of the SAW. The YAG laser had a specially designed long pulse width of around 100 ns that allowed a long interaction time between the fringes and the generated SAW.

The SAW propagated along the equator of the ball perpendicular to the fringes in many turns. They were detected at each turn by the optical knife edge method with an argon ion laser probe focused at a position 3–4 mm away from the fringes. After the detection, frequency components lower than 20 MHz were filtered out, resulting in suppressing the detection of bulk waves in the steel ball generated by the laser irradiation. Figure 8.2 shows the output signal with an unexpectedly large number of tone burst. The number of round-trip propagation was as large as 20, corresponding to a total propagation length of 50 cm. Figure 8.3 shows the expanded signals at the first and twelfth turns, respectively, for the as-received, 50 nm-thick Ag-coated and 150 nm-thick Ag-coated balls, respectively. The signals were averaged over 10 measurements

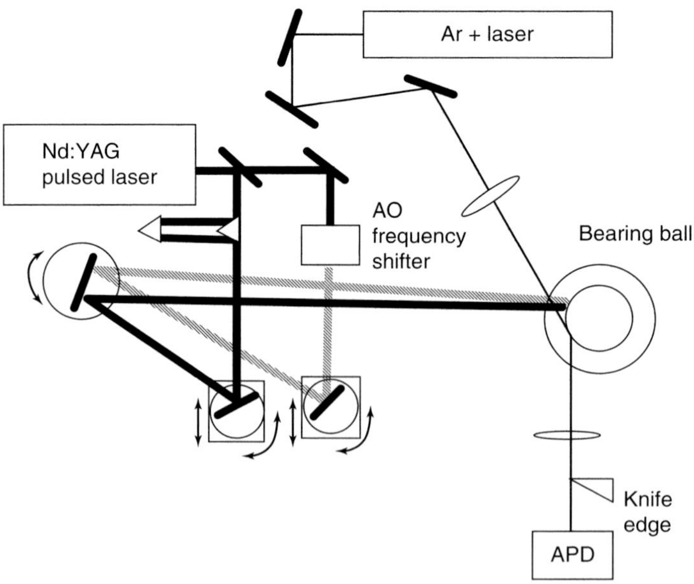

Figure 8.1 Apparatus for generating and detecting SAW on a bearing ball by scanning interference fringes method. Source: Yamanaka et al. 2000 [3]. Reproduced with permission of AIP Publishing.

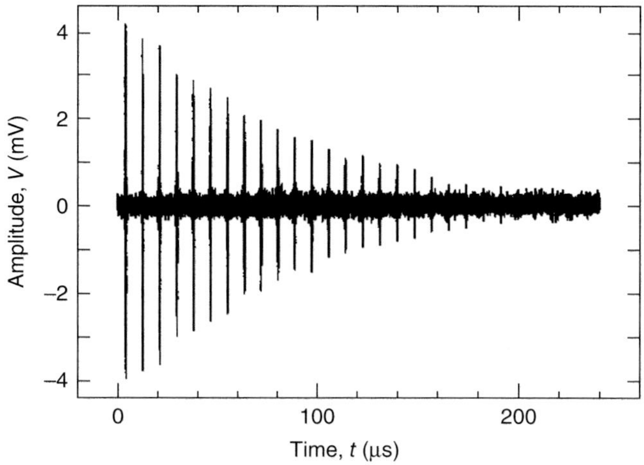

Figure 8.2 Detected signal of SAW with a large number of monotonically decaying periodic pulses. Source: Yamanaka et al. 2000 [3]. Reproduced with permission of AIP Publishing.

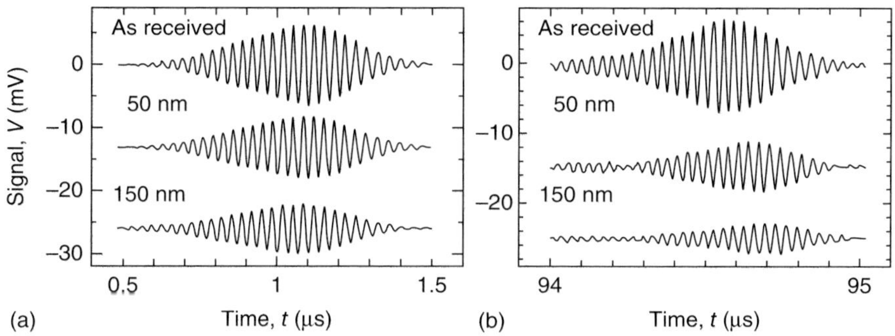

Figure 8.3 Expended waveforms of SAW at (a) the first and (b) the twelfth turn. Source: Yamanaka et al. 2000 [3]. Reproduced with permission of AIP Publishing.

and digitized with the sampling interval of 2 ns. Although the temporal separation of signals in Figure 8.3a,b was large (93 μs), their waveforms hardly changed. That made it easy to apply the pulse-echo-overlap method to deduce the delay time between the first and the twelfth turns. For the as-received steel ball, it was 93.474 μs ± 0.002 μs. This indicates that the phase velocity measurement was as accurate as 0.002%.

8.2 Invention of a Ball SAW Device and Application to Gas Sensors

Any waves would diverge by diffraction of wave propagation. Hence, it was a mystery for a while why the SAW generated on the ball bearing by SIF propagated for such a long distance. The other one of the present authors, YT, realized that it was due to the balance between the diffraction and the geometrical focusing by spherical shape of the surface [4].

Let us suppose that a line source of a finite length is located along the equator on a spherical object as shown in Figure 8.4. It will generate two beams of Rayleigh wave propagating away from the line source and converging toward the either poles because the surface is not flat but curved. However, the Rayleigh wave does not converge into a single point because a wave from a finite aperture spreads by diffraction. The two effects would balance when the length of the line source is appropriately chosen, and a diffraction-free propagation of a finite beam width would be realized. In order to validate the conjecture, the following experiment was designed. A glass ball 80 mm in diameter was prepared and its surface was optically polished. The density, and the longitudinal, shear and Rayleigh wave velocities of the glass were 2510 kg m^{-3}, 6019 ms^{-1}, 3655 ms^{-1}, and 3334 ms^{-1}, respectively. An ultrasonic transducer of one to three composite made of epoxy and lead zirconate titanat (PZT) was cut into a strip, 1.5 mm × 20 mm, with thickness of 1.4 mm. The transducer was glued on the glass ball and connected to an electric pulse generator, which fed an electrical pulse with a rise time of 1 ns. Longitudinal and shear waves were generated in the glass and reverberated within the glass sphere. A Rayleigh wave with frequency about 1.1 MHz was excited, too, with a high efficiency because the width of the transducer was half the wavelength.

Figure 8.5 shows a waveform received by the same transducer that was used for the transmitter. The peak marked with R1 corresponds to a round-trip of the Rayleigh wave around the sphere. R2 and R3 correspond to the twofold and threefold round trips. The fourfold round-trip, R4, is barely discernable. The peak marked with L corresponds to a longitudinal wave that traveled from the transducer to the opposite surface in the glass and was then reflected back. The peak S is a shear wave reflection. Other small peaks are multiple reflections of bulk waves within the sphere.

The authors observed the propagation of the Rayleigh wave along the surface using a pin transducer, which consisted of a circular disk of polyvinylidene

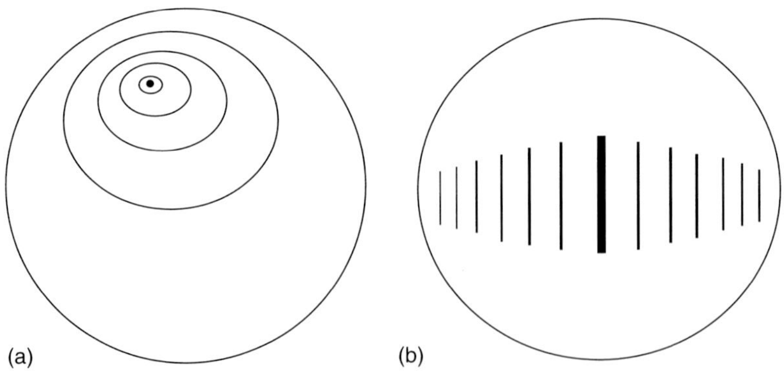

(a) (b)

Figure 8.4 Schematics of Rayleigh wave propagation on a spherical solid: (a) point source excitation, (b) line source excitation. Source: Tsukahara et al. 2000 [4]. Reproduced with permission of AIP Publishing.

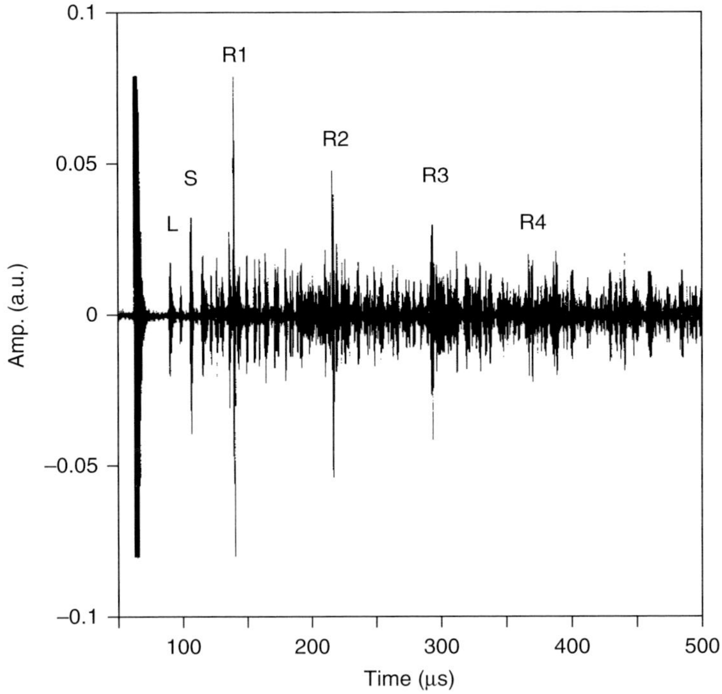

Figure 8.5 Waveform received by the same transducer used as a transmitter. Source: Tsukahara et al. 2000 [4]. Reproduced with permission of AIP Publishing.

difluoride (PVDF) film, 2 mm in diameter. The pin transducer was placed on the spherical surface with glycerol as a coupler, and the waveform at that position was recorded. Figure 8.6 shows a series of waveforms obtained at different lateral positions, 5 mm apart from each other, at 90° from the transducer. The peaks C1, CC1, and CC2, corresponding to Rayleigh waves, were collimated in a beam width between 5 and 10 mm, regardless of their propagation distances. Figure 8.7 shows the waveforms obtained at 180° from the transmitter. The beam width of the Rayleigh wave was about 20 mm, which was the size of the source transmitter located at the opposite end of the sphere. The waveforms of Rayleigh waves did not change at all when the surface area outside of the propagation zone was touched by fingers. These measurements of the beam profile validated the model of diffraction-free propagation of the Rayleigh wave around a solid sphere.

Now, it is extremely interesting to explore the possibility of fabricating surface acoustic wave devices on spherical solids by preparing interdigital transducers (IDTs) with piezoelectric thin films on small ceramic balls, or by preparing IDTs on piezoelectric balls such as single crystal quartz [5]. It should be noted that the same mechanism of diffraction-free propagation could be applied to any waves such as the light waves and wave functions in quantum mechanics.

The principle of the ball SAW gas sensor is illustrated in Figure 8.8a. A sensitive film was deposited on the surface area around the equator of a ball. A

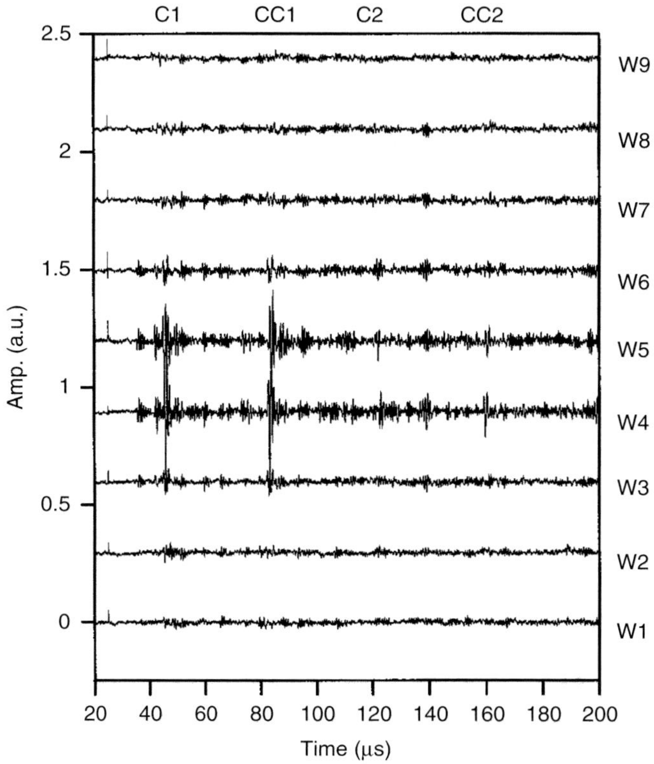

Figure 8.6 Waveforms obtained by a pin transducer at positions 90° from the transmitter. Lateral positions measured from the center of the beam for W1–W9 are +20, +15, +10, +5, 0, −5, −10, −15, and −20 mm. Source: Tsukahara et al. 2000 [4]. Reproduced with permission of AIP Publishing.

collimated beam of SAW was generated by an IDT and detected by the same IDT after many turns of propagation. The velocity and attenuation of SAW were changed by absorption of molecules into the sensitive film. Although the change was small for a very low concentration, it was amplified after multiple round trips, more than 50 turns. If the attenuation coefficient variation multiplied by the circumference of the sphere is $\Delta\alpha L = 10^{-3}$, the relative amplitude after one turn is $\exp(-\Delta\alpha L) = \exp(-10^{-3}) = 0.999$, which is practically indistinguishable from unity in experiments. However, the relative amplitude in ball SAW sensors after 100 turns is $\exp(-100\Delta\alpha L) = \exp(-0.1) = 0.905$, whose difference from unity is easily detected in experiments.

We verified the advantages of the ball SAW gas sensor taking H_2 as a target gas. Two sensors were fabricated: one with a single crystal quartz ball, 10 mm in diameter, with an acoustic frequency of 45 MHz, and the other with 1 mm in diameter with the frequency of 156 MHz. A photograph of the latter sensor is shown in Figure 8.8b. The IDT was fabricated by photolithography. The films of Pd or PdNi were deposited by thermal evaporation under the pressure of 10^{-4} Pa at a rate of 0.01 nm s^{-1} resulting in thicknesses of 20 nm or 40 nm. The

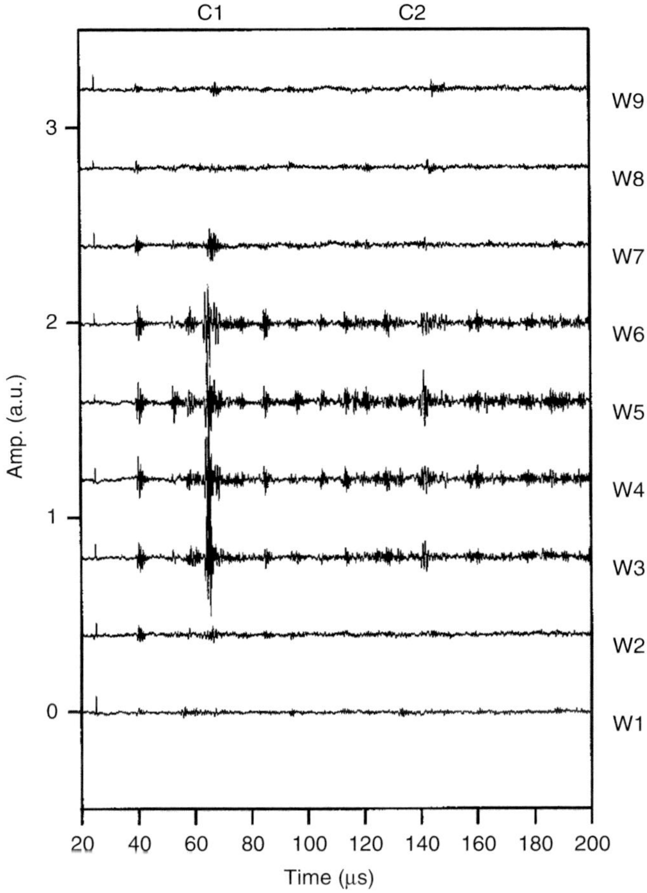

Figure 8.7 Waveforms obtained by a pin transducer at positions 180° from the transmitter. Lateral positions measured from the center of the beam for W1–W9 are +20, +15, +10, +5, 0, −5, −10, −15, and − 20 mm [4].

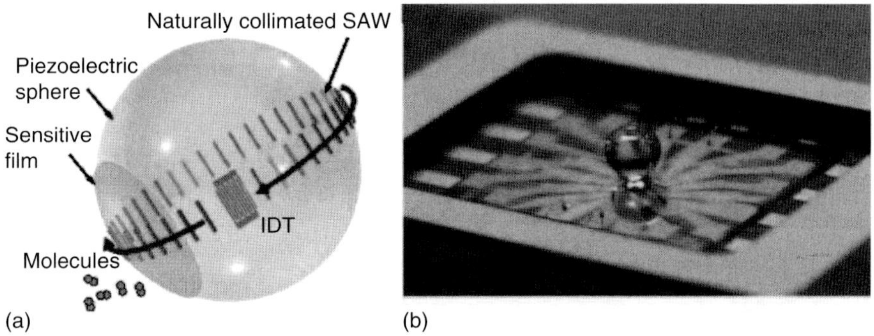

Figure 8.8 Ball SAW sensor: (a) the adsorption of gas molecules to the sensitive film causes a change in delay time and amplitude of SAW propagating around the piezoelectric ball, (b) A $\phi = 1$ mm quartz ball SAW H_2 gas sensor with 40-nm-thick Pd30% Ni film installed in package [5].

performance of the ball SAW sensor was evaluated in a flow cell at a gas flow rate of 5 l min^{-1}. Initially, the carrier gas of 100% Ar was introduced. The gas was subsequently changed to 0.1% H_2 in Ar by adjusting gas concentration with a mass flow controller. After 10 min, it was changed back to 100% Ar. H_2 concentration varied from 0.1% up to 3.0% with a step of ~0.6%.

The delay time and amplitude of SAW were measured by an ultrasonic pulser/receiver. Figure 8.9a shows a multiple round-trip signal on the 10 mm device at center frequency of 45 MHz. The signal was visible after 50 turns, which corresponds to the total propagation length of 1600 mm. Considering the width of IDT (0.7 mm) and the wavelength of SAW (0.07 mm), the Fresnel distance was 1.75 mm. The propagation length of 1600 mm was extremely large compared to the Fresnel length which was absolutely impossible for the wave propagation on planar or cylindrical surfaces because of diffraction loss. The delay time τ to reach the amplitude of $1/e$ was 200 μs and the frequency was 45 MHz, so that the Q factor, $\pi f\tau$, was 28 000. Figure 8.9b shows the effect of the hydrogen content to the delay time of the SAW at 3% H_2 in Argon carrier gas. The hydrogen gas concentration was changed and a clear response with a signal to noise ratio of four was obtained even at 10 ppm H_2 concentration. This was the highest sensitivity at the time with SAW H_2 sensors. In the other extreme, the 100% H_2 concentration that meant the 100% hydrogen gas without the carrier gas, gave a clear delay time response. Hence, it was experimentally validated that the ball SAW hydrogen gas sensor was able to detect the H_2 concentration ranging from 10 ppm to 100% as shown in Figure 8.10.

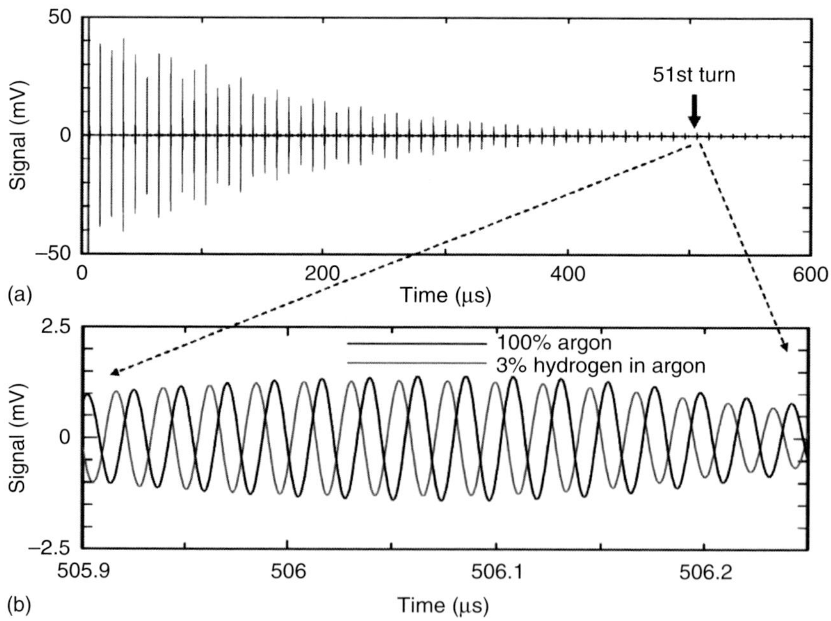

Figure 8.9 Waveform of ball SAW sensor. (a) Forty-five megahertz SAW signal observed on *Z*-axis cylinder of ϕ = 10 mm quartz ball, (b) effect of 3% H_2 in Ar gas producing significant difference in delay time at 51st turn [5].

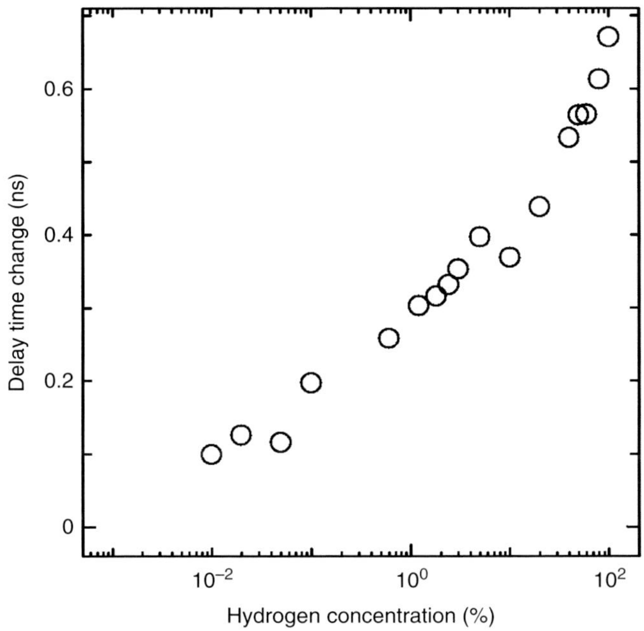

Figure 8.10 Delay time responses for hydrogen concentration from 0.01% to 100% [5].

8.3 Unexpected Fluctuations in the Output Signal of the Gas Sensor Leading to the Development of Trace Moisture Sensors

A novel semiconductor/MEMS manufacturing technologies developed for a spherical silicon substrate with 1 mm diameter was applied to the fabrication of ball SAW sensors by one of the present authors, NT. Especially, a key technology to fabricate an IDT was the so-called maskless exposure technology that enabled the fabrication of precise patterns on a spherical surface without photomasks. The material of the spherical substrate was single crystal quartz, and the SAW was excited and detected by an IDT fabricated in the Z-axis cylinder. Figure 8.11a shows the entire body of the ball SAW sensor, and Figure 8.11b shows an enlarged view of the IDT made by Au/Cr with 5 μm line width and spacing for 160 MHz excitation. A number of sensors were thus fabricated and used for the detection of specific hydrocarbon molecules in natural gases. While doing so, he noticed that one of the sensors was especially susceptible to the variation of the environment. The phase shift of the output signal fluctuated and it seemed that the fluctuation corresponded to the amount of water vapor content in the natural gases. Now he decided to develop a ball SAW hygrometer [6].

The sensor head and a detecting circuit of the ball SAW hygrometer are shown in Figure 8.12a,b, respectively. The sensor head consisted of the ball SAW sensor, radio frequency (RF) switches, and the first-stage amplifier. The gas inlet and outlet were VCR joints, and a cavity with the ball SAW sensor was provided between

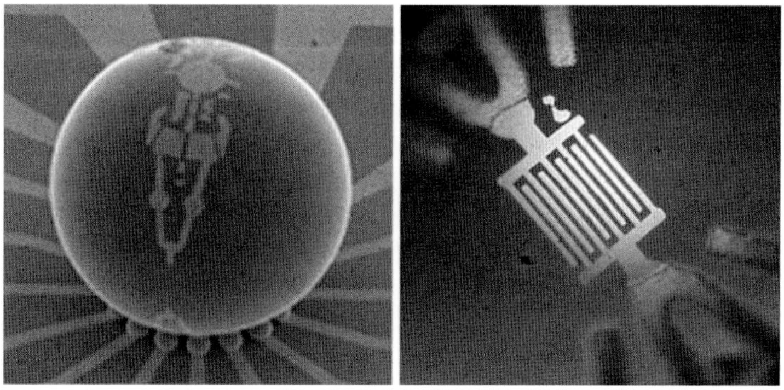

Figure 8.11 Ball SAW sensor with 1 mm diameter: (a) entire body of the sensor, (b) enlarged photograph of IDT. Source: Takeda and Motozawa 2012 [6]. Reproduced with permission of Springer Nature.

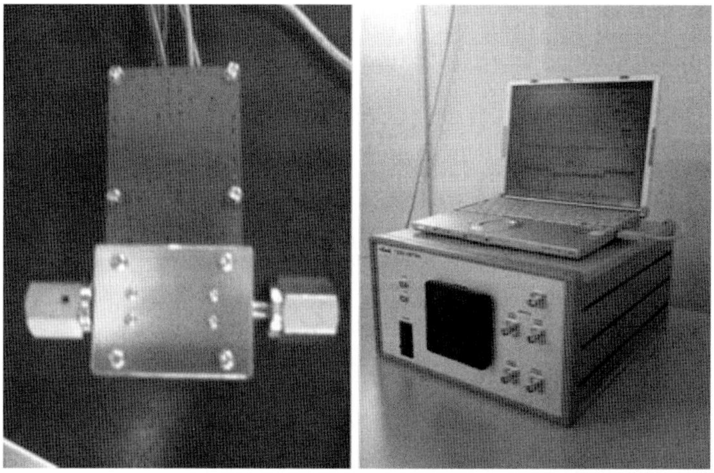

Figure 8.12 Appearance of the ball SAW hygrometer: (a) sensor head, (b) detection circuit box. Source: Takeda and Motozawa 2012 [6]. Reproduced with permission of Springer Nature.

them. The cavity was only about 0.2×10^{-6} m^3 in volume because of the small size of the sensor. The ball SAW sensor was driven by a 0.5 μs burst signal with a center frequency of 160 MHz. After the excitation, the sensor was switched to receiving amplifiers by RF switches. Figure 8.13 shows a typical output waveforms of the amplitude. A cycle in the waveform shows one turn of the burst signal, and it took about 1 μs for each turn. Turn-by-turn, the amplitude decayed, and after 50–100 turns, the amplitude and phase were compared to a reference signal from the RF oscillator.

To evaluate the sensitivity of the ball SAW hygrometer, a 1 μmol mol^{-1} water vapor generation system was provided by using a diffusion tube

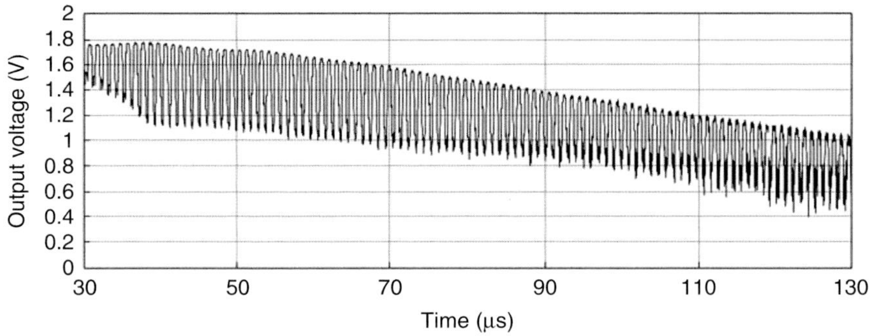

Figure 8.13 Typical output wave form of amplitude from the ball SAW hygrometer. Source: Takeda and Motozawa 2012 [6]. Reproduced with permission of Springer Nature.

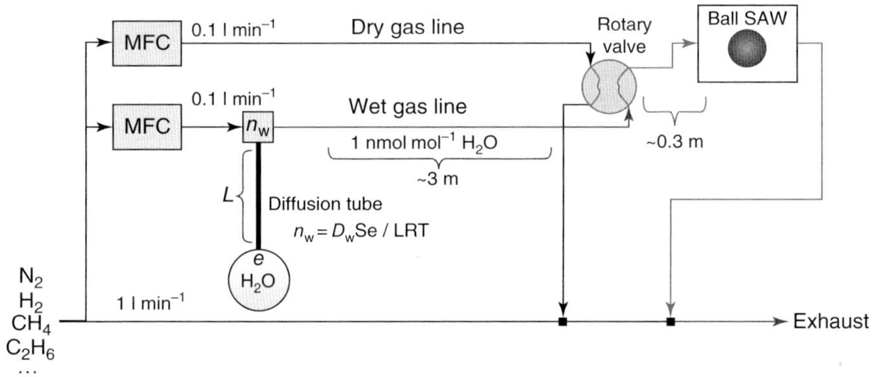

Figure 8.14 Diffusion method of generating gases including 1 μmol mol^{-1} moisture. Source: Takeda and Motozawa 2012 [6]. Reproduced with permission of Springer Nature.

method as illustrated in Figure 8.14. The supplied nitrogen was dried down to <14 nmol mol^{-1} H_2O, which is approximately equivalent to −100 °C in frost point, by an adsorption-type purifier. A wet gas line of 1 μmol mol^{-1} water vapor (approximately equivalent to −76 °C in frost point) and a dry gas line as a reference were switched alternatively to achieve a step-wise injection of water vapor. A rotary valve was used to switch the lines with minimum dead volume. The gas flow rates of both lines were controlled by mass flow controllers. Figure 8.15 shows an alternative injection of 1 μmol mol^{-1} water vapor into nitrogen at four-minute intervals. The rise time from 10% to 90% of the ball SAW sensor against a step-wise change of the gas from dry N_2 to 1 μmol mol^{-1} H_2O in N_2 was ~15 s. Figure 8.16 shows the relationship between the phase shift of SAW and the water vapor concentration. Below 100 μmol mol^{-1} water vapor concentration (approximately equivalent to −42 °C in frost point), it seems that the phase shift was proportional to the square root of the water vapor concentration, which should reflect a physics underneath the observed phenomena. The detection limit of this current proto-type ball SAW hygrometer was 10 nmol mol^{-1}, which was approximately equivalent to −102 °C in frost

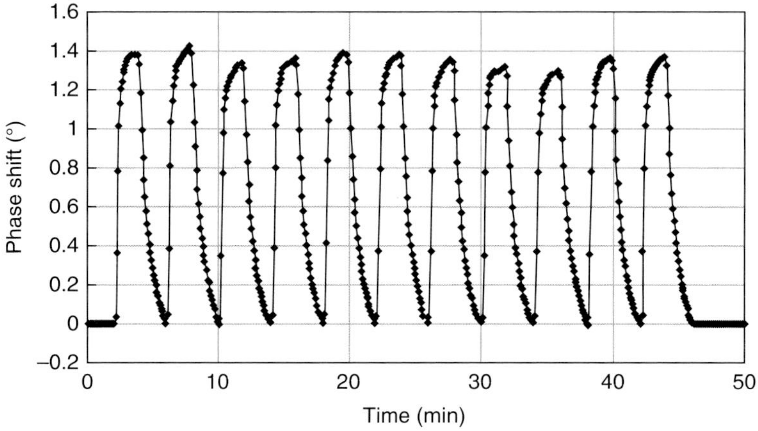

Figure 8.15 Change in delay time measured by the phase shift during the alternative injection of 1 μmol mol^{-1} moisture into N_2 carrier gas. Source: Takeda and Motozawa 2012 [6]. Reproduced with permission of Springer Nature.

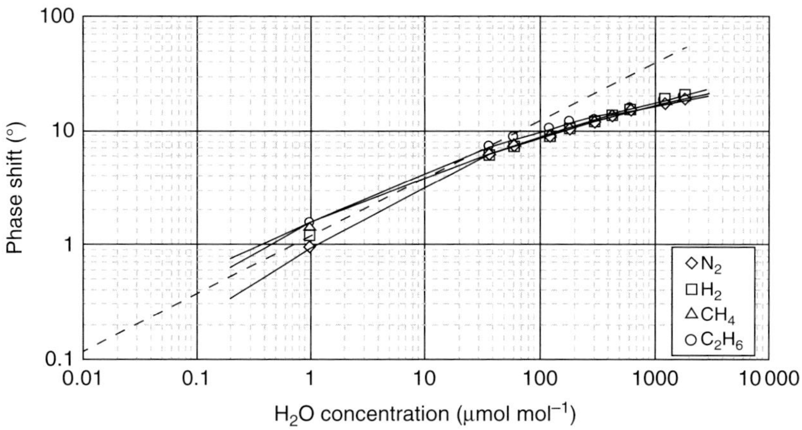

Figure 8.16 Phase shift versus moisture concentration in various carrier gases. The dashed line indicates the square root dependence of the phase shift on the moisture. Source: Takeda and Motozawa 2012 [6]. Reproduced with permission of Springer Nature.

point, because the noise level of the detecting circuit was around 0.1° in phase shift.

It was experimentally proved that the specific ball SAW sensor with a spherical single crystal quartz 1 mm in diameter was good for the very sensitive hygrometer, but more importantly, the same sensitivity was not obtained with any other ball SAW sensors albeit every possible effort was made to reproduce the good result.

8.4 Sol–Gel Silica Film for the Trace Moisture Sensors

It took almost a year, with many experiments and materials characterization, for the present authors to come up with a hypothesis to explain why that specific single crystal quartz ball was sensitive to the water molecules in the carrier gas. It was that a damaged layer of amorphous silicon oxide was formed during the lapping of the surface of the specific quartz ball, and the damaged layer was especially susceptible to water vapor. However, its mechanism was not clear, and it was still difficult to reproduce the "magic" damaged layer by lapping. Hence, we made an assumption that the deposition of amorphous silicon oxide could replace the "magic" damaged layer to detect the trace moisture. As a means of deposition, the sol–gel method was employed [7].

An IDT of 150-nm-thick Cr layer was formed on the surface of $\phi = 3.3$ mm quartz sphere by using subtract method. The IDT was designed so as to generate a fundamental 80 MHz SAW as well as a third harmonic SAW with 240 MHz. By using the two frequencies at once, temperature compensation was achieved as shown in the following.

Amorphous silicon oxide (SiOx) was synthesized by sol–gel method and coated on the surface of ball SAW sensor; 3.47 g tetraethoxysilane (TEOS), 0.75 g isopropanol (IPA), and 1.50 g 0.1 N hydrochloric acid were mixed and stirred by sonication (27, 45, 100 kHz, 60 min). TEOS was polymerized by hydrolysis and resulted in Si*Ox*. After sonication, the mixture was diluted with IPA and 0.5 mass% Si*Ox* solution was obtained. The area of the surface covering the propagation route of the ball SAW was coated with the Si*Ox* solution by spin coating. The condition of the spin coating was 3000 rpm, 20 s. The thicknesses of the SiOx layer were 10–29 nm by interference microscopy. N_2 gas was fed into a bubbler and humidified to −95, −70, −60, −50, −40, −30, −20, −10, 0 °C in frost point. The humidified N_2 gas was supplied to the ball SAW sensor and, in parallel, to a chilled mirror hygrometer used for a reference.

Figure 8.17a,b shows the delay time of 80 and 240 MHz SAWs, respectively. Note that the two curves look similar, specifically with a periodic variation of about 20 min. This periodic variation was the effect of temperature fluctuation due to the air conditioning of the laboratory. The SAW with frequency of 240 MHz had a three times shorter wavelength compared to that of 80 MHz. Hence, it stayed in the three times shallower surface of the sphere during its propagation and experienced about three times stronger effect of the water vapor to the SiOx layer, while the effect of the temperature variation to the delay time was almost same to the both frequencies. By subtracting the delay time for 80 MHz from that for 240 MHz, we obtained a temperature compensated delay time as shown in Figure 8.17c. The humidity range from −95 to −40 °C in frost point was measured with a very small noise level. This experimental result supports the assumption that the deposition of amorphous silicon oxide could replace the "magic" damaged layer to detect the trace moisture. It was shown in Figure 8.17c that the polarity in the delay time change against the moisture

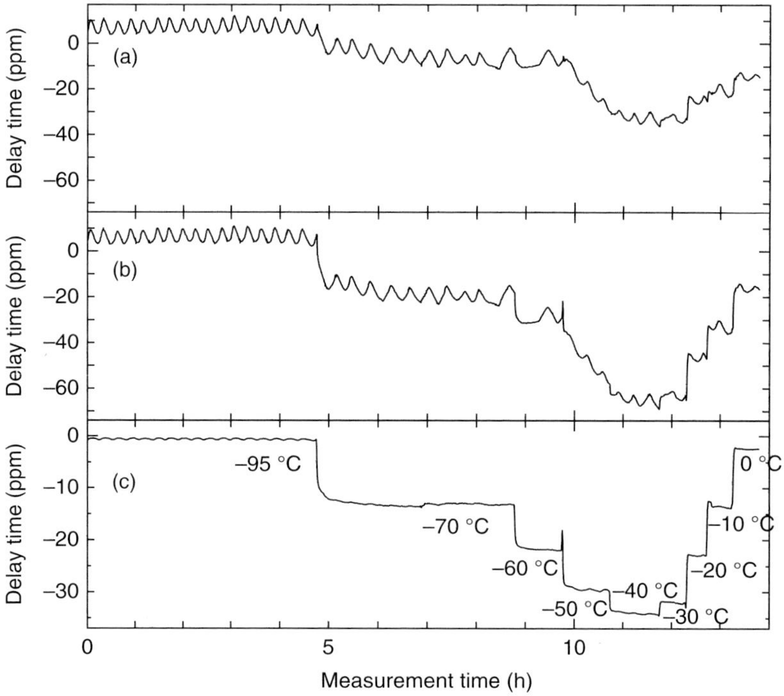

Figure 8.17 Delay time of ball SAW sensor when the frost point was changed from −95 to 0 °C: (a) with 80 MHz, (b) with 240 MHz, and (c) the difference between (a) and (b) [7].

was reversed in the humidity range from −30 to 0 °C in frost point. This may be considered to be a mass loading effect on the SAW propagation by a large amount of water molecules adsorbed in the SiO*x* layer.

8.5 A Thermodynamic Model of Interaction of Water Vapor with Amorphous Silica Glass

Now that the coating of a thin layer of amorphous silica was established to reproduce the accidental invention of a ball SAW hygrometer, it remains an unanswered question how the water molecules interact with the amorphous silica layer at room temperature. Remember that the moisture density is low as 1 $\mu mol\,mol^{-1}$, the phase velocity of the surface acoustic wave increases with the moisture density, which means that the amorphous silica film stiffens with moisture. In the following, we concentrate on this phenomenon, stiffening of the amorphous silica layer due to the ambient moisture in the low-density limit.

It has been known that the water molecules dissociate and the resultant H^+ and OH^- chemically react with the network in silica glass at an elevated temperature as

$$\text{(bulk)} \equiv \text{Si—O—Si} \equiv \text{(bulk)} + H_2O \leftrightarrow 2 \cdot \text{(bulk)} \equiv \text{Si—OH} \tag{8.1}$$

It was pointed out recently [8, 9] that the process could take place even at a room temperature. As an elementary process in the reaction described by the Eq. (8.1), we start with a simple model of dissolving the "solutes" of H^+ and OH^- into a "solvent" of amorphous silica and analyze the thermodynamics of the system. Let H_2O in the ambient gas and the dissociated H^+ and OH^- dissolving in the matrix of silica are in the equilibrium as

$$H_2O \leftrightarrow H^+ + OH^- \tag{8.2}$$

where the chemical potentials, μ, for the constituents satisfy the equation

$$\mu(H_2O) = \mu(H^+) + \mu(OH^-) \tag{8.3}$$

Since we assume that the reaction is a simple dissociation, we propose that for each constituent

$$\mu = \mu_0(T) + kT\ \log([x]) \tag{8.4}$$

where T and $[x]$ are the temperature and the density, respectively, of the constituent x, and k is the Boltzmann constant. T has the same value in all constituents in the equilibrium. By taking the Eq. (8.4) in to the Eq. (8.3), and taking the logarithm assuming

$$[H^+] = [OH^-] \tag{8.5}$$

we have the following Eq. (8.6),

$$[H^+] = \mathrm{Sqrt}\{[H_2O]\}\cdot \exp\{\Delta\mu_0/2kT\} \tag{8.6}$$

where

$$\Delta\mu_0 = \mu_0(H_2O) - \mu_0(H^+) - \mu_0(OH^-) \tag{8.7}$$

Let us look at the dependence on the moisture density. Maintaining the temperature at a constant value and changing the density of H_2O, $[H_2O]$, in the Eq. (8.6), the dependence of $[H^+]$ on $[H_2O]$ is

$$\log\,[H^+] \sim (1/2)\ \log[H_2O] + \mathrm{Const.} \tag{8.8}$$

It seems reasonable to assume that the variation in the phase velocity of the surface acoustic wave is proportional to $[H^+]$ in its low-density limit, then the dependence demonstrated in the previous experiments as shown in Figure 8.16 is in good agreement with the Eq. (8.8) at small values of $[H_2O]$. When $[H_2O]$ becomes larger, the experimental data deviates from the relationship in the Eq. (8.8) and starts to behave according to the familiar mass-loading model. This suggests that we are observing a new phenomenon of penetration of H^+ and OH^- dissociated from water molecules into the amorphous silica layer at room temperature in the very low level of concentration, thanks to the very high sensitivity of ball SAW sensors.

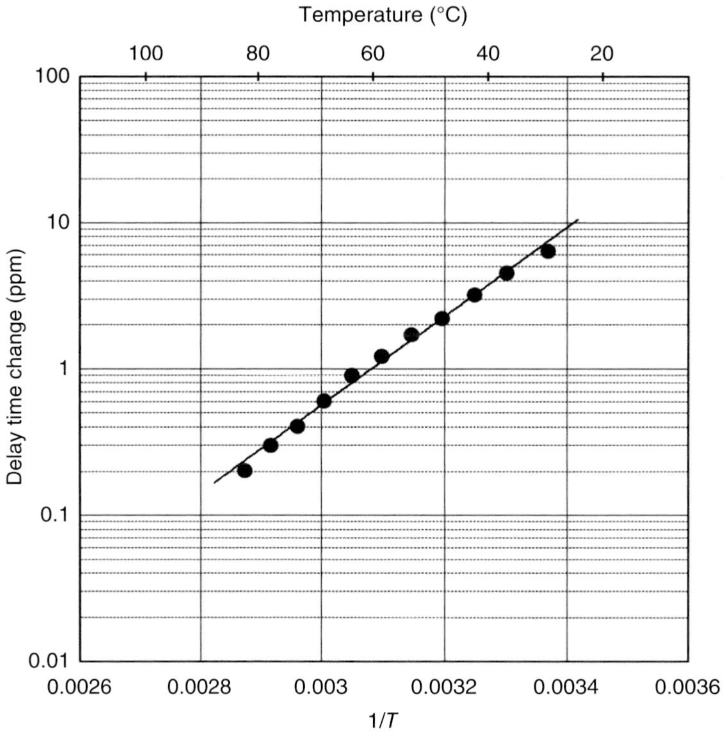

Figure 8.18 Phase velocity change vs temperature at a constant moisture density, 1 μmol mol^{-1}. The *x*-coordinate is T^{-1} (K) and the *y*-coordinate is log (change in the delay time) in parts per million. The change in the delay time is proportional to the phase velocity change when it is small enough.

Next, we examine the dependence on temperature. Maintaining the density of H_2O, $[H_2O]$, and changing the temperature, T, the dependence of $[H^+]$ on T is

$$\log [H^+] \sim \Delta\mu 0/2kT + \text{Const.} \tag{8.9}$$

An experimental plot in Figure 8.18 shows a good agreement with a fitted line according to the first term in the right-hand side of the Eq. (8.9) assuming that $\Delta\mu 0$ has a weak dependence on T. It should be noticed that the experimental data are consistent with a positive value for $\Delta\mu_0$ in the Eq. (8.9). This indicates that H^+ and OH^- would like to be in the amorphous silica layer rather than in the ambient gas when the temperature goes down.

It was shown that a simple model of dissociation of H_2O and dissolution of the resultant H^+ and OH^- into the matrix of amorphous silica leads to a strikingly good agreement with experimental dependences on the moisture density as shown in Figure 8.16 and on the temperature as shown in Figure 8.18.

The simple thermodynamic model presented here is a phenomenological one where we have neglected the mechanism how the H^+ and OH^- react with the network of amorphous silica, which is an important process for the stiffening of the amorphous silica layer. The simulation [8] and experiments [9] suggested that the chemical reaction described by the Eq. (8.1) does occur even at a room temperature. The combination of these studies with our theoretical and experimental work will be an important step toward the understanding of the reaction of water molecules and protons with the amorphous silica surface at room temperature in the very low water molecule density.

8.6 Concluding Remarks

It is remarkable that we have encountered with a puzzling phenomenon and an exciting discovery at each crossroad on a long and winding path from a single experiment in the university laboratory to the commercialization of ball SAW sensors. We are convinced by this experience that one has to be prepared for the unexpected. Do not dismiss something apparently wrong. Think about it many times until finally coming up with a correct explanation. Sometimes it takes years, as in our case of the mechanism of phase velocity change due to the water molecules. We need to propose a hypothesis and validate it by numerical simulation or experiment in order to understand the unexpected phenomena. In many cases, alas, the hypothesis turns out to be ultimately wrong. Then, do not give up and dig it to the level of atomistic models. In taking this painstaking route, it is always important to look into the cumulative knowledge such as reference papers, databases, and simulation codes.

Some reference papers related to this chapter are listed in the following section: on the basic of the laser ultrasonic [10, 11], on the mode analysis of SAWs [12, 13], on the phase velocity scanning of laser for the excitation of SAW [14–16], on the oscillation and wave propagation of elastic sphere [17–19], on the gas sensors by SAW devices [20–26], on the optical phenomena on a sphere [27], on the hydrogen sensors with Pd/Ni alloy [28–32], on the trace moisture sensor based on the laser interferometry [33], on the deposition of amorphous silica [34, 35], on the units of trace moisture [36], on the standard of trace moisture measurements [37], on the novel technologies to manufacture circuitry on spherical objects [38, 39], on the generation of trace moisture [40], on the mass-loading effect on SAW propagation [41], on the chemical reaction of H_2O with silica glass [42], and on the dissolution of gas molecules into metals and amorphous silica [43, 44].

We hope this chapter to be helpful for the researchers exploring a path from discovery to product.

References

1 Strutt, J.W. (1877). *The Theory of Sound*. London: Macmillan and Co.
2 Nishino, H., Tsukahara, Y., Nagata, Y. et al. (1993). Excitation of high frequency surface acoustic waves by phase velocity scanning of a laser interference fringe. *Appl. Phys. Lett.* 62: 2036–2038.
3 Yamanaka, K., Cho, H., and Tsukahara, Y. (2000). Precise velocity measurement of surface acoustic waves on a bearing ball. *Appl. Phys. Lett.* 76: 2797–2799.

4 Tsukahara, Y., Nakaso, N., Cho, H., and Yamanaka, K. (2000). Observation of diffraction-free propagation of surface acoustic waves around a homogeneous isotropic solid sphere. *Appl. Phys. Lett.* 77: 2926–2928.
5 Yamanaka, K., Ishikawa, S., Nakaso, N. et al. (2006). Ultramultiple roundtrips of surface acoustic wave on sphere realizing innovation of gas sensors. *IEEE Trans. UFFC* 53: 793–801.
6 Takeda, N. and Motozawa, M. (2012). Extremely fast 1μmol·mol^{-1} water-vapor measurement by 1 mm diameter spherical SAW device. *Int. J. Thermophys.* 33: 1642–1649.
7 Takayanagi, K., Akao, S., Yanagisawa, T. et al. (2014). Detection of trace water vapor using SiOx-coated SAW sensor. *Mater. Trans.* 55: 988–993.
8 Ewing, C.S., Bhavsar, S., Veser, G. et al. (2014). Accurate amorphous silica surface models from first-principles thermodynamics of surface dehydroxylation. *Langmuir* 30: 5133–5141.
9 Warring, S.L., Beattie, D.A., and McQuillan, A.J. (2016). Surficial siloxane-to-silanol interconversion during room-temperature hydration/dehydration of amorphous silica films observed by ATR-IR and TIR-Raman spectroscopy. *Langmuir* 32: 1568–1576.
10 Scruby, B. and Drain, L.E. (1990). *Laser Ultrasonics Techniques and Applications.* Bristol: Adam Hilger.
11 Royer, D., Dieulesdaint, E., Jia, X., and Shui, Y. (1988). Optical generation and detection of surface acoustic waves on a sphere. *Appl. Phys. Lett.* 52: 706.
12 Viktorov, I. (1967). *Rayleigh and Lamb Waves*, 33. New York: Plenum.
13 Auld, B.A. (1990). *Acoustic Fields and Waves in Solids*, 278. Malabar, FL: Krieger.
14 Yamanaka, K., Nagata, Y., and Koda, T. (1991). Selective excitation of single-mode acoustic waves by phase velocity scanning of a laser beam. *Appl. Phys. Lett.* 58: 1591.
15 Yamanaka, K., Kolosov, O., Nishino, H. et al. (1993). Analysis of excitation and coherent amplitude enhancement of surface acoustic waves by the phase velocity scanning method. *J. Appl. Phys.* 74: 6511.
16 Yamanaka, K. (1997). Precise measurement in laser ultrasonics by phase velocity scanning of interference fringes. *Jpn. J. Appl. Phys. Part 1* 36: 2939.
17 Lamb, H. (1882). On the vibrations of an elastic sphere. *Proc. Lond. Math. Soc.* 13: 189.
18 Gilbert, F. (1971). Excitation of the normal modes of the earth by earthquake sources. *Geophys. J. R. Astron. Soc.* 22: 223.
19 Aki, K. and Richards, P.G. (1980). *Quantitative Seismology: Theory and Methods.* New York: W.H. Freeman.
20 Mcgill, R.A., Chung, R., Chrisey, D.B. et al. (1998). Performance optimization of surface acoustic wave chemical sensors. *IEEE Trans. Ultrason. Ferroelect. Frteq. Contr.* 45 (5): 1370–1380.
21 Grate, J.W., Nelson, D.A., and Skaggs, R. (2003). Sorptive behavior of monolayer-protected gold nanoparticle films; implications for chemical sensing. *Anal. Chem.* 75 (8): 1868–1879.

22 Rapp, M., Binz, D., Kabbe, I. et al. (1991). A new high-frequency high-sensitivity SAW devices for NO_2 gas-detection in the sub ppm range. *Sensors Actuators B Chem.* 4 (1–2): 103–108.

23 D'Amico, A., Palma, A., and Verona, E. (1982). Hydrogen sensor using a palladium coated surface acoustic wave delay line. 1982 IEEE Ultrasonics Symp. Proc. (San Diego, CA, 1982), New York: IEEE, p. 308.

24 Jakubik, W.P., Urbanczyk, M.W., Kocowski, S., and Bodzenta, J. (2002). Bilayer structure for hydrogen detection in a surface acoustic wave sensor system. *Sensors Actuators B Chem.* 82 (2–3): 265–271.

25 Anisimkin, V.I., Ktelyanskii, I.M., Verardi, P., and Verona, E. (1995). Elastic properties of thin film palladium for surface acoustic wave (SAW) sensors. *Sensors Actuators B Chem.* 23: 203–208.

26 Ymanaka, K., Ishikawa, S., Nakaso, N. et al. (2004). Ball SAW devices for hydrogen gas sensor. *Proc. Ultrason. Symp.* 299–302.

27 Tsai, C.S., Chen, W., Le, P., and Tsai, C. (2001). Acousto-optic interactions and devices in a spherical waveguide. *J. Opt. A* 3 (4): S46–S53.

28 Lewis, F.A. (1967). *The Hydrogen-Palladium Systems*, 138. London: Academic.

29 Hughes, R.C. and Schubert, W.K. (1992). Thin films of Pd/Ni alloys for detection of high hydrogen concentrations. *J. Appl. Phys.* 71 (1): 542–544.

30 Okuyama, S., Umemoto, K., Okuyama, K. et al. (1997). Improved response time of Al-Al_2O_3-Pd tunnel diode hydrogen gas sensor. *Jpn. J. Appl. Phys.* 36 (11): 6905–6908.

31 Leisure, R.G., Nygren, L.A., and Hsu, D.K. (1986). Ultrasonic relaxation rates in palladium hydride and palladium deutride. *Phys. Rev. B* 33 (22): 8325–8328.

32 Richards, P.M. (1986). Vacancy and labeled-particle hopping with interactions. *Phys. Rev. B* 33 (5): 3064–3075.

33 Funke, H.H., Grissom, B.L., McGrew, C.E., and Raynor, M.W. (2003). Techniques for the measurement of trace moisture in high-purity electronic specialty gases. *Rev. Sci. Instrum.* 74: 3909.

34 Pfeifer, K.B. (1995). Heat of adsorption and thin film surface area studies of a silica sol-gel film exposed to HCl and H_2O. *Langmuir* 11: 4793.

35 Pfeifer, K.B., Kelly, M.J., Guilinger, T.R., Peterson, D.W., Sweet, J.N., and Tuck, M.R. (1994). Development of solid state moisture sensors for semiconductor fabrication applications. Proceedings of the International Conference on Microcontamination, San Jose, CA (October 4–6), 87–97.

36 The Society of Polymer Science, Japan (1968). *Handbook of Materials and Moisture*, 144. Tokyo: Kyoritsu Shuppan Co. Ltd.

37 Abe, H. (2009). A marked improvement in the reliability of the measurement of trace moisture in gases. *Synthesiology* 2: 223.

38 Ishikawa, S. (1999). Spherical-shaped semiconductor integrated circuit, US Patent 5, 955,776.

39 Ishikawa, S. (2001). Maskless photolithography system that digitally shifts mask data responsive to alignment data. US Patent 6, 251,550.

40 Abe, H. and Kitano, H. (2006). Development of humidity standard in trace-moisture region: characteristics of humidity generation of diffusion tube humidity generator. *Sens. Actuators A* 128: 202.

41 Skeie, H. (1970). Electrical and mechanical loading of a piezoelectric surface supporting surface waves. *J. Acoust. Soc. Am.* 48: 1098.
42 Tomozawa, M. (2006). Water in glasses. *New Glass* 21 (3): 9. [in Japanese].
43 Takeda, N., Oizumi, T., Tsuji, T. et al. (2013). Deep sub micoro mol/mol^{-1} water vapor measurement by dual ball saw sensors for temperature compensation. *Proc. TEMPMEKO 2013* HMC-1: 173.
44 Sieverts, A. (1929). Absorption of gases by metals. *Z. Metallk.* 21: 37. [in German].

9

From Sensors to Low-cost Instruments to Networks: Semiconducting Oxides as Gas-Sensitive Resistors

David E. Williams

School of Chemical Sciences, University of Auckland, Private Bag 92019, Auckland 1142, New Zealand
MacDiarmid Institute for Advanced Materials and Nanotechnology, PO Box 600, Wellington 6140, New Zealand

This chapter illustrates not only the path from basic science to commercial application but also the reverse and, indeed, the intimate and interesting interaction of the two, using semiconducting oxides as gas-sensitive resistors as an example of the journey. This chapter is about a solid-state device interacting with a gas, and not at first sight about electrochemistry or electrochemical engineering. However, concepts from these disciplines underpinned and enabled the journey. This chapter discusses the basic science of reactions at an interface and the applied science of a practical device and its utilization in the field. Themes that recur throughout electrochemistry as a discipline are found here: charge transfer at oxide surfaces; surface vs bulk composition and consequences for reactions and measurement; consequences of minor phases and details of microstructure; interaction of microstructure and mass-transport-determining consequences for the measured signal; and instrument design and measurement technique.

This chapter does not pretend to be a comprehensive review. Rather, it describes a personal journey that illustrates how these basic scientific threads unite along the path from discovery to product.

9.1 Overview

Gas-sensitive resistors based on heated semiconducting oxides (Figure 9.1) are in many ways "cheap and cheerful": they are relatively easy and inexpensive to make and the signals are large and easily measured. In the literature, the lack of selectivity, interfering effects of water vapor, and the lack of signal stability over time have repeatedly been remarked, from the earliest reports [1]. In recent times, because the devices can be obtained very inexpensively, the signal is easy to measure and now they can be made with very low power consumption using micro-machined substrates, they have become popular for "citizen science" deployments to indicate variations in air quality (e.g. http://airqualityegg.com/). The signals obtained are indicative but not reliable or quantitative. This chapter,

Electrochemical Engineering: The Path from Discovery to Product,
First Edition. Edited by Richard C. Alkire, Philip N. Bartlett, and Marc T. Koper.

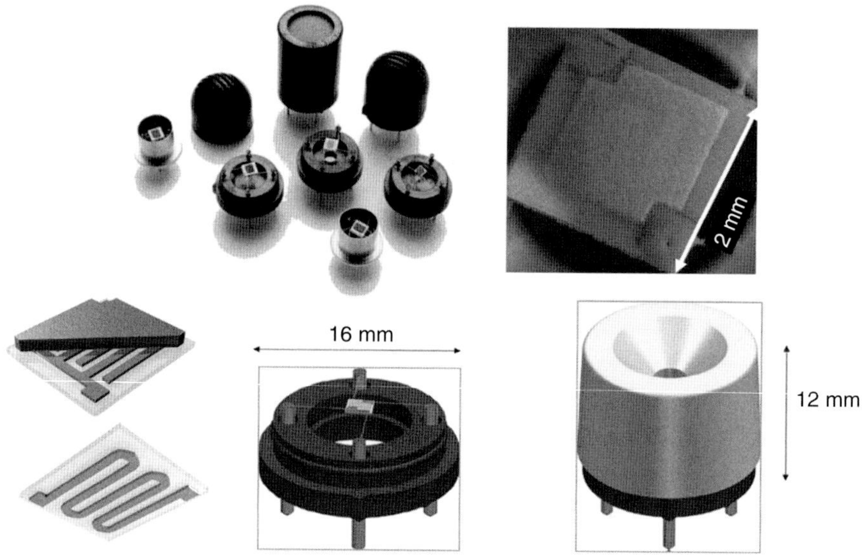

Figure 9.1 Gas-sensitive resistors based on semiconducting oxides. A Pt heater is printed on one side of an alumina ceramic tile and an interdigitated electrode on the other side. A porous layer of oxide is printed over the electrodes. Sensors are mounted suspended on their connecting wires and protected by a suitable housing. In other forms, sensors with low power consumption and fast thermal response are prepared using micro-hotplates by deposition of heater, electrodes, and sensing material onto a thin membrane of Si_3N_4 prepared by microfabrication on a Si wafer.

however, shows how to obtain reliable and quantitative information from these simple devices, by proceeding from the fundamental science through the science of device operation and instrument construction, optimized to take advantage of the outstandingly useful characteristics while not being swamped by the disadvantages, and finally fundamental study of the characteristics of air quality networks to develop tools that build confidence in data reliability and that optimize management of the network for cost and reliability. Following this overview, there are three sections: multiscale modeling [2] and its use to optimize device construction, instrument design and sensor operation, and network design.

The basic ideas concerning charge transfer between a semiconductor surface and species adsorbed from the ambient atmosphere resulting in a change of surface potential with consequent effects for the kinetics of adsorption were described in the 1950s [3–5]. In 1962, Seiyama [6] described the use of a heated thin film of ZnO as a gas chromatography detector. Simultaneously, Naoyoshi Taguchi discovered the properties of heated porous pellets of SnO_2 as sensitive detectors of the presence of trace explosive gases in air, through changes in electrical conductivity. Taguchi's study [7] is an absorbing epic of a lone inventor pursuing an idea through systematic, careful, and comprehensive empirical experimentation until a useful and commercially successful product

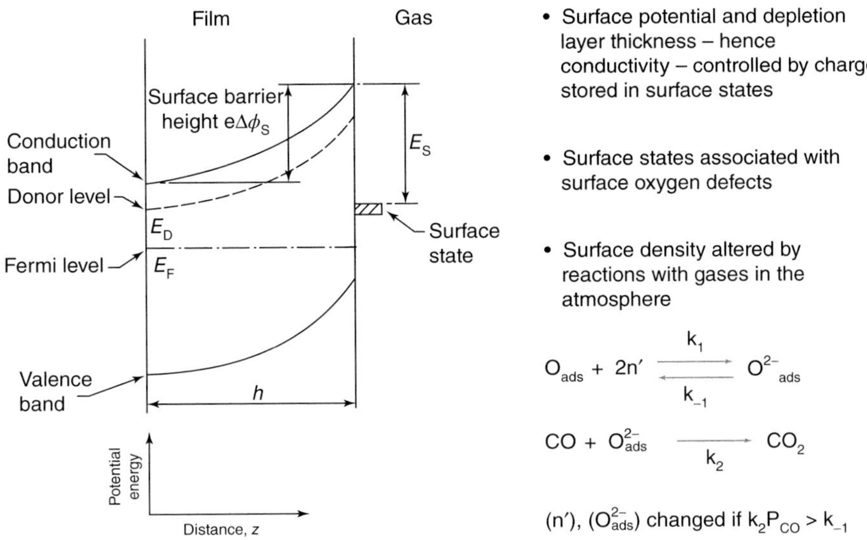

Figure 9.2 Schematic of an electronic band structure near the gas–solid interface of a semiconducting oxide. The basic concepts used to describe the response are listed on the right, including a phenomenological reaction scheme. Source: McAleer et al. 1987 [9]. Reproduced with permission of Royal Society of Chemistry.

is obtained. Windischmann and Mark [8] developed the idea that the charge carrier concentration near the surface of an oxide is controlled by charge transfer to adsorbed oxygen, whose surface coverage was changed by reaction with other gases in the ambient (Figure 9.2). This phenomenological model, an electrochemical steady-state formalism of changes in occupancy of a surface state causing changes in mobile carrier density, continues as the basic understanding of the response mechanism. The general characteristics of the response (Box 9.1) were established early on and the continuing concerns of stability, cross-sensitivity, and effects of changing humidity were noted [1].

Box 9.1

General Features of Gas-Sensitive Resistors Based on Semiconducting Oxides

- Porous body or thin film
- Temperature typically at 400 °C
- In an environment containing oxygen (air)
- Large conductivity changes caused by the presence of trace gases in the atmosphere.
  - Initially discovered for propane, carbon monoxide, hydrogen, and methane, soon extended to SO_2, H_2S, NO_2, and many others.
- Magnitude of conductivity change depends on the gas present, oxygen partial pressure, and temperature.

Our sensor development was driven by a perception from users that there was a market need for an alternative to SnO_2 for toxic and inflammable gas detection, which was less sensitive to the interfering effects of water vapor, more selective, and less susceptible to poisoning by silicones. It started with a simple question: was the behavior of SnO_2 and ZnO – the only examples in the literature at that time – in some sense unique; were the surface reactions such that by alteration of oxide composition, would it be possible to confer selectivity? Because there was no a priori basis for answering these questions, a screening experiment was set up, designed to test the conductivity response of as many oxides as possible to a range of important gases in low concentration in air over a range of temperature. The important gases should also include the traces of those of commercial interest, a reduced oxygen partial pressure and a switch between wet and dry air; that required an automated experiment to be designed to uncover phenomena rather than making detailed studies. The Commodore PET® personal computer had just become available. It had an IEEE interface, an RS232 serial interface, and a built-in BASIC programming language. With this, it was easy to set up the required measurement and control system. We accumulated the data from a number of rigs running simultaneously on a DEC PDP 11/34 computer that had a time-sharing operating system, by making the PET pretend to this that it was a keyboard. We scanned simple oxides and then complex oxides chosen from different structural families, particularly families that were known for catalytic properties or which had a rich defect chemistry within which we could explore substitutions of variable valency metals: perovskite, pyrochlore, spinel, for example [10]. We had a simple standardized preparation, just pulverizing the constituent oxide powders together then firing ("beat and heat" method) and tested everything, as pressed, sintered pellets, regardless of whether the product was a single phase or not. After several hundred oxides, we concluded that gas sensitivity of conductance was a common property: all that was required was that in the temperature range 200–500 °C, the oxide had a resistivity that was not too small so that changes were easily seen, and not so large that measurement was not possible. Oxides differed in detail in the relative response to the different gases [11–29]. The general classification of "n" and "p" type was established as an organizing principle ("n" type giving conductivity increase in reduced partial pressure of oxygen, and in response to small concentrations in air of gases such as hydrocarbons, H_2, NH_3, H_2S, and CO generally considered as reducing, and conductivity decrease in response to small concentrations in air of gases generally considered as oxidizing, such as Cl_2 and NO_2, "p" type giving the opposite) [30]. Some materials, however, switched between "n" and "p" type with change of gas concentration or change of temperature: Figure 9.3. We showed that the effects could be rationalized using a simple effective medium model for the porous system [2, 31, 32] (Section 9.2.1). We deduced that this common pattern of behavior implied a surface reactive species that was common to all oxides: consistent with the simple ideas that the response was kinetically determined by reaction of the target gas with a surface adsorbed oxygen that functioned as a surface trap for electrons [30]. These ideas have found recent support in infrared spectroscopy studies [33].

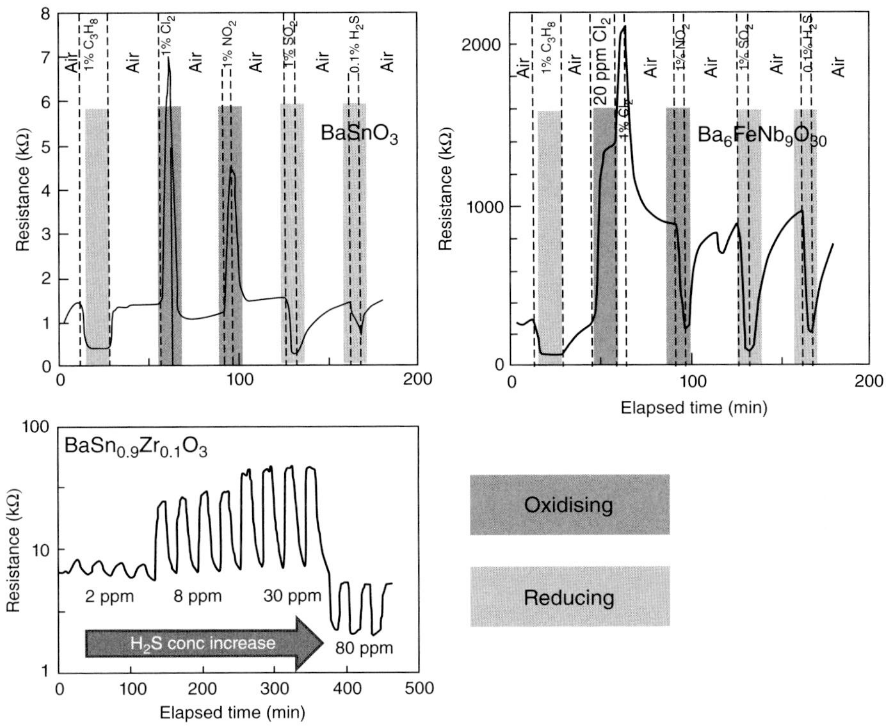

Figure 9.3 Screening of oxides for gas sensor behavior. Oxides show regular patterns of behavior. Gases in air may be classified as reducing or oxidizing toward the surface and oxides as "n" or "p" type. Behavior may switch with change of temperature or gas concentration. Source: Williams and Moseley 1991 [31]. Adapted and reproduced with permission of Royal Society of Chemistry.

In parallel, in 1987–1989, we published detailed studies of the sensor response of porous SnO_2 and other oxides [9, 10, 28, 29, 34–38] that built on the ideas of Windischmann and Mark, of Morrison and of Heiland, in which we described the effects of surface-trap- and surface-barrier-controlled conductivity in porous sensors, the importance of microstructure and how it might be altered by dopants that act as sintering aids or sintering inhibitors, and the effects of surface decoration by catalysts, including the effects of charge exchange with the catalyst, the effects of heterojunctions between oxides in composites of different oxides (recently extended by others [39, 40]), and the effects of changes of gas composition within a porous pellet on the way electrical current flows through the pellet and hence on the observed signal [9, 36, 37, 41] (Section 9.2.1). In contrast to classical surface science studies on well-defined single crystal surfaces (e.g. [42, 43]), we focused on porous polycrystalline materials. By systematic variations of composition, e.g. [44, 45], and of heat treatment temperature, and by titrating the surfaces against "poisons," we showed that sites that mediated the electrical response to water vapor could be distinguished from those that mediated the response to other gases [46]. The relative response to water vapor and other target gases could be hugely modified by surface segregation of dopants in the oxide [47] (Section 9.2.2). In this way, the issues of humidity response – the signal due to variation of water vapor pressure is in practice, one of the major interferences limiting the quantitative application of these devices – were resolved; $Cr_{1-x}Ti_xO_3$ (CTO) [48, 49] was introduced into commercial production for gas sensors (Section 9.3).

During this development, the exceptionally large conductance response to very low concentrations of ozone of heated porous films of WO_3 was noted [2, 50, 51] (Section 9.3.2). Ozone is widely used as a disinfectant, with associated health and safety consequences. Measurement of ozone using low-cost, light-weight devices is also of importance for balloon-borne measurements of the atmosphere [52] and for urban air quality determination. Hence, there was a market driver for sensor and instrument development. The issues for practical development were drift of the zero ozone signal, humidity effects, and the slow response. The first key observation was that the behavior of the porous-sensing element could be well described by a simple parallel combination of two resistors, one representing the "bulk," which contributed to the drift, and the other representing the "surface," which provided the signal – an old idea [1], which we had elaborated to describe the effects of microstructure on response quite generally [53]. If a zero ozone signal could be obtained, then the drift could be canceled, as could the humidity dependence [50]. We deduced and proved that, if the sensor temperature was stepped to a sufficiently high temperature, then the surface was "reset" such that a step back to a lower temperature in the presence of ozone gave a reproducible transient response. Thus, a measurement at a fixed time after the temperature step – essentially a measurement of the initial kinetics of the response – gave a reliable measurement of the ozone concentration. The signal is sensitive to the flow rate over the sensor element because ozone decomposes on heated surfaces. Hence, a zero signal could be obtained by simply dropping the air flow rate to zero [54]. In typical urban air, the signal to ozone is so much larger than that due to other gases present in the atmosphere that the effects of interferences are negligible. As a consequence, highly reliable, low-cost, small ozone measurement instruments are now manufactured for monitoring of disinfection and high-quality, low-cost instruments for urban air quality measurement are widely sold [54–57]. Methods for management of large-scale networks of air quality monitoring instruments, and particularly for data validation automatically to detect instrument drift, are being developed [58] and have led to further commercial development [59] (Section 9.3.5).

9.2 Basic Science of Semiconducting Oxides as Gas-Sensitive Resistors

9.2.1 Multiscale Modeling of Gas-Sensitive Resistors

9.2.1.1 Introduction

In gas-sensitive resistors, there are phenomena at a range of different size scales that interact to result in the observed signal. These are as follows:

- The surface reactions that develop the signal, where the size scale is that of the charge depletion or accumulation layer in the surface and of surface crystallographic defects that function as active sites both for developing the electrical response and for mediating the reactions of species from the gas phase.
- Current flow, percolation of the charge carriers in the solid, diffusion, and reaction in the gas phase, where the size scale is that of the microstructure – the particles making up the sensor and their size relative to the surface charge depletion/accumulation layer. At this size scale, agglomerates of particles and the size of any junctions between them in relation to the surface charge depletion/accumulation layer are important, as is the porosity.
- Averaged effects of the composition, where the device can be treated as a uniform effective medium and the size scale is that of the device as a whole – the thickness of the sensing layer and the geometry of the measurement electrodes, with the effective medium parameters dependent on the details of the microstructure.

For understanding the phenomenology and developing effective sensor devices, the most productive approach is to start at the most coarse scale. Thus, at the first level, the whole device is considered as an effective medium: that is, the effects of the surface reactions are represented as lumped parameters and the details of the microstructure are simply represented as effects upon these lumped parameters (Section 9.2.1.2). At this level, the effect of the gas on the conductivity is represented as a change in a volumetric trap state density: no details of the variation of charge carrier concentration with position within individual grains of the materials are considered. Formally, this is equivalent to considering the behavior as a surface-trap-limited conductivity rather than a surface-barrier-controlled conductivity [9] and is clearly an approximation. However, it turns out to be very useful in helping to systematize observations. At this level, also, reactions on the internal surface of the device are considered as uniformly distributed. The reactions change the local composition and hence the local conductivity. How this is reflected in the measurement depends on the construction of the device: specifically, for a printed film, on the spacing of the electrodes in relation to the thickness of the film, on the presence of other layers printed on top of the sensor element, and on the distribution through the sensing layer, or on top of the sensing layer of any gas reaction catalysts [60–66]. The lumped parameters include an average gas diffusivity and an average surface reaction rate constant: the details here connect to the details of modeling of the microstructure. Because the measured result depends on these averaged parameters, this modeling reflects the effect of changes in the averaged parameters. It led to the invention of device structures that are self-diagnostic, detecting changes in microstructure over time or detecting surface poisoning through their effects on gas diffusion and reaction through the structure changing the way current flows through the device. These changes in current flow can be detected by the electrode geometry dependence of the measured signal [60, 61] (Section 9.2.1.3).

At the second level of modeling (Section 9.2.1.4), the detailed effects of the microstructure of the material are considered: how does the current flow through the microstructure; what is the effect of agglomerates and particularly of the grain size of the material in relation to the surface space charge dimension [67]; can the behavior be represented by simple equivalent circuits [53, 68]; what is the effect of porous catalytic additives? At this level, the effect of microstructure – porosity and pore size distributions, particle agglomeration, and internal surface area – on the diffusion and reaction of gases, and hence on the measured signal, can be considered.

At the third level, the details of surface reactions are explored, particularly how changes in surface composition might modulate the response to different gases relative to one another [44, 47, 49, 69] (Section 9.2.2). The signal to water vapor is of particular interest. Understanding the effects of changes in surface composition deriving from surface segregation of components of the sensor material leads to improvements in performance through improvements in processing. The effect of heterojunctions – between particles of different composition – and the effect of surface decoration by nanoparticles of different materials come in at this level. Particularly interesting phenomena that can lead to new types of device have been found. The final level is consideration of the atomistic details of surface defect and reaction models (Section 9.2.3). Computational models of surface structure appear at this level [49, 70–79] as a means of exploring hypotheses concerning the surface reactions. They are satisfying in helping to understand the basic science of the surface reactions and their connection to the electrical signal.

9.2.1.2 Effective Medium Model 1: Rationalization of Composition Effects on Response

In this model, the oxide is treated as a uniform medium. All effects of microstructure are blurred out. Charge carriers are in equilibrium between the conduction and valence bands, donor and acceptor states within the band gap, and states on the grain surfaces. These surface states are associated with chemisorbed oxygen and hydroxide, but all details to their nature are ignored, other than general statements concerning their occupancy and energy within the band gap of the oxide. The surface states are treated as uniformly distributed throughout the solid: this is the effective medium approximation, ignoring the details of the microstructure. It is assumed that the grains are so small that the bands are essentially flat within the grains of the porous solid, a condition that applies when the grains are sufficiently small but which is clearly an approximation. The gas sensitivity is assumed due to a variation in the concentration of surface states. With the bulk donor states just below the conduction band, the bulk acceptors just above the valence band, and the surface states in the gap well separated in energy from these, the gas-sensitive conductance change relative to the baseline is [2]

$$S = \frac{d(\sigma/e)}{dN_S}\left(\frac{1}{\sigma_0/e}\right) \approx \frac{1}{N_S}\left(\frac{\mu_p K_2 N_S/(N_A+N_D) - \mu_e K_1((N_A+N_D)/N_S)}{\mu_p K_2 N_{S,0}/(N_A+N_D) - \mu_e K_1((N_A+N_D)/N_{S,0})}\right) \tag{9.1}$$

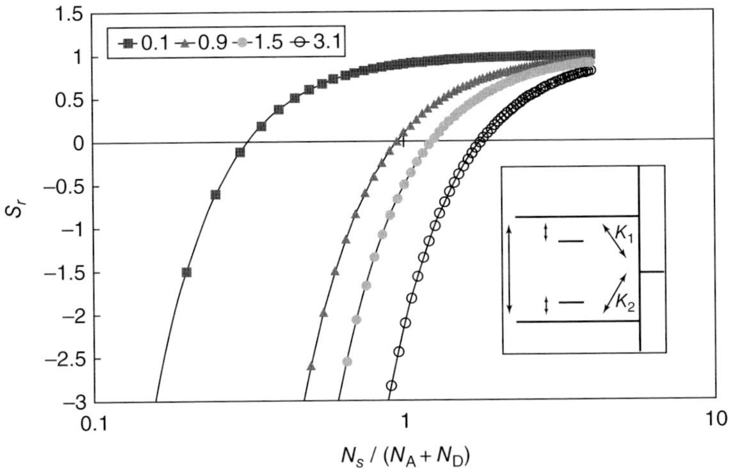

Figure 9.4 Effective medium model; surface trap-limited conductivity. Dependence of relative response, S_r on the ratio of surface trap state concentration, N_S, to bulk donor, N_D, and acceptor, N_A state concentration, for different values of $\alpha = \mu_e K_1/\mu_p K_2$ (legend). The inset defines the equilibrium constants, K_1 and K_2, of electrons between the surface states and the conduction and valence bands of the solid. S_r positive corresponds to "n"-type response to a reducing gas and S_r negative to "p"-type response. Here μ_e and μ_p respectively denote the electron and hole mobility.

Symbols are defined on figure 9.4. The subscript 0 denotes the baseline values. The form of the behavior (Figure 9.4) is shown by

$$S_r = \left(1 - \frac{\alpha}{\overline{N}_S^{\,2}}\right) \tag{9.2}$$

where $\overline{N}_S = N_S/(N_A + N_D)$ and α is defined in Figure 9.4. This formulation rather nicely describes the patterns of behavior for a wide range of different oxides responding to gases like reactive hydrocarbons, hydrogen, or carbon monoxide [31]. The basic model is the same for all: reaction with a gas decreases the surface density of states, as formulated phenomenologically in Figure 9.2. The result is a decrease or increase in conductivity dependent on the magnitude of α, or a change in the sign of conductivity response from a decrease to an increase as gas concentration increases and hence N_S decreases. The magnitude of $\overline{N}_S$ can be manipulated by altering the bulk donor or acceptor state densities, and that of α by altering the position of these levels within the band gap. Systematic studies of the effects of cation substitutions in solid solution series thoroughly justified this formulation [44, 45, 69, 80]. An example is given in Figure 9.5 for a series of solid solutions between wolframites, MWO_4, and SnO_2, on the one hand, and TiO_2, on the other hand, prepared and tested in the form of porous pellets. The substitutions here are charge-neutral, so the only effect is the effect of localized states introduced into the band gap from the *d*-electron states of the metal M. The transition from n-type (conductance increase) response to CO

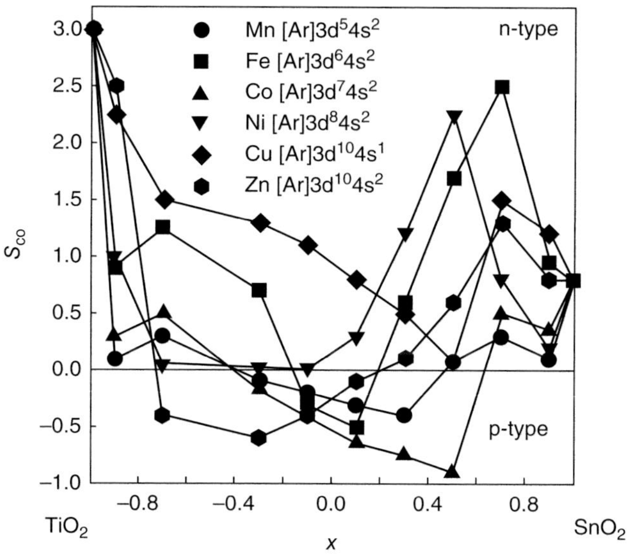

Figure 9.5 Sensitivity $\Delta\sigma/\sigma_0$ of conductivity of sintered pellets of compounds in the solid solution series $(MWO_4)_{1-x}$ $(M'O_2)_x$ to 1% carbon monoxide in dry air, measured at 400 °C M = Mn, Fe, Co, Ni, Cu, Zn; M′ = Sn, Ti. Source: Dusastre and Williams 1999 [44]. Reproduced with permission of Royal Society of Chemistry.

in air to p-type (conductance decrease) varies systematically with composition, and further systematically as the d-state occupancy changes.

Another interesting result of these studies is that alteration of the cation has a marked effect on the response to water vapor, which is quite different to the effect on the gas response. The substitutions with Fe, Co, and Ni significantly diminished the signal to change of water vapor pressure relative to the effect caused by the presence of the reactive gas [44]. A reasonable deduction is that the surface states mediating the response to water vapor are different from those mediating the response to the gas and can be manipulated independently by changing the material composition. This is a deduction of significant practical interest.

9.2.1.3 Effective Medium Model 2: Diffusion–Reaction Effects on Response; Effects of Electrode Geometry and "Self-Diagnostic" Devices

Theoretical Development An oxide surface at an elevated temperature in air can be a combustion catalyst for traces of oxidizable gases such as hydrogen, carbon monoxide, and reactive hydrocarbons. The oxide surface may also be decorated with other oxides or metal catalysts such as nanoparticulate Pt. The result might be a complete combustion of the target gas or a transformation into partial oxidation products. The oxide itself has local conductivity that is dependent on the local composition of the gas within the porous solid, which varies with position away from the exterior interface of the sensor as a consequence of the surface reactions. The interaction of diffusion and reaction gives rise to a position-dependent conductivity. Variations in microstructure of the device also give rise to a spatially variable conductivity. If the gas concentration profile changes – for example, due to changes in microstructure or changes in surface reactivity that might be caused by poisoning of reactive sites on the surface – or if the microstructure changes – for example, cracks develop in the material – then the position dependence of the conductivity will change. This position-dependent conductivity can be sampled by appropriate electrode geometry. Modeling is accomplished by the following:

1. Specifying the surface chemical reactions and their rates. The rate constants are treated as volume-averaged, although their dependence on the internal surface area of the porous solid needs to be acknowledged;
2. Specifying the dependence of the conductivity, κ, of the material upon the concentration of each of the gaseous species, and how the contributions add. The conductivity is volume averaged;
3. Solving the diffusion–reaction equation for the chemical species (gas partial pressure, P, reaction rate, $R(P)$:

   $$D\nabla^2 P - R(P) = 0 \tag{9.3}$$

   The diffusion coefficient is treated as volume averaged, but its dependence on the porosity and tortuosity of the pore space of the solid needs to be acknowledged;
4. Solving the Laplace equation for the electric potential:

   $$\nabla \cdot (\kappa \nabla V) = 0 \tag{9.4}$$

5. Calculating the electric field at the electrodes, hence the current, hence the conductivity.

The procedure is general and numerical solution is straightforward [81].

However, simple model cases are particularly useful for showing the important characteristics [81]. They are chosen because the electric field can be calculated analytically and are evaluated by ignoring the effect of the finite thickness on the current flow. Figure 9.6 shows one implementation of the idea, for a planar device comprising a porous sensing layer on an impervious insulating substrate, with electrodes on the substrate underneath the sensing layer that are semi-infinite. The length and width of the device are much higher than the sensing layer thickness, so the electric field varies only with position $X = x/h$ above the substrate. For semi-infinite planar electrodes of negligible thickness, the electric field, E, across the plane bisecting the electrode gap, as a function of X depends on the ratio $A = a/h$, of the half-width of the electrode gap to the thickness of the layer [82]:

$$E \approx \frac{V}{\pi X\sqrt{1 + (A/X)^2}} \tag{9.5}$$

The current density across this bisecting plane is

$$j = \kappa E \tag{9.6}$$

The ratio of the current density at the base of the layer ($X = 0$) to that at the top of the layer ($X = 1$) for uniform conductivity through the layer is thus

$$\frac{j(0)}{j(1)} = \frac{\sqrt{(1 + A^2)}}{A} \tag{9.7}$$

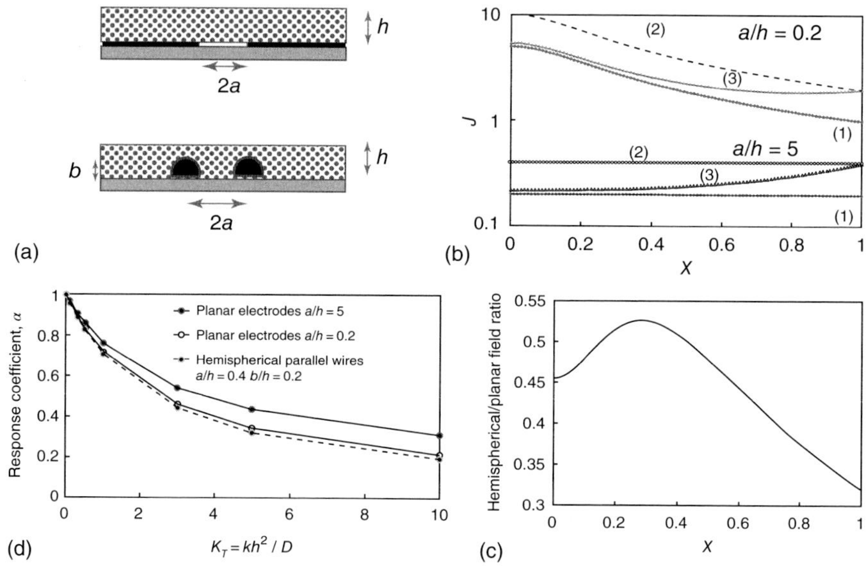

Figure 9.6 (a) Two model geometries; porous oxide thickness *h* on impervious substrate: semi-infinite coplanar spacing 2*a*, and hemispherical electrodes radius *b* and center distance 2*a*; (b) current density through the layer against position, *X*, along the symmetry plane between the electrodes, coplanar electrodes, two spacing with (1) no gas; (2) gas with no concentration profile and $K_p = 1$; (3) gas with concentration profile for first-order combustion, $K_T = 10$, $K_P = 1$; (c) electric field ratio for planar/hemispherical electrodes against position, *X*, along the symmetry plane between the electrodes; (d) response coefficient against rate constant for first-order decomposition, illustrating the effect of electrode geometry.

The ratio is 1.4 for $A = 1$ and 10 for $A = 0.1$. The current becomes strongly focused at the base of the layer as the ratio a/h decreases. The measured conductance is the integral:

$$\sigma \approx \frac{1}{\pi} \int_0^1 \frac{\kappa}{X\sqrt{1 + (A/X)^2}} \, dX \tag{9.8}$$

The measured result thus depends strongly on A. If $A = a/h$ is small, then the measurement is dominated by the contribution of the part of the layer close to the electrodes, at the base of the sensing layer while if a/h is large then the measurement samples the whole layer. The result, in the presence of a gas, depends on whether the gas (or its reaction products) causes the conductivity to increase or decrease and on the rate constant for the surface reaction rate relative to the characteristic timescale for diffusion of gas into the porous electrode structure, which determines the composition profile through the layer. A simple case that as an example allows easy exploration of the effects is if the reaction assumed is a first-order decomposition of the gas into a product that has no effect on the local conductivity: $R(P) = kP$. The rate constant, k, would depend on the internal surface area of the porous material and upon the composition of the pore surface since this could significantly affect the gas decomposition rate (e.g., the surface-catalyzed combustion rate of gases such as CO or ethanol in air). The steady-state concentration profile through the layer is then

$$\frac{P}{P_0} = \frac{\cosh[X(kh^2/D)^{1/2}]}{\cosh[(kh^2/D)^{1/2}]} \tag{9.9}$$

where D denotes the gas diffusivity through the porous layer, dependent on the porosity and tortuosity of the assembly. If the response of conductivity to change of gas partial pressure is

$$\frac{\Delta\kappa}{\kappa_0} = K_P P \tag{9.10}$$

Then, Figure 9.6d shows how the measured conductance response, $\Delta\sigma/\sigma_0$, would vary with the kinetic parameter $K_T = kh^2/D$, and how it would differ depending on the electrode spacing. The calculated signal drops sharply with an increase in K_T.

Another simple model is the one which has hemispherical electrodes, radius b, in the form of parallel strips spacing $2a$, on an impervious substrate. Here, with $B = b/h$ and $A = a/h$, the conductance is

$$\sigma \approx \frac{1}{A\ \ln\left(\frac{2A-B}{B}\right)} \int_0^1 \frac{\kappa}{\left(1 + \frac{X^2}{A^2}\right)} \mathrm{d}X \tag{9.11}$$

Calculation for this geometry illustrates the sensitivity to the details of the geometry of the electrodes, which determine the geometry of current flow through the layer. Figure 9.6c shows the electric field ratio for the two geometries, as a function of position above the substrate along the symmetry plane between the electrodes. Subtle changes of geometry change the weighting of different depths within the sensor. In practice, it is straightforward to vary the thickness and spacing of electrodes, by printing.

More generally, for planar devices – prepared, for example, by printing – one might expect characteristics to vary through the device thickness: the particle size might vary, giving rise to a variation in the pore structure and surface area. The consequence would be a variation in the conductivity in air alone, κ_0, in the sensitivity to the target gas, K_P, and in the decomposition rate, K_T. The composition profile of the gas, $P/P_0 = f\ (X, K_T)$, would be altered as a result. Indeed, these characteristics might be deliberately varied, by printing successive layers with differently modified materials [65, 83].

Semiconducting oxides show a dependence of response on gas concentration of the form:

$$G = \alpha_g P^{\beta} \tag{9.12}$$

where $G = \Delta\sigma/\sigma_0$ for conductivity increase or $G = \Delta R/R_0$ for resistivity increase (R denotes electrical resistance). Taking the conductivity increasing case, for the symmetrical electrode geometries used here for illustration, the effect of diffusion–reaction on the response impacts the response coefficient, α_g,but not the functional form of the response:

$$\alpha_g = \int_0^1 \kappa_0 K_P f(X, K_T) E\ (X, \text{electrode geometry}) \mathrm{d}X \tag{9.13}$$

where the integral is evaluated on the plane bisecting the electrodes. A device with two different electrode gaps – for example, one large with respect to the layer thickness, designed to probe the average behavior of the whole layer, and one small, designed to probe the base of the layer – will have two response coefficients, α_L and α_s. They will be different and hence report differently the variations of κ_0, K_P, and K_T. As the gas composition in the atmosphere being sensed varies, the signal from the large gap plotted against that from the small gap will move along a line of slope α_L/α_S. This is the "operating line" for the sensor. If the spatial variation of any of κ_0, K_P, and K_T changes, then the response will move off the operating line. Spatial variations could be caused by alteration of the microstructure by local cracking or change of thickness of the device if sensor material sublimes, or by poisoning of the sensor (for example, by reaction with silicones [46, 60] or H_2S [61]), or by the presence of an uncalibrated interfering gas. This is a powerful concept for a self-diagnostic sensor [81]. If two gases are present, whose signals are additive but which have different values for K_T, then the two different electrode responses could be used to calculate the concentrations of the two different gases [84]. In an unknown gas mixture, the pattern of variation of α_L against α_S is in some sense characteristic of the gas composition: the device functions as a two-element array that can greatly increase discrimination in an "electronic nose" [85, 86].

Experimental Demonstration This concept was first demonstrated using cylindrical devices, comprising a pellet clamped between electrodes, one of which was an outer ring and the other a central disc, such that gas diffusion into the pellet was possible only from the outer rim [87]. The response to CO and propane on SnO_2 and on a complex niobate were studied. The signals from the two electrodes were very different, simply because the gas composition was different in the interior of the pellet compared to the outside. Given a model for the gas concentration dependence of the conductivity and for the kinetics of the surface decomposition, a self-consistent solution could be found for the rate constant for surface reaction and the response coefficient.

To fit the data required a change from first-order to zeroth-order kinetics of combustion with an increase in gas concentration and the fit shown in Figure 9.7 confirmed this deduction. One specific and important deduction from this work based on the different variation with gas composition of the electrical response and the surface-catalyzed combustion rate was that the sites responsible for the surface-catalyzed combustion were not the same as those responsible for mediating the electrical response. Another was that the differences in response in different regions of the pellet reflect variations in microstructure. The size scale of the pellet (13 mm diameter) was such that the gas concentration in the interior could be reduced to very low levels so that the effects of gas concentration gradients in the interior of the device were very clear.

Significant effects are also clearly observable in printed devices, of thickness scale 100 μm. Poisoning of methane sensors by silicones was unambiguously detected as well as interference due to H_2S [46, 61]. SnO_2 gives a useful signal to trace concentrations of methane in air, and sensors are widely deployed for leak detection. Silicones derived from polishes or lubricants are a major

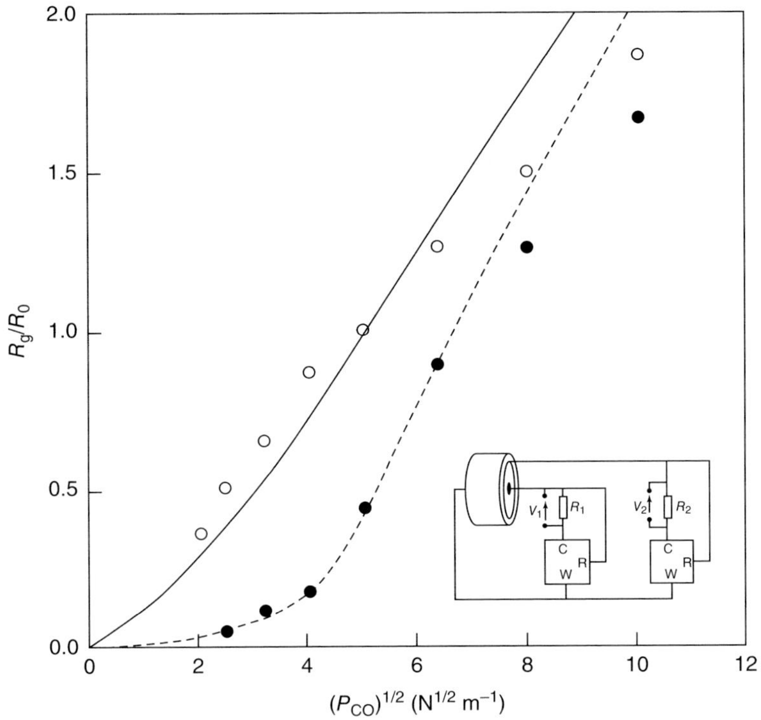

Figure 9.7 Dependence of gas response on electrode geometry: cylindrical pellet of SnO_2 with gas access from the circumference only. Response of inner ring (open circles) and outer disc (closed circles) to CO in air at 430 °C. Source: Williams et al. 1995 [87]. Adapted and reproduced with permission of Royal Society of Chemistry.

problem affecting the sensors because the silicone can decompose to silica on the hot sensor surface, essentially coating the active surface with a layer of glass. The effect is detectable with a multielectrode device because the poisoning changes the surface-catalyzed decomposition rate of methane and hence the gas concentration gradient through the device. Figure 9.8 shows the device construction and the signal on the different electrode gaps. The signal followed Eq. (9.12) with $\beta = ½$. An accurate numerical calculation gave $E(X)$. Assuming a first-order decomposition of the gas, and assuming that the surface is poisoned uniformly through the thickness of the device – that is, that K_T and K_P were uniform through the thickness – the response on the different electrodes could be fitted to Eq. (9.13), assuming the concentration profile given by Eq. (9.9) to derive the values of K_P and K_T, as shown in Figure 9.9 [46].

The use of electrode geometry to probe composition gradients within a porous sensor body is also very useful in practical application to detect the presence of uncalibrated interferences. Figure 9.10 shows operating lines for response to ethanol and to CO on a dual-gap sensor fabricated from chromium-titanium oxide (CTO). For this sensor material, the combustion rate for ethanol on the sensor surface is significantly higher than that of CO, so the gradient of the operating line is less (the signal on the narrow gap is relatively lower).

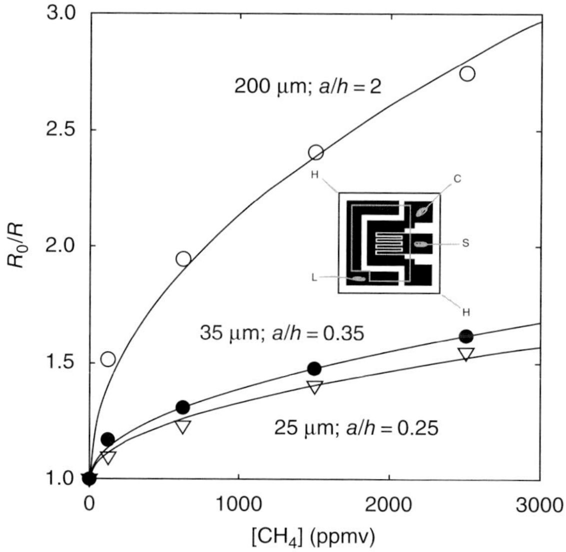

Figure 9.8 Methane response of multi-gap SnO_2 sensor operating at 625 °C. Printed SnO_2 on an impervious alumina substrate with gold electrodes. SnO_2 thickness 100 μm. Three different electrode gaps. Inset shows a multi-gap electrode arrangement. Source: Williams and Pratt 1998 [46]. Adapted and reproduced with permission of Royal Society of Chemistry.

Thus, signal variations due to the presence of CO can be distinguished from those due to the presence of ethanol, a common interference in domestic applications.

The gas composition gradient could be very different for different gases, depending on the rate of surface-catalyzed combustion, which can be altered by surface-decorating the device with a catalyst or by mixing a catalyst into the device or layering it on top of the device [84, 86]. For single gases, therefore, a comparison of the signals on wide and narrow gaps can be used qualitatively to

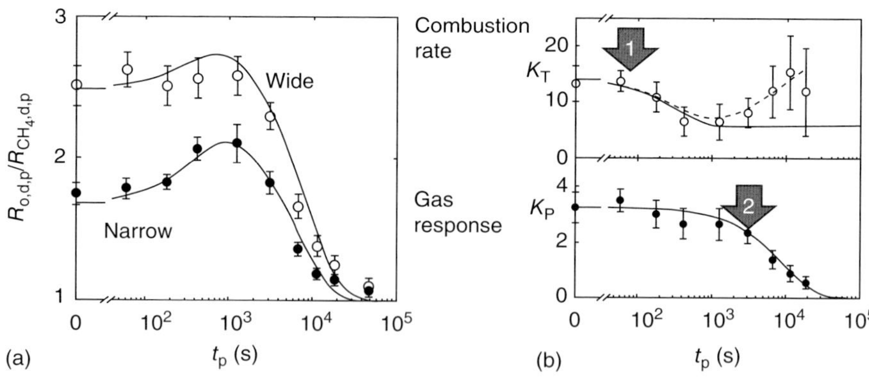

Figure 9.9 (a) Variation of signal, $R_0/R = \sigma/\sigma_0$ to 0.25 vol.% CH_4 in dry air. (b) Derived surface combustion rate constant, $K_T = kh^2/D - k$, is the first-order rate constant for surface catalyzed combustion, h the porous oxide layer thickness, and D the gas diffusivity within the oxide layer – and conductivity response constant, K_P, as a function of total time of previous exposure to 25 ppm hexamethyldisiloxane in air, t_p. The arrows mark the titration of two different types of surface site: one mediating surface-catalyzed combustion and one mediating the electrical response. Source: Williams and Pratt 1998 [46]. Adapted and reproduced with permission of Royal Society of Chemistry.

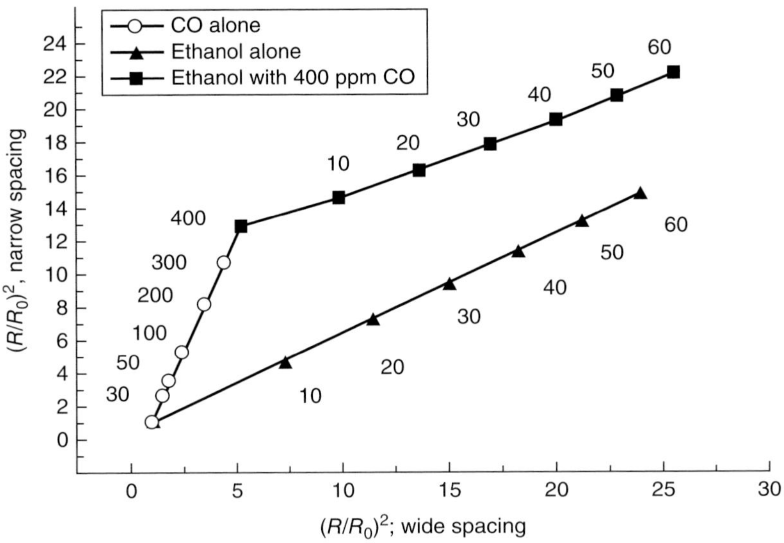

Figure 9.10 Illustration of operating line principle using multi-gap electrodes and CTO as the sensing material: signal on the narrow gap against that on the wide gap. The operating line slope is smaller in the presence of ethanol than in the presence of CO because the surface-catalyzed combustion rate constant for ethanol on CTO is significantly larger than that for CO. Labels: gas concentration in air/ppm. Source: Williams 1999 [2]. Reproduced with permission of Elsevier.

discriminate one gas from another – carbon monoxide from ethanol or acetone, or acetone from toluene, for example [84]. We showed that the multielectrode array devices could be used to derive the combustion kinetics [84, 87]. We also showed that different gases in a binary mixture could be resolved and independently measured, provided that the composition gradient was significantly different for the different gases. When the effects of different composition gradients are factored out, the signals due to different gases in a mixture are in fact additive [84].

9.2.1.4 Microstructure Model: Percolation and Equivalent Circuit Representation

The microstructure of porous oxide gas sensors is very different at different size scales (Figure 9.11): there are large open voids and connected porosity on a range of size scales, fine particles associated into agglomerates which are themselves nanoporous, and very fine size-scale coatings on the surface of larger crystals, especially toward the outer surface of the device. These variations arise during manufacture and are associated with the input particle size of the powder used to prepare the device, the way in which the powder of the sensing material is dispersed into the printing ink that is used to make the device, the way in which the ink dries before firing, the way the ink components burn off during firing and effects on the ceramic powder of the locally reducing environment that may be present during this process, and sintering and sublimation processes of the ceramic after the ink components are removed. Close control of the input powder,

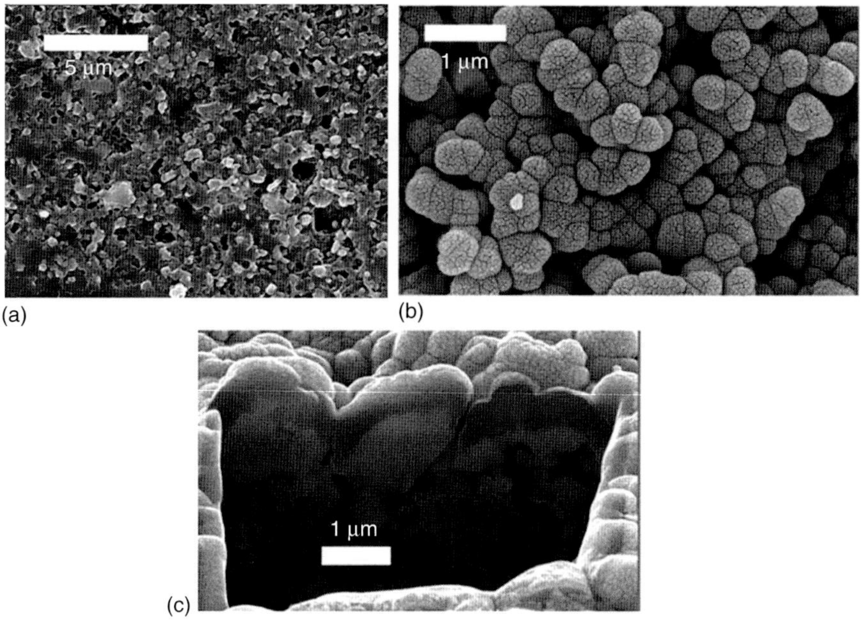

Figure 9.11 Sensor microstructure at different scales (chromium-titanium oxide). (a) Cross section showing micrometer scale voids. (b) Surface view showing microstructure within each agglomerate. (c) Focussed-ion beam section through the surface, showing very fine-scale porosity on the immediate outside surface of the sensor.

the mixing and the printing and firing, is the essential art to be mastered for fabrication of reliable and repeatable devices. The microstructure is the factor above all others that controls the response of the device [53].

The porosity and particle size of the sensor material are clearly important. If the material is "n" type – electrons as majority charge carriers – then (as noted above) there will be a charge carrier depletion layer near the gas interface of each particle. If the material is "p" type – holes as majority carriers – then there will be an accumulation layer. A significant effect on gas sensitivity of reducing the particle size toward the size scale of the depletion or accumulation layer is to be expected, which should be different according to whether the material is "n" or "p" type or, more specifically, different according to whether the gas causes an increase or decrease in the charge carrier concentration near the interface. The effects are very marked for "n"-type materials where a conductivity decrease is caused by the gas [88]. As might be expected, the effects depend on the bulk donor density, which determines the size scale of the surface depletion layer (Figure 9.12). The ideas describing these effects are straightforward: charge transport from one gas-exposed crystallite (grain) to another may be controlled by a Schottky barrier at the intergrain contact, or by a charge-depleted neck joining the two grains, or grains may be fully depleted, causing a surface-trap-limited conductivity (in fact, the simple approximate model used above to rationalize composition effects), Figure 9.12. Transition between these regimes is dependent on the grain size in relation to the surface depletion layer width [9, 90, 91].

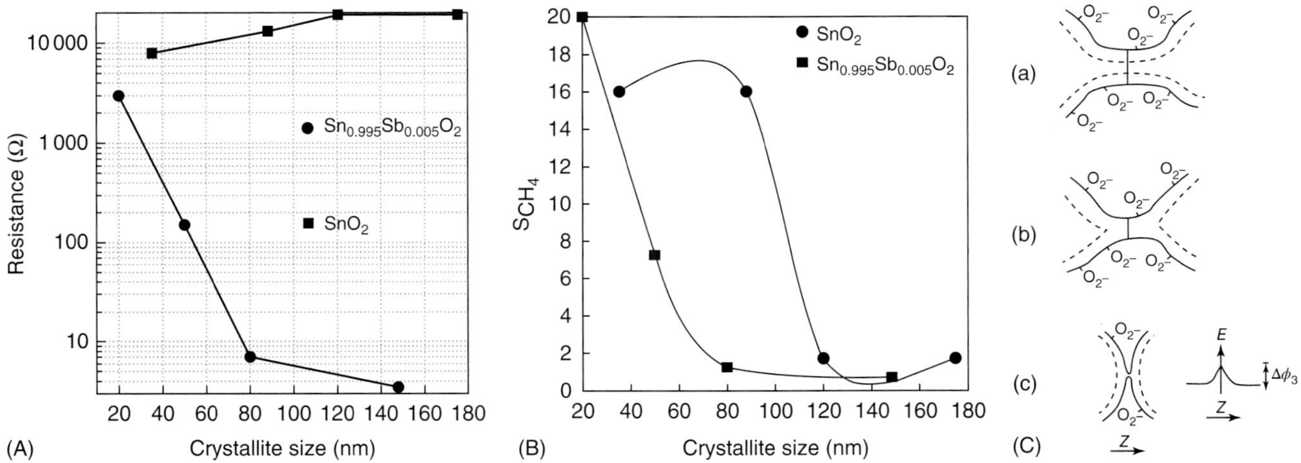

Figure 9.12 Crystallite size effects on conductivity and gas response. Resistance (A) and sensitivity $S = \sigma/\sigma_0$ to 1% methane in air (B) of pressed pellets of $Sn(Sb)O_2$ (from Dusastre [89]). (C) Three models for conductance limited by intergrain constrictions: open neck, closed neck, and Schottky barrier. Source: McAleer et al. 1987 [9]. Reproduced with permission of Royal Society of Chemistry.

There are more subtle effects of microstructure. Evidently, there is a strong effect on composition gradients and hence conductivity gradients within a device, as noted above. There are also variations of microstructure through the thickness of a device, which can be either adventitious or deliberately achieved, for example, by screen printing. These effects, which can be detected (as noted above) by devices using different electrode spacings to probe different physical regions [53], will also have subtle and interesting consequences. Here, the behavior that we seek to explain and hence modify is the irrational power law response (Eq. (9.12): β can take values ranging from 0.2 to 1.3, depending on the microstructure [53, 92]). Some simple modeling approaches are helpful to understand the effects. We show that it is a consequence of the electrical connectedness through the structure.

A general model that might take account of the effects of agglomeration, voids and porosity of different size scale, is of a randomly connected resistance network, some elements of which are gas dependent (those parts exposed to the gas within the pores) and some of which are not (those parts in the interior of large agglomerates, or the interior of crystallites of the material). A simplification is to model all elements and connections of nodes of the network as identical. There is obviously an approximation involved in averaging properties in this way. Kirkpatrick [93] has given the conductance sufficiently far from the percolation limit of a random resistor network in which each resistor connects adjacent points of a cubic lattice as

$$\sigma = \sigma_N \left(1 - \frac{3}{2}p\right) \tag{9.14}$$

where $(1 - p)$ denotes the probability of a connection between adjacent points and σ_N is the conductivity of the individual network elements. Using this idea, the sensor model is as an array of "domains" (which might be single particles or particle agglomerates) separated by "connections" [9], in which case $(1 - p)$ is interpreted as the probability of the presence of a domain and p as the porosity. The properties of the whole random network can therefore be obtained simply by examining the properties of the series combination of R_1 (the connections) and R_2 (the internal resistance of a domain), Figure 9.13. The most general approach would be to represent both R_1 and R_2 in Figure 9.13 as parallel combinations of a gas-sensitive and a gas-insensitive resistor. Different simplifications have been found applicable to different circumstances. A simple response law for the gas sensitivity of the electrical conductivity ("n"-type) or resistivity ("p"-type) of the gas-sensitive elements gives an apparent power law response, dependent on the range of gas concentration explored, where the exponent depends on the relative values of the gas-sensitive and gas-insensitive resistors [53]. Figure 9.13B shows an example for a conductance-increasing (resistance-decreasing) response.

For example, simulation of a conductance-increasing response requires at least a gas-insensitive element in series with a gas-sensitive element. Figure 9.13 shows an example of a simulated conductance-increasing response for a two-element series circuit, where σ_2 is independent of the gas concentration and σ_1 has the variation with gas partial pressure, P_g:

$$\sigma_1 = \sigma_1^0 (1 + K_P P_g^{1/2}) \tag{9.15}$$

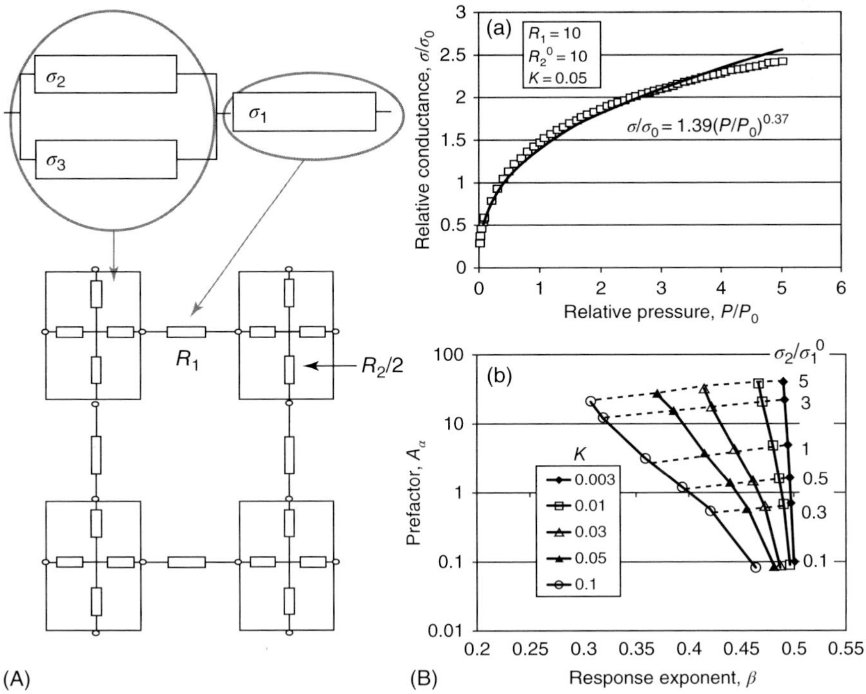

Figure 9.13 Equivalent circuit approximation for microstructure effects on conduction and gas response. (A) "Agglomerates" and "connections" represented as resistors, R or conductances, σ. A series–parallel circuit represents the agglomerates as a parallel combination of "surface" and "bulk" elements, connected by a series element representing the inter-agglomerate connection. (B) Results of simulation appropriate to an "n"-type (surface depletion layer) material, using a reduced two-element network, of gas-sensitive "connection" and gas-insensitive "bulk": (a) conductance response of the network; (b) correlation of pre-factor and exponent in Eq. (9.12). Symbols are defined in the text. Source: Reproduced with permission from Ref. [53].

Equation (9.15) is the form of variation found for the response to carbon monoxide of thin films of tin dioxide and can be rationalized with a simple model for the surface reactions [5, 23]. This network would be an approximation for the case of agglomerates whose internal resistance was much less than that of the gas-sensitive surface of the agglomerate, so that the signal came from the connections between agglomerates [9]. Figure 9.13 shows the computed response of the series network empirically fitted to the form of Eq. (9.12). Eq. (9.12) is indeed a reasonable approximation for the behavior, with β an irrational number dependent on the parameters of the network. Specifically, if K_P is fixed (i.e., the gas being detected is fixed), then β may vary widely depending on the ratio σ_2/σ_1^0. However, the empirical response parameters, α_g and β, should be correlated, $\log(\alpha_g)$ varying roughly linearly with β. This is exactly what is found, Figure 9.14. We prepared near-ideal microstructures as inverse opals formed from SnO_2 nanoparticles [92] and also [53] SnO_2 devices with a very wide range of microstructure, by screen printing, altering the particle size of the input powder, the degree of aggregation in the input powder, and the

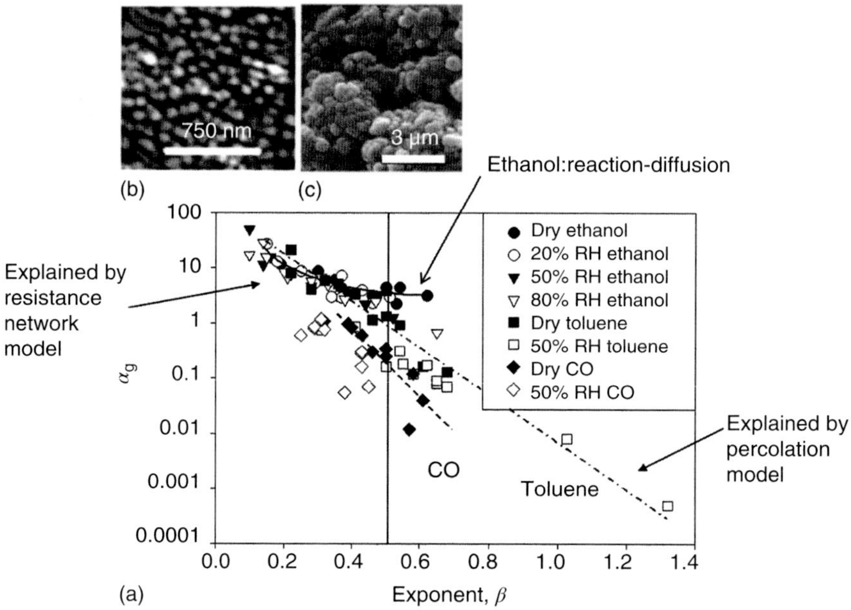

Figure 9.14 Effects of microstructure variation on response parameters of SnO_2 sensors explained by microstructure models. (a) Correlation of log(α_g) with β, for different gases. (b,c) Example of microstructure variation obtained by altering processing parameters. Scale bar: (b) 750 nm; (c) 3 μm. Source: Reproduced with permission from Ref. [53].

amount of surfactant and ratio of powder to printing vehicle in the printing ink, and measured the apparent response law, Eq. (9.12), to three different gases: ethanol, CO, and toluene. Figure 9.14 shows that the response exponent, β, varied widely and could be very different for the different gases on the same sensor. The particles were prepared by an emulsion sol–gel route [94]. An increase in surfactant concentration in the emulsification stage (which tended to decrease the particle size of the oxide) tended to lead to higher values of response exponent; an increase in the proportion of vehicle in the printing stage (which ought to tend to favor higher porosity and disaggregation in the finished sensor) also tended to lead to higher values of response exponent. A given gas on different sensors is equivalent to moving along a line of constant K_P in Figure 9.13 and different gases on the same sensor is equivalent to moving along lines of constant σ_2/σ_1^0 (Figure 9.13). The correlation of log(α_g) varying roughly linearly with β is generally as expected from the very simple resistor network model. Interestingly, the correlation is the same regardless of whether the gas is wet or dry, meaning that K_P should be the same: that is, water vapor does not compete on the same sites as the gas, further consolidating the ideas that there are distinctive sites for the water signal, first elucidated from the studies of systematic substitution described in Section 9.2.1.2 and developed further in the studies described in Section 9.2.2.1. The correlation for CO and toluene is log-linear as expected from Figure 9.13 but that for ethanol is curved. The explanation is that ethanol burns on the surface of SnO_2 at a rate significantly higher than that of CO or toluene, so the results for ethanol are strongly affected by the reaction–diffusion effects discussed in Section 9.2.1.3. However, there is one notable subtlety remarked here and discussed later: the response exponent, β, can be significantly >0.5, which the simple resistor model would not predict.

The resistor network model also worked well for conductance-decreasing responses. For ozone on WO_3, an "n"-type material, a simple parallel model, of a gas-sensitive "surface" element with a gas-insensitive "bulk, " described the results and is consistent with the ideas above, where the "bulk" has a higher conductance than the "surface." These ideas proved very useful for the development of the low-cost air quality measurement instruments that use these devices [50] (see Section 9.3.2). For "p"-type materials where the observed conductance is controlled by a carrier accumulation layer at the surface, a more complete equivalent circuit is needed since the bulk conductance would in general be less than the surface conductance. The ideas can be illustrated by the study of the effects of microstructure on the response of CTO to reducing gases [68, 95]. The series–parallel model shown in Figure 9.13A accounted for the response over a very wide range of gas concentration and microstructure, both to ethanol [95] and to CO [68], with both the surface of the agglomerates and the connections contributing to the gas sensitivity:

$$\sigma_1 = \sigma_1^0/(1 + K_\mathrm{P} P_\mathrm{g}); \qquad \sigma_2 = \sigma_2^0/(1 + K_\mathrm{P} P_\mathrm{g}) \tag{9.16}$$

Figure 9.15 shows two things: firstly, the measured response law is approximately to the square root of gas concentration, but the assumed response for the model is linear – an important distinction when it comes to deducing mechanisms of surface reaction – and secondly, that the variation is subtly nonlinear, as different elements of the network become dominant over different gas concentration ranges: only measurements over a large concentration range, not normally attempted in the literature, will show these effects.

The simple resistor network model does not describe possibly interesting behavior that might appear if the porosity and agglomerate size were such that the percolation limit for current passing through the network was approached. A simple model construction explored that case [67]. An approximation for the surface depletion layer is just a simple slab of fixed thickness, δ, whose conductivity varies with the gas composition. So now, a model microstructure can be constructed by assembling cubes of side δ (Figure 9.16). Those cubes that are adjacent to void space can be identified and their conductivity varied. The conductivity of the whole structure in response to variation of the conductivity of these surface elements can then be examined.

A power law variation indeed represents the variation of the conductivity of the assembly with respect to the conductivity of the surface elements. For "p"-type (conductivity decreasing) responses, the behavior is well represented by the simple resistance network model. The model essentially reproduced the experimental results. For "n"-type (conductivity-increasing) responses, subtleties appeared. The observed response exponent was always >1. This is the effect of approaching the percolation limit for current flow. In this case, the boundary elements that connect agglomerates become particularly important: percolating paths can be switched off or turned on with alteration of the conductivity of the boundary elements. The result should be a very strong dependence of the conductivity of the

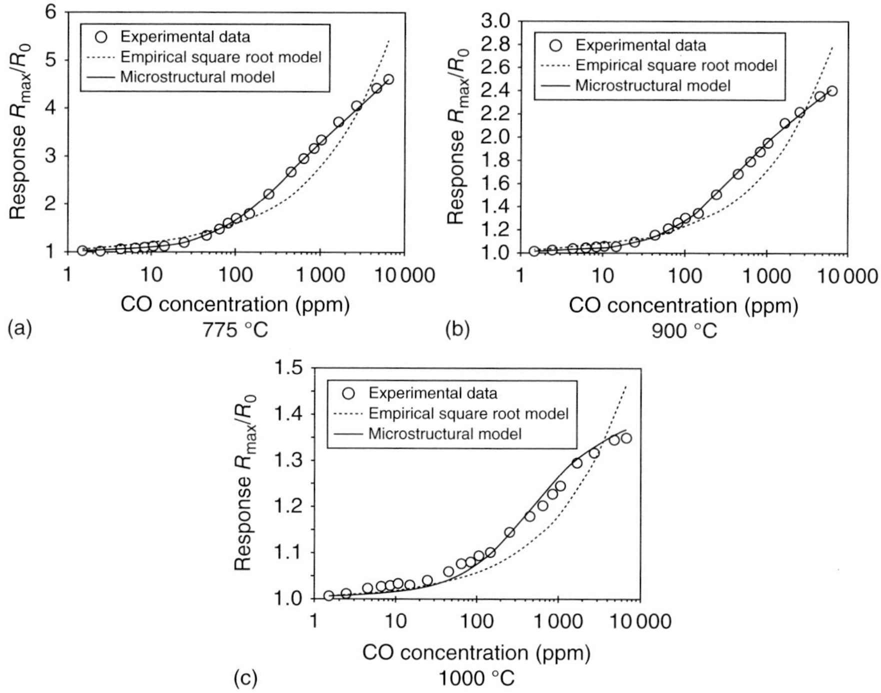

Figure 9.15 Response to carbon monoxide in air, R_g/R_0, for CTO sensors fired at different temperature to achieve different microstructure. The average particle size, determined by measurement on scanning electron micrographs, increased from 0.4 to 0.6 μm with increasing firing temperature. The solid line is the fit to the equivalent circuit model shown in Figure 9.13, using Eq. (9.16) for the "surface" and "connection" conductances (linear response law). The dotted line is an empirical fit to Eq. (9.12) with $\beta = 0.5$. Source: Naisbitt et al. 2006 [68]. Reproduced with permission of Elsevier.

assembly on the conductivity of the boundary elements, reflected in a response exponent > 1. This is then the explanation for the large response exponents that can be observed experimentally (Figure 9.14). The results presented in Ref. [67] did not include calculations varying the ratio of "zero-gas" conductivity ratio of the boundary elements to the bulk elements, and so the full range of the experimental behavior seen in Figure 9.14 was not captured. Even so, the combination of modeling using a simulated microstructure, representation using simple resistance networks, and modeling of reaction–diffusion effects through the device represented as an average structure, has clearly captured in total the critically important effects of microstructure on the response of gas-sensitive resistors, as noted in Figure 9.14.

9.2.2 Surface Segregation and Surface Modification Effects

9.2.2.1 Surface Modification by "Poisoning"

The surface sites that mediate both signal generation and catalytic oxidation are susceptible to modification by reaction with H_2S or SO_2 or siloxanes (from

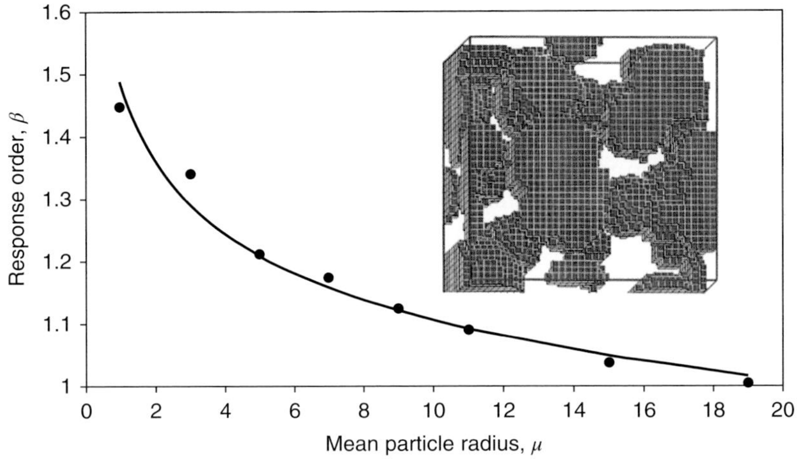

Figure 9.16 Model microstructure and effect of current percolation on power law response order, β, for conductance-increasing response. The inset shows a typical model microstructure of spheres with randomly placed centers and a Poisson distribution of radius, constructed from elementary cubes. Dependence of β on mean radius of the spheres for structures of porosity $\approx$0.54. Source: Williams and Pratt 2000 [67]. Reproduced with permission of Elsevier.

sprays of polishes or lubricating oils for example) present in trace amounts in the atmosphere. These effects are frequent causes of signal "drift" and in the absence of periodic recalibration would result in erroneous alarms, or failures to alarm. Poisoning is expected to develop progressively from the outside to the inside of the sensing layer, and thus influence the signal from a wide electrode gap before one from a narrow electrode gap. The movement of the operating point for the device on the space defined by the signals from the two electrode gaps would also be changed. The expected effects are easily simulated and the principle was demonstrated for sensors utilizing both SnO_2 and CTO [60, 61]. A specific practical example studied was the detection of both methane and H_2S from sour gas, compromised by sensor poisoning by siloxanes.

The effects of reaction with H_2S or SO_2 are to cause a sulfation and hydroxylation of oxide sensor surfaces [96], which then result in a change in response to other gases. These effects can be reversed by an increase in sensor temperature [96]. Reaction with siloxanes causes a permanent silication of the sensor surface. The effects can be detected using multiple-gap sensors because the reactions not only cause a change of the electrical signals but also of the surface combustion rate. Progressive poisoning with siloxane has been used to reveal the different general types of surface site present on polycrystalline tin dioxide, as these react with siloxane at different rates and are thus removed from the surface at different doses of siloxane [46]. The electrical response can be derived separately from the combustion rate constant. Four different classes of surface site were distinguished: an electrically neutral oxygen site that catalyzes methane and hexamethyldisiloxane (HMDS) combustion, two electrically charged oxygen sites that respectively mediate methane response and dissociative chemisorption of water, and a site for molecular chemisorption of water.

9.2.2.2 Surface Modification by Segregation

As noted above, and as might be expected, the gas-sensitive behavior of semiconducting oxides can be significantly altered by differential surface segregation of different cations substituted in the oxide. Two studies highlighted these effects in detail [47, 49].

Substitution of Sb into SnO_2 is well known as a means to increase the conductivity (Figure 9.12). Sb(V) on Sn sites is charge balanced by reduction of Sn(IV) to Sn(II), which provides the donor states for the conductivity. However, Sb is surface segregated, to a degree dependent on the heat treatment temperature: the enthalpy of segregation is ≈ -20 kJ mol^{-1}. At the surface, Sb is reduced to Sb(III), which acts as a surface trap state for electrons. It forms an electrically active surface state that interacts with water and has a significant effect on the surface-catalyzed combustion rate of carbon monoxide, but only in the presence of water vapor. The response of the electrical conductance to change of water vapor pressure correlates directly with the surface segregation of Sb [47]. This work complements that described above, where the study of the effects of sensor poisoning identified different types of site that were electrically active, that were responsive to water vapor differently than reactive gases, and were differently active for surface-catalyzed combustion.

A different system where the effect of surface segregation on sensor response has been very clearly demonstrated is that of the titanium-substituted chromium oxide, $Cr_{2-x}Ti_xO_3$ [49] (CTO). Pure Cr_2O_3 prepared as a porous pellet and heated in air has a relatively low resistivity. There is a small resistance increase on exposure to traces of reducing gases such as CO in air: the material is a "p"-type conductor. Substitution of a small amount of Ti (e.g. $x \sim 0.01$) causes a significant increase in resistivity and a significant and practically useful resistance-increasing response to gases such as CO in air (Figure 9.17). The material is easy to fabricate, has a significantly lower response to changes in water vapor pressure than SnO_2, and is significantly more resistant to poisoning by silicones: that made it an attractive material for commercial development. It became the first alternative material to SnO_2 to be introduced for mass-manufactured sensors. The key to the behavior was the surface segregation of Ti. The surface conductivity of pure Cr_2O_3 was attributed to Cr(VI), charge balanced by Cr vacancies: Cr(VI) was detected by X-ray photoelectron spectroscopy (XPS), and the signal significantly diminished upon substitution of Ti. Hence, the mechanism by which the signal was developed was by the removal of the surface acceptor states associated with Cr(VI) that masks the effects of states associated with adsorbed oxygen. The defect was formulated as a pair: $(V_{Cr}{}^{///}Cr_{Cr}{}^{\bullet\bullet\bullet})$. The vacant d-states associated with Cr provide the electron acceptor states above the valence band edge, for the conductivity. Again, XPS showed a clear signature for Cr-3d states at the valence band edge. The ionic lattice model for the defect state upon Ti substitution that minimizes the electrostatic energy has three Ti(IV) ions substituted on Cr sites, arranged symmetrically around the Cr vacancy. Whether the model showed the defects to be surface segregated depended on the crystal face that was examined. In the optimal configurations, Ti(IV) and Cr(VI) were immediately subsurface. A significant distortion of the arrangement of oxygen anions around the Cr vacancy and around the higher valent metal ions was calculated. Oxygen anions

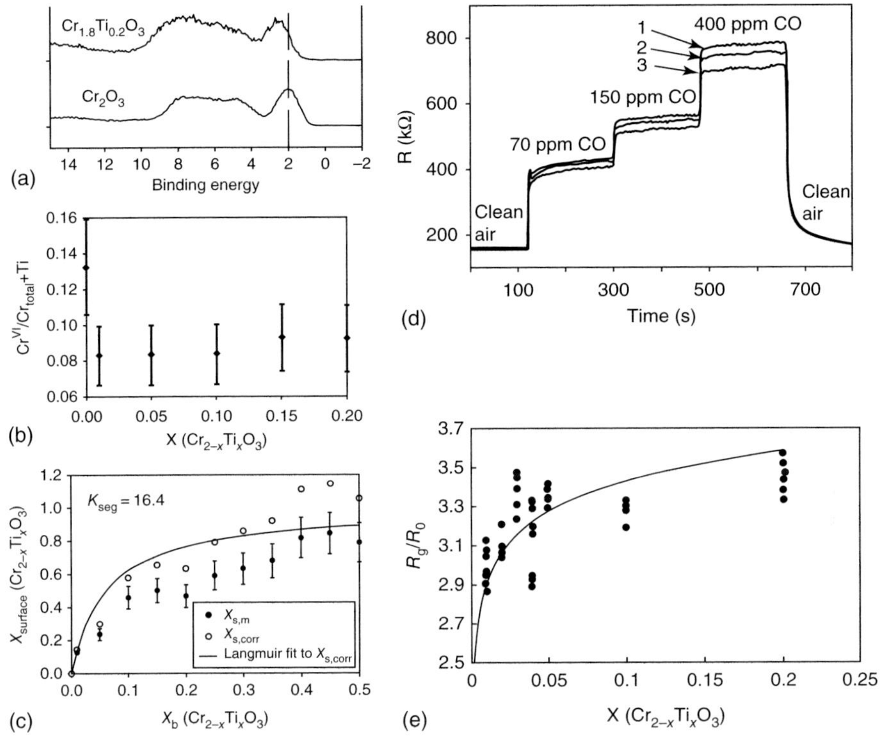

Figure 9.17 Effects of Ti substitution in chromium titanium oxide. (a) X-ray photo electron spectroscopy, valence band edge: shift of Cr-d states, $x = 0.2$ compared with $x = 0$; (b) diminution in Cr(VI) – 2p signal in XPS with Ti substitution: Cr(VI)/(Cr(total) + Ti) against x; (c) surface segregation of Ti: $x_{surface}$ determined by XPS against x_{bulk}; (d) Resistance change in response to CO in air of a screen-printed $Cr_{1.8}Ti_{0.2}O_3$ sensor at 400 °C, showing the effects of change of relative humidity: 1 is 50%, 2 is 25%, 3 is 10% RH at 20 °C. Source: Reproduced with permission from Ref. [49]. (e) Resistance response to 400 ppm CO in air increases sharply with increase of x from zero to 0.01, after which the variation is quite small: screen printed sensors at 400 °C in air at 100% RH at 20 °C. Each measurement is on a different sensor. Source: Data taken from Ref. [48].

surrounding the chromium vacancy moved away from the vacancy and slightly up out of the surface. The defect model of Ti(IV) on Cr sites, charge compensated by Cr vacancies, has been supported by X-ray absorption measurements [97, 98]. These surface defect models led to a rationalization of the gas response using a model analogous to the Mars–van Krevelen model for catalysis: in this case, the idea was adsorption of the reactive gas (CO) at the surface Cr vacancy and of oxygen at sites over the high-valent metal ion – Cr(VI) or Ti(IV).

An unusual aspect of the behavior of CTO was that, while there was a significant signal to CO, there was no significant catalysis of combustion of this gas, at temperatures where the electrical response was high, which is in contrast to observations on pure Cr_2O_3, which shows high catalytic activity for CO oxidation at moderate temperatures. More complex gases such as ethanol were efficiently catalytically oxidized on CTO. In practice, this is important because it means that the signal due to common interfering gases like ethanol and acetone can be distinguished from that due to CO by controlling the microstructure and thickness of sensors fabricated from the material, and the geometry of the electrodes (as described above, in Section 9.2.1.3). The loss of catalytic activity toward CO combustion can be attributed to the displacement of Cr(VI) species from the surface by segregation of Ti. The practical success of the CTO material rested on the relatively small signal to water vapor. An interpretation is that, unlike SnO_2, there are no sites equivalent to the bridging oxygen vacancies on which water will easily dissociate (see Section 9.2.3). The electrical signal in response to CO increased sharply with an increase in Ti substitution, up to $x \approx 0.05$, then leveled off. The signal to water vapor increased more gradually, up to $x \approx 0.2$. An interpretation is that the CO response is mediated on sites different to that mediating the response to water vapor. Practically, it means that the composition can be adjusted to maximize the response to CO relative to water vapor [99].

9.2.2.3 Surface Grafting as a Means for Altering Response

The idea is simply stated: is it possible to provide by deliberate surface modification an electrically active gas-sensitive surface state on a semiconducting oxide such that the charge carrier concentration near the oxide surface, and thus the oxide conductivity, is controlled by charge transfer to this deliberately created state? Then, if such a state can be formed, can the electron occupancy be altered by interaction with a gas? Essentially, the surface oxygen states mediating the response are to be replaced by the deliberate surface modification. We first deduced that this was possible from studies of SnO_2 surface decorated with Pt and Pd [36]. Later, we systematically explored the effect of surface decoration with Pt [100, 101]. We showed that a room temperature response to CO in air occurred and that very sensitive devices could be fabricated by screen printing [2]. The conductivity of the oxide depended on the oxidation state of the Pt surface clusters, as did the gas response. Oxidation of the clusters removed the gas response. The concept was generalized: a Ru complex was adsorbed onto $(Sn,Sb)O_2$ and then decomposed to form very small Ru clusters [102]. Again the conductivity of the oxide was strongly modified by the adsorbate and dependent on the oxidation state of the Ru. In principle, this seems like a powerful general method for creating a chemical sensor with deliberately engineered surface sites. However, the signal was very sensitive to the chemical state of these engineered sites: particularly, the interaction with oxygen and with water [101, 102], both of which will adsorb strongly on isolated transition metal centers.

9.2.3 Surface Defect and Reaction Models

We suggested that the surface defect states that mediate both the electrical behavior and the surface-catalyzed combustion are best formulated as an association complex of an oxygen vacancy with Sn(II) or Sb(III) and were then able to propose a simple model that unifies the interpretation of the behavior of SnO_2 as both a gas sensor and a combustion catalyst. We postulated that a correct formulation of the “absorbed oxygen” surface species mediating the electrical response is an oxygen molecule trapped in or on a surface oxygen vacancy; that the combustion reaction proceeds partly through these species and partly through lattice oxygen at the surface; that water competes with oxygen for the surface vacancies, blocking this route; and that the binding energy of water to the Sb(III)Vo surface defect complex is less than that to the Sn(II)Vo complex. First, an elementary reaction between oxygen and a surface defect complex can be written as follows:

$$[Sn_{Sn}^{//} \bullet V_O^{2\bullet}] + O_2 \rightarrow [Sn_{Sn} \bullet O_O] - O \xrightarrow{2n^/} [Sn_{Sn} \bullet O_o] - O^{2-} \quad (9.17)$$

The overall stoichiometry of this reaction can be represented as

$$4n^/ + V_O^{\bullet\bullet} + O_2 \rightarrow O_O + O^{2-}{}_{ads} \quad (9.18)$$

in which two electrons are trapped for each surface oxygen atom, as the interpretation of the electrical behavior requires. Interestingly, the mechanism as formulated here does not require dissociation of molecular oxygen. A dissociative adsorption of water onto the surface defect complex was proposed:

$$[Sn_{Sn}^{//} \bullet V_O^{\bullet\bullet}] + O_O + HOH \rightarrow [OH^{\bullet}{}_O \bullet Sn_{Sn}^{//} \bullet OH^{\bullet}{}_O] \quad (9.19)$$

In comparison with Eq. (9.17), the surface electron trap in Eq. (9.19) may lie at higher energy in the band gap, so replacement with water of oxygen adsorbed at the defect results in an increase in conductivity. Where Sb is surface segregated, the surface defect complex could be written as $[Sb_{Sn}{}^{/}V_O{}^{\bullet\bullet}Sb_{Sn}{}^{/}]$. A similar formulation for the surface reactions can then be made. If the energy difference between OH state and oxygen state is bigger on this site, then the effect of water vapor on the conductivity would be larger. Such a statement is consistent with water being less strongly bound at this state than at the tin defect state represented in Eq. (9.19).

Computational modeling has supported and further developed these ideas. First, static lattice modeling considers the oxide as fully ionic. In its simplest form, the potential energy function describing the ionic interactions is adjusted to give a good fit to the lattice parameters, given the crystal structure. Then, the effect on local structure of introducing different ionic species substituted into the lattice, or of different types of defect – interstitial or vacancy – can be computed. These calculations supported the idea that defect pairs or defect complexes were favored structures that were surface segregated [70]. Next, computations using density functional theory (DFT) develop more detail. These computations necessarily involve "scenarios" [72]: ideas concerning the placement of the adsorbed gas and model structures for the oxide (SnO_2), which are taken either as slab models [72, 74–79] or as cluster models [71, 102] for a particular crystal face, usually (110), in vacuum. A favored model is of the surface oxygen vacancy formed by removal of the bridging oxygen on the (110) face (Figure 9.18).

Manassidis and Gillan [75] described the electronic structure resulting from removal of all of the bridging oxygen (the "fully reduced surface"). They show the electrons left by removal of bridging oxygen localized largely above the bridging Sn (labeled Sn_{6c} in Figure 9.18) in the bridging oxygen vacancy (~30%) with the rest distributed toward the surrounding oxygen. The tin atom below the vacancy can be described as Sn^{2+}, but with a strongly asymmetrical electron distribution, that can be considered consistent with the polarizability of this ion. Subsequent

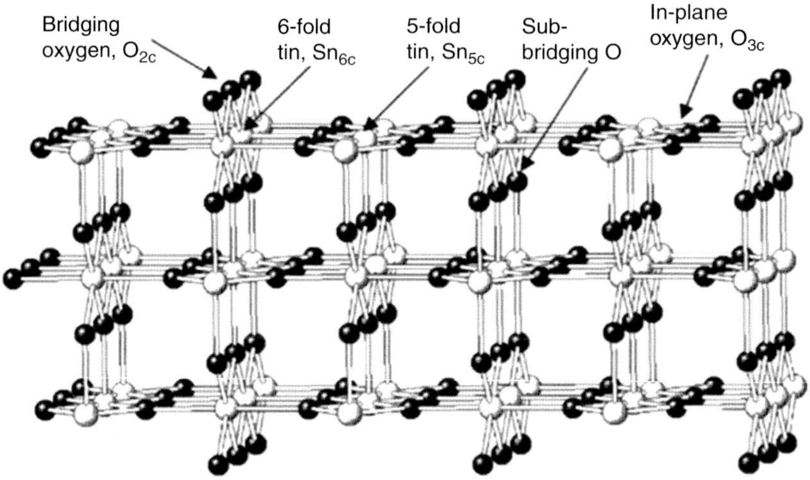

Figure 9.18 Model for the (110) face of SnO_2. Source: Sensato et al. 2002 [78]. Reproduced with permission of Elsevier.

studies have essentially confirmed these ideas. This surface oxygen vacancy is identified as a reactive site for gas adsorption. Both experimentally and computationally, however, the most stable fully reduced surface does not have just the bridging oxygen removed: it has both bridging and in-plane oxygen vacancies, with a structure similar to SnO [43].

DFT modeling studies agree on the strong adsorption of an oxygen molecule at the bridging oxygen vacancy. Our first attempt at a model indicated that oxygen adsorbed at such a site would be dissociated [103], but this involved an approximation in the computation that was later shown to be invalid [77]. Oviedo and Gillan [77] describe a number of adsorption modes. The most stable was found to be a "straddled," non-dissociative mode: one oxygen atom inserted into the vacancy and the other coordinated to the adjacent Sn_{5c} site. Oxygen here has a singlet ground state, in contrast to the triplet ground state of the gas-phase molecule. This configuration has a shorter Sn—O bond length than in the crystal and a longer O—O bond length than in the molecule. The phenomenological model given above (Eq. (9.17)) is very similar. Modeled with initial surface coverage of vacancies of 0.5 (a "half-reduced" surface), this configuration spontaneously dissociated to the filled vacancy and a singlet oxygen atom bonded to Sn_{5c}. However, at lower vacancy coverage, 0.33, this dissociation did not occur. Repulsion between the adsorbed species is an important factor. Another state, "twisted" molecular adsorption with the molecule centered over the vacancy but twisted with respect to it, is also favorable. These adsorbed states result in a broad distribution of electronic states within the band gap and a charge transfer toward the surface. The results have been confirmed by others, with differences in the calculated adsorption energies [72, 78, 79]. The oxide can also be represented as a small cluster of atoms, taken to represent a particular crystal face [71, 102]. It is then possible to compute activation energies between different proposed oxygen states during reaction with a gas and hence deduce from a straightforward analysis of reaction kinetics the steady state response of electronic charge on the cluster to change of gas atmosphere [71]. The computations showed that water would dissociate spontaneously on the bridging oxygen vacancy and that the competition between water and oxygen adsorption at this site was finely balanced [74, 104]. The results illustrate that the phenomenological kinetic models used in earlier literature are essentially consistent with these more sophisticated ideas [2, 9, 35].

A different approach has been presented by Ponce et al. [73]. Here, a 3 nm diameter spherical nanoparticle is constructed, as a rutile core fitted into a sphere of the defined diameter. Then, the sphere is allowed to relax in an environment of defined oxygen chemical potential, minimizing the surface Gibbs energy, using DFT to calculate the structure. A disordered zone about 0.2 nm across formed around the rutile core. The reduced particle ($SnO_{1.89}$) showed the excess electronic charge accumulating on Sn within this zone. Oxygen dimer and trimer species were noted at the interface. These studies illustrate the complexity of the interface but do not allow a simple interpretation of the different types of surface site that were identified by the surface titration experiment [46]. They do, however, focus attention on the details of the surface defects and the atoms surrounding them as a way of understanding the behavior of multicomponent oxides.

9.3 Commercial Development of Sensors and Instruments

9.3.1 Introduction

Chemical sensors are big business, and much of it is based on electrochemical principles. The scientific literature is vast. Much of this literature is driven by the perceived practical application. However, the successful commercial development of a chemical sensor device requires much more than just a good idea. First, the science has to be solid because exposure to the market is a brutal test that rapidly exposes weaknesses in the measurement principle or the implementation in a measurement device. Second, there has to be a market or potential market that is large enough to justify the significant development costs that will be incurred in taking a good idea and turning it into a robust device. Third, there is significant conservatism to overcome with a new sensor device. It has to be thoroughly established that the measurement result is completely reliable, since the consequence of a measurement is a decision or action, and the consequences of a wrong decision or sensor failure can be disastrous: not just from considerations of health and safety but also from loss of market confidence even if the measurement is not particularly safety critical. There are, therefore, significant regulatory and financial barriers to the introduction of a new sensor. Further, if there is an existing technology, then any new technology has to demonstrate a very large advantage in order to persuade a customer to change. As a consequence, the number of distinct commercially successful devices is quite small (Box 9.2). Each of the successes has been characterized by a good idea, a solid scientific base, and a significant market need that could pay the development costs. Most have been characterized by the entrepreneurial flair of just a few individuals who recognized a market need and who recognized the sound science that would underpin the market solution. In most cases, also, there was a government research funding base from which the science sprang, which underpinned the development in its early stages, which identified causes of failure and which drove improvements.

Box 9.2

Chemical Sensors are Big Business

- There are about 10^{10} blood glucose tests using electrochemistry sold per year.
- Every modern car has an electrochemical oxygen sensor in the exhaust to control the fuel/air ratio in the engine.
  - Many have two: one to control the engine and the other to monitor the gas coming out of the catalyst that cleans up the exhaust.
- Coal mines, oil and gas rigs, and petrochemical plants are monitored by chemical sensors for explosion and toxic gas hazards: electrochemical, semiconductors or catalytic beads (pellistors).
- There are about 10^8 pregnancy tests sold per year.
- Measurement of pH using glass electrodes is the most frequently performed chemical measurement in industry.
- Ion-selective electrodes are widely used in hospitals, in Industry, and for environmental measurement.
- A breathalyzer has an electrochemical sensor measuring the concentration of alcohol in the breath.

As noted earlier in this chapter, our sensor development started with a perception from users that there was a market need for an alternative to SnO_2 for toxic and inflammable gas detection. It started with a simple question: was the behavior of SnO_2 and ZnO – the only examples in the literature at that time – in some sense unique. The screening program that was established to answer this question, and the associated program of detailed measurements on SnO_2, was the basis for all our subsequent commercial development, and was rich in scientific results, as described above. The results emphasized the importance of microstructure, developed ideas concerning the surface states responsible for the signal and the theory connecting surface reaction kinetics and conductivity response, and uncovered the room temperature response of surface-decorated SnO_2 leading to the ideas of the conductivity of the oxide acting as a transducer of adsorption at the surface modifier [9, 23, 36–38, 100]. We recognized that additives modified microstructure, which modified the current distribution through the sample and hence the signal [35, 105].

In the course of this work, CTO was noted as a material with a small effect of water vapor and a useful signal to NH_3, which appeared to be selective and also very good for detection of H_2S [96, 106, 107]. The useful signal to CO was at first missed. $FeNbO_4$ and WO_3 were identified as useful materials for detection of

Cl_2 [108]. The perception of market need, the lack of interest from established sensor manufacturers, and the ethos of the time led to the creation of a new company, Capteur Sensors and Analysers Ltd., with venture capital funding. We decided first to go for niches in the market, which were not well served and where our materials seemed to have particular advantage, and drive the manufacturing technology – hence NH_3 and Cl_2. Our work had shown us that one of the major reasons for fluctuating signals was poor temperature control of the sensor element. We had worked on screen printing using standard inks formulated for the production of hybrid electronic circuits as a fabrication method in other projects on electrochemical sensors [109, 110] and decided to use this methodology to fabricate the devices. A Pt track on one side of a ceramic tile functioned both as heater and Pt resistance thermometer [52, 111]; operating the track at constant resistance by using the balance signal from a Wheatstone bridge incorporating the heater as the control for the heater drive maintained the tile at a temperature constant to perhaps 0.1 °C (deduced by observing the baseline stability of the semiconducting oxide). After exploring other fabrication methods for depositing the sensor material, we settled on screen printing for this, also, and through experience and the science described earlier invested considerable effort in the control of the sensor microstructure. CTO and WO_3 were the materials chosen. The market pull for CO detectors for domestic use was, however, very strong, driven by fatalities caused by faulty ventilation of natural gas burners for water heating. We re-checked CTO and now noted that indeed it was an excellent sensor for CO, as noted earlier in this chapter. This became one of the main planks of the company.

Serendipity then struck. At a trade show, Peter McGeehin was asked about a sensor for trace ozone in the atmosphere, for controlling ozone generators sold in the United States to domestic consumers for suppressing odors. We naively reasoned that, as ozone was an oxidizing gas similar to chlorine, we should try our WO_3 sensor. The signal to ozone was spectacularly large [22, 112] (Figure 9.19). The signal was surprising because it was large with a sensor temperature of 500 °C: one might have expected ozone to be completely decomposed on the sensor surface at this temperature. The speed of response was slow, becoming even more so with decreasing temperature. The speed of response obtained at high operating temperature was, however, adequate for the domestic application. We were soon making $\sim 10^6$ per year sensors. The company shortly after executed a successful trade sale.

The availability of the sensors prompted new collaborations in fields in which we had had no previous connection. Tony Cox and Rod Jones at Cambridge used the sensor to make a device for measuring atmospheric ozone. They used a chemical scrubber to generate a zero signal. Discussions led to new proposals to build balloon-borne instruments for measuring ozone in the stratosphere [113]. These proposals funded the basic science, which led to understanding the chemistry of the response and to ways of operating an instrument [50, 51, 112]. Further development on measuring ozone in the troposphere led to a useful prototype instrument. The problems to solve were the slow response and the need to continually check the zero, to apply the corrections for baseline drift and cancellation

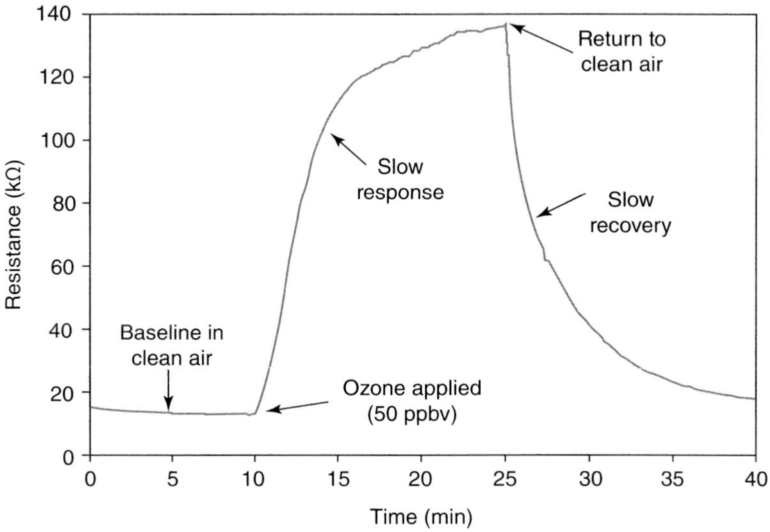

Figure 9.19 Resistance response to trace ozone in air of a printed WO_3 sensor at 500 °C; rapid air flow.

of humidity effects. The signal was extremely large at the measurement temperature (450–500 °C) but went to zero rapidly with temperature increasing to 600 °C. The consequent idea was that at 600 °C, the sensor surface was "reset" to a reliably reproducible state. If the temperature was then stepped over a relatively small range – from 600 to 550 °C, for example – the sensor signal would start evolving from this initial state at a reproducible rate determined by the ozone concentration. This proved to be the case. It also proved that a sufficiently large signal was developed in less than 60 s at 550 °C and that less than 60 s at 600 °C was sufficient to "reset" the surface [50]. Adjustment of the two temperatures allows a compromise between sensitivity, overall response time, and sensor stability. The signal was very sensitive to air flow rate because ozone decomposed on the warm plastic surrounds of the sensor. Developing a reliable signal required a very high flow rate, which was achieved with a pump connected to a pipe on the rear of the sensor housing drawing air in and a nozzle molded into the cap to direct the inlet air to the center of the sensor chip. A "zero ozone" signal could be obtained simply by controlling the pump to drop the flow rate to a low level. Figure 9.20 shows an early prototype, illustrates the measurement sequence, and gives a typical result.

These connections and discussions stimulated thoughts about using the new sensors in instruments for use in spatially dense networks to measure air quality. Could we measure the atmosphere without needing a truck to carry around the equipment? Conventional (spectroscopic) instruments are bulky, expensive, and require significant power, large temperature-controlled enclosures, and weekly or often daily calibration. These considerations limit their deployment in the field both in terms of the number of instruments and choice of monitoring site. The resulting networks of regulatory instruments may not adequately resolve the strong gradients in vertical and horizontal pollutant concentration, which are significant for assessment of personal exposure and the identification of

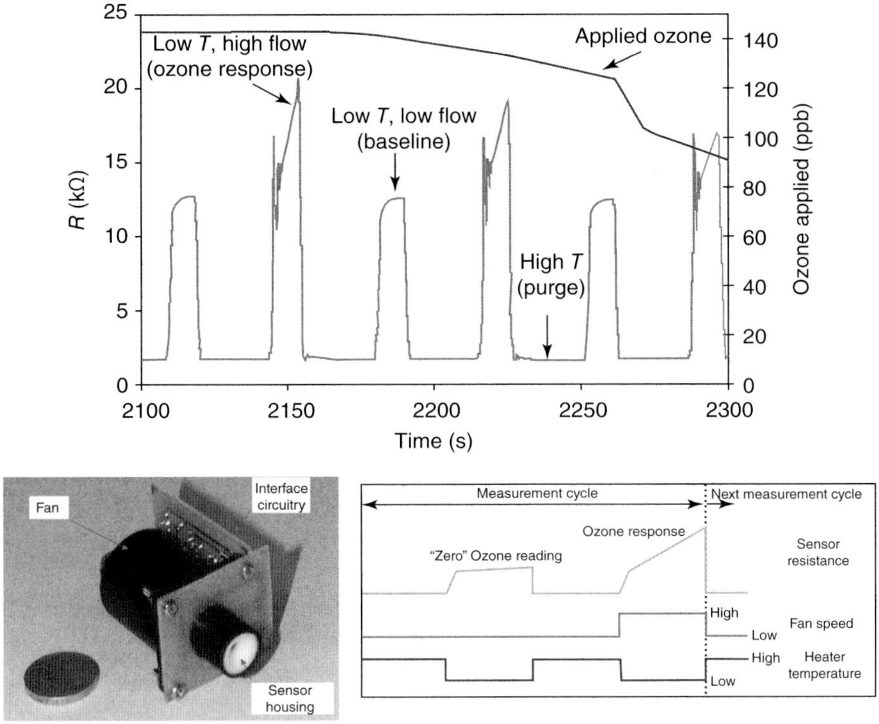

Figure 9.20 Early prototype ozone measurement module, with measurement cycle alternating fan speed and sensor temperature, and sensor resistance variation in response to the temperature and fan speed steps.

health or environmental impacts [114]. It seemed that we had a way to deal with issues like that and that there was a market opportunity. Again, there seemed little interest from established manufacturers, so in 2001, we formed a new company, Aeroqual Ltd., to make low-cost air quality instruments using the new sensors [57]. We considered that the timescale for return on investment would be too long for venture capital funding and decided to start with relatively small funding from private investors, aiming to sell sufficient instruments within a short period to continue to fund the development of the company. In 2001, the US Food and Drug Administration approved ozone as a sterilizing agent and that prompted a change in direction. The prototype ozone instrument derived from the earlier scientific work was particularly good. In a short time, the crude prototype was converted into a polished commercial instrument (Figure 9.21) and this led to a small but promising business providing hand-held ozone monitors to the growing ozone industry in the United States. Instruments for air quality measurement networks followed, driven by a growing world-wide market in less-developed countries with poor air quality and limited air-monitoring infrastructure. These instruments (Figure 9.22) utilize a variety of low-cost sensors for different parameters, including sensors based on semiconducting oxides. They build upon the design and operating experience gained from

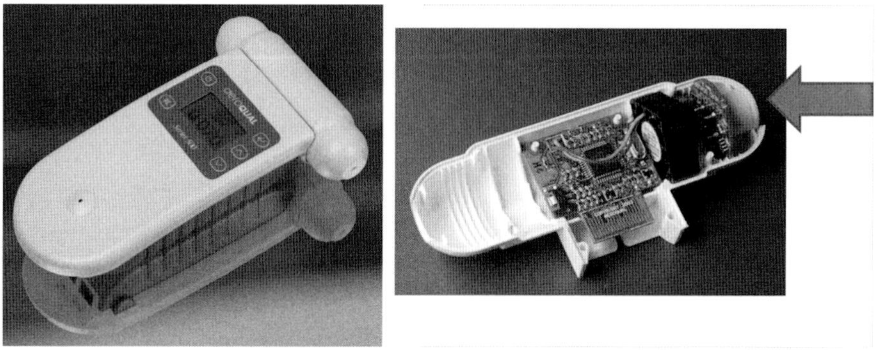

Figure 9.21 Hand-held instrument for measurement of ozone at part-per-billion levels in the atmosphere. The base unit has power, display, and communications. The detachable head unit holds the sensor ID, the sensor calibration, and the measurement result computation. Heads are calibrated as complete units. The arrow points to the sensor, mounted on its heater control, and resistance measurement circuitry, in front of the fan. Source: Reproduced with the permission of © Aeroqual Ltd.

extensive commercial deployment of the ozone instruments. They are designed to be significantly less expensive and significantly more compact than standard reference instruments (UV spectrophotometers for ozone; chemiluminescence analyzers for NO_2). They can include careful air handling, systems for automated checking of span and zero, and environmental temperature control, which we found greatly improved instrument stability. They have been extensively evaluated in field trials and thousands are now in operation. The data were very comparable in precision and accuracy with that delivered by standard reference instruments [54], showing that when correctly handled the sensors were stable and accurate devices. Still, the goal was to develop true low-cost instruments and demonstrate them in dense networks [56, 115].

The first trial of a dense sensor network for ozone was carried out in Vancouver, British Columbia [55], a city that has an extensive high-quality regulatory monitoring network with which the results of the low-cost devices could be compared. The instruments were a modification of the hand-held unit and were mounted on the reference stations and were also deployed across the city, mounted on private houses, schools, and also free-standing, powered by solar panels (Figure 9.23). A network was also deployed around Auckland, New Zealand [116, 117]. The networks have provided a rich source of data and insights into factors that affect the distribution of ozone across the city [117] and the way in which pollutants disperse in streets [118], leading to new research in environmental science and epidemiology of air quality effects on health, as well as into practical matters such as how to mount the instruments, protect them from the weather, and communicate the data [116].

The major cost in operating a network of instruments is maintenance and regular calibration. If regular calibration were to be needed, then a low-cost sensor network would not be low-cost at all. So, an intriguing question is: if a device is calibrated when it is put out in the field, how do you know that the calibration remains stable? Trivial occurrences such as spikes or complete failures or

Figure 9.22 Compact air quality measurement station utilizing low-cost sensors. The station is configured simply by choosing modules for the parameter of interest. The housing is temperature controlled. The unit can be remotely operated. Source: Reproduced with the permission of © Aeroqual Ltd.

spectacular drifts of signal are easy to pick. However, if the data are to be useful, the measurements have to be reliable: meaning that subtle drifts in span or zero need to be detected. A specification is needed for the calibration drift and methods need to be devised to check whether the measurement results can be relied on: unreliable data are worse than no data at all. The users of the data, whether a regulatory body or a citizen scientist, need to know that they can trust the results. How can this be achieved in the absence of expensive periodic calibration? That is a challenging question that has turned out to be a real frontier question and which is now the final step in the journey [58, 119]. The key turns out to be the correct formulation of the problem. It is in part a problem in statistics: what is the likelihood of a particular reading given all the other information that is available; is the low probability that may be deduced a consequence of some local event or is it indicative of an instrument issue? It is also in part a problem of understanding

Figure 9.23 Low-cost ozone measurement instrument configured to operate in an air quality measurement network.

how the various signals that can be derived from a sensor can indicate issues, given expectations based on experience and models, so linking back to the basic sensor science discussed earlier. These thoughts led to the perception of a market need. It is quite demanding to install and operate a sensor network and deliver reliable data, and in general people are interested only in the data. They need to be assured that the data are reliable and fit for purpose, and do not really care exactly how they have been obtained. This perception of a market need and our approach based on our own challenges operating low-cost sensor networks has led to the formation of another company: this time directed at providing comprehensive air quality measurement and monitoring services to clients around the world: installing and operating dense sensor networks and delivering reliable data [59].

The next sections step through the journey in detail.

9.3.2 Development of a Low-Cost Instrument for Measurement of Ozone in the Atmosphere

The challenge to overcome for a low-cost instrument is to sustain a reliable signal, without calibration, for long periods of time, with minimal, simple maintenance. Following the discovery of the strong response of WO_3 to ozone and its commercial application in the control of domestic disinfection ozone gas generators, it was evident that there was a long-term drift in sensor signal that would limit use in more challenging applications. The drift was partly due to a baseline drift and partly due to the slow response and recovery following ozone exposure. The study of the response to temperature steps led to the resolution of these issues. First, we hypothesized that the conductance of the device could be represented as a parallel combination of a "bulk" conductance and a "surface" conductance – following the ideas developed in detail above for different representations of microstructure effects on the conductance of semiconducting oxide sensors. We hypothesized that the drift was due to a time variation of the "bulk" element, which could be caused by injection of oxygen vacancies at the interface and their slow transport into the bulk. If this were the case, then the drift should be canceled by simple subtraction of the zero ozone signal. We also supposed that one element in the variability was a variability in the surface oxygen vacancy concentration, which a simple response model proposed was dependent on the ozone concentration. As we knew that the response to ozone disappeared at high enough temperature, and also that the oxygen vacancy mobility increased dramatically with increasing temperature, we thought that we could both anneal any oxygen vacancy concentration profile and establish a defined initial state for the surface by holding the sensor first at a sufficiently high temperature then stepping down to a temperature at which the ozone signal was significant. This method proved extremely successful and the theoretical model solidly justified [50].

We used the temperature step method to study in detail the response to ozone as a function of oxygen partial pressure [51]. Typical resistance transients of a screen-printed WO_3 sensor in response to a small temperature step in the presence and absence of ozone are shown in Figure 9.24. We achieved a satisfyingly complete description of the results using the simple effective medium model. The idea is that the conductivity is controlled by oxygen vacancies that are generated and destroyed at the gaseous interface by reaction with oxygen and which also react with ozone at the interface. The ozone reaction is more rapid than the oxygen reaction:

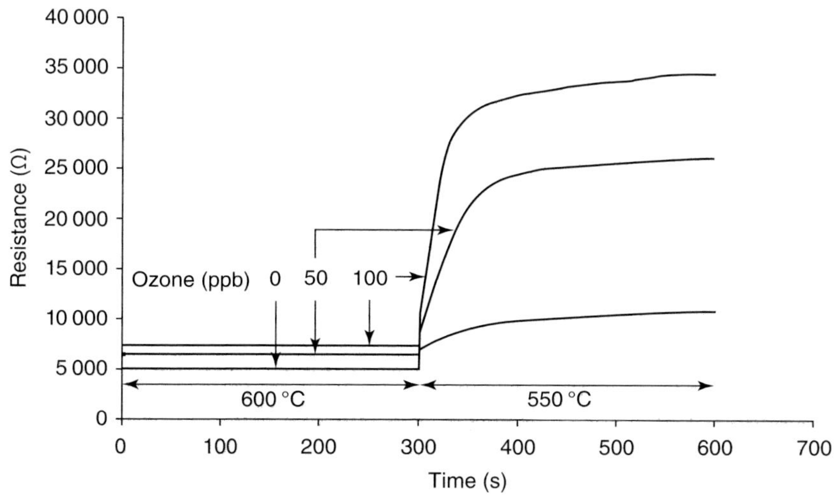

Figure 9.24 Resistance transients of a screen-printed WO_3 sensor in response to temperature steps, in the presence and absence of ozone in air. Source: Reproduced with permission from Ref. [51].

$$X \underset{k_{-1}}{\overset{k_1}{\longleftrightarrow}} V + \frac{1}{2}O_2 \qquad V + O_3 \xrightarrow{k_2} X + O_2 \tag{9.20}$$

where X denotes the unperturbed lattice. V is used as a shorthand notation for a neutral species that can be considered as an ion pair between an oxygen vacancy and a reduced lattice tungsten ion: oxygen vacancies are charge balanced by reduction of tungsten ions. Charge carriers are produced by thermal excitation of electrons from reduced tungsten ions into the conduction band. The charge carrier concentration, and hence the conductivity, changes in response to changes in the oxygen vacancy concentration at the interface. Assuming that the charge carrier concentration is much smaller than the concentration of impurities in the lattice determines that the concentration of the ionized electron trap states is essentially fixed by charge balance with the impurities. The steady-state response varied from linear to quadratic between different sensors (Figure 9.25). The equivalent circuit model for the microstructure that was described in detail earlier (Figure 9.13) accounted for the variability. The variation of the interface element with ozone concentration is given by

$$\sigma_2 = \frac{a_0 P_{O_3}}{a_1 P_{O_3} + 1} \tag{9.21}$$

where the parameters a_0 and a_1 depend on the two temperatures, the time after the step that the measurement is taken and the oxygen partial pressure. The response time constant, τ, is

$$\tau = (k_2 P_{O_3} + k_{-1} P_{O_2}^{1/2})^{-1} \tag{9.22}$$

with the rate constants defined in Eq. (9.20). It varied as expected, linearly with ozone concentration. The dependence of τ on oxygen partial pressure was very small, consistent with the model, which requires that for a significant response $k_2 P_{O_3} \gg k_{-1} P_{O_2}^{1/2}$. The response time constant also varied with the thickness of the printed device: an effect that was successfully modeled, using the equivalent circuit representation, as an effect of variation of the microstructure. The use of multielectrode devices (as described in Section 9.2.1.3) indicated that there was little gradient of ozone concentration through the sensor thickness: that is, the decomposition rate of ozone on the heated WO_3 at these elevated temperatures was small, which is the unusual effect that leads to the response. This work also showed the importance of an initial "burn-in" period for the sensors: ideally heated in ozone. We assumed that this was necessary to scavenge traces of organic material left from the manufacture. It also showed that after exposure to a humid atmosphere at room temperature for some time, the sensors had to be heated to 600 °C for a short period to develop a reproducible response. We assumed that this was due to hydroxylation of the surface after long exposure to water vapor at room temperature, similar to observations on SnO_2.

Practically, the variation of response with microstructure meant that it was possible to adjust microstructure to reduce as far as possible the response time and to prepare sensors that were optimized for different concentration ranges (Figure 9.25). That meant that sensors could be prepared that were useful for disinfection monitoring (concentrations up to parts-per-million) as well as others

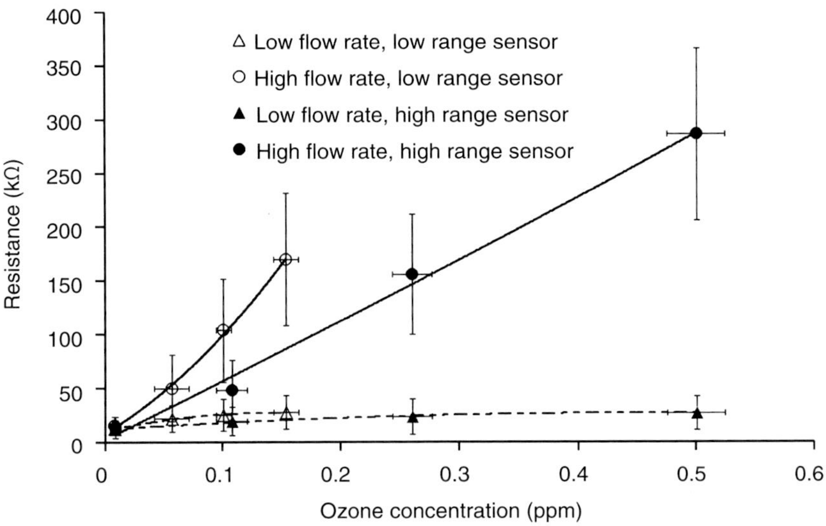

Figure 9.25 Variation of WO_3 sensor resistance with ozone concentration, measured for a large number of sensors of two different microstructures (mean ± 1SD). The resistance is measured 5 s and 35 s after the step from high to low temperature. Source: Reproduced with permission from Ref. [54].

that were useful for atmosphere monitoring or for monitoring environments useful for preventing mold growth on food during transportation [120], opening up a wide range of useful niche markets for the instruments. The equivalent circuit representation has proved to be a very robust description of manufacturing variability, emphasizing yet again the importance of microstructure control at every size scale of the device. Comparison of results in the literature shows the evolution in performance of these devices [50–52, 54, 113].

As noted above, the instrument design used an air flow rate step to determine the zero. The previous work had indicated that the ozone concentration could be derived from measurement of the conductance difference, $\Delta\sigma$, measured as the difference between high air flow and low air flow conductance, each determined a fixed time after the temperature step. If σ_1 (Figure 9.13) is small, then

$$P_{O_3} = \frac{a\Delta\sigma}{1 - b\Delta\sigma} \tag{9.23}$$

with a and b constants to be determined by calibration.

Prediction of the ozone concentration from Eq. (9.23) is in practice problematic because the "zero" conductance can be very significantly larger than the "signal" conductance and so the result can become very sensitively dependent on the accuracy of measurement of the "zero". Alternative prediction formulae using the sensor resistance that are less sensitive to measurement errors can be derived that preserve the drift correction idea and that are robust against the variability in the sensor response due to manufacturing variability.

The instrument (Figure 9.21) is in two parts. A detachable head houses the sensor, fan or air pump, air inlet filters, and sensor control electronics. The linearization function is loaded into this unit, which delivers a digitized, linearized output.

The second part of the unit comprises the power, communications, and display. The pump draws air through the housing by an outlet directly beneath the sensor. The top of the housing is formed as a conical nozzle that directs air onto the face of the sensor element at a rate precisely controlled by the pump or fan. The nozzle dimensions and flow rates depend on the sensor and housing dimensions and were empirically adjusted to achieve the desired performance. Calibration is a critical part of the manufacturing procedure. Because ozone is so reactive, considerable care is needed. The detachable heads are calibrated as complete units. The units are mounted in a rack inside a large Perspex box though which filtered ambient air (thus containing ozone dependent on ambient conditions) is drawn. The air inside the box is stirred by a fan. Ozone is generated inside the box by a controllable, shielded UV source. The air inside the box is sampled to a spectrometric reference analyzer that is itself periodically calibrated against standard instruments ultimately traceable to NIST, Washington DC. This arrangement means that all the parts of the instrument that could affect the reading are incorporated in the calibration stage, including the inlet system. The instrument samples a large, uniform mass of air that is simultaneously sampled by the reference device, just as it would when deployed. In the first stage of the procedure, linearization, the sensor resistance for each head is read at a sequence of increasing ozone concentration. The resistance is fitted to the linearization function and the parameters of this function are loaded into the head. In the second stage, carried out independently, the ozone concentration in the box is again cycled and the linearized output from each head unit is compared to the reference analyzer measurement. For over 3000 instruments constructed to cover two different ranges (0–150 ppb and 0–500 ppb), this final comparison gave slope 1 ± 0.05 and offset 0 ± 0.004 ppb [54].

In the atmosphere, running alongside reference analyzers, the new ozone instruments had impressive performance. However, the slope could be significantly different from unity. Field studies with instruments over a wide variety of locations, correlating hourly averaged ozone over 48 h periods showed slope of instrument predicting reference analyzer of 0.9 ± 0.2 and offset 6 ± 6 ppb. Figure 9.26 shows an example. The standard error of estimate for a given instrument was 3 ± 2 ppb over the range 0–80 ppb. For the lumped data of a group of instruments, it was 6 ppb.

The initial linearization uses just three or four different ozone concentrations. The number of different concentrations used is simply determined by the time, hence the cost. The linearization fit involves a least-squares fit to a nonlinear function. The estimated ozone concentration derived from such fitting is P_{est}, a function of the measured low-flow and high-flow resistances, R_0 and R_g, and the parameters of the fitting function (the coefficients a and b in Eq. (9.23), for example):

$$P_{est} = f(R_0, R_1, a, b). \tag{9.24}$$

The estimating function would, in general, not be exact, being derived from a least-squares fit of the assumed response function to the calibration data. The parameters would themselves have an error estimate that reflected the goodness-of-fit. The data, R_0, R_1, would also be subject to some measurement

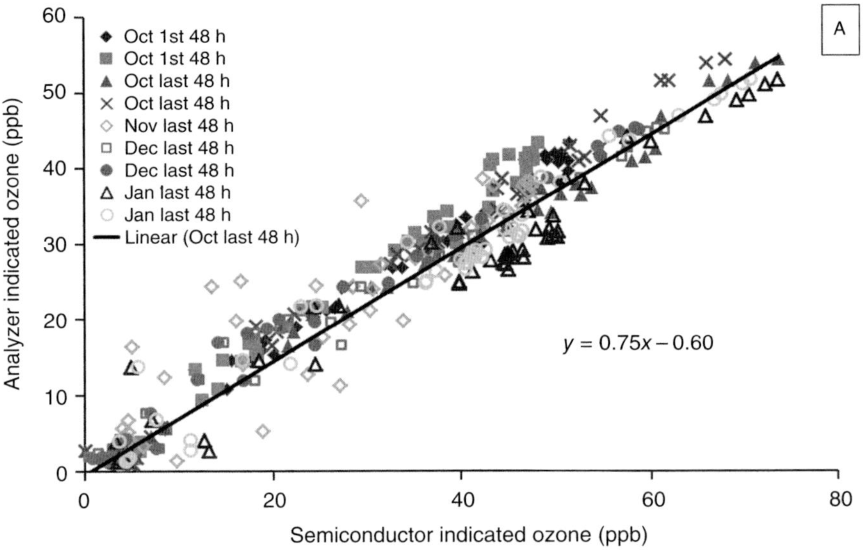

Figure 9.26 Response of two different WO_3-based ozone instruments in the atmosphere (unit similar to that shown in Figure 9.22) correlated with the measurement result of a reference analyzer with which they were co-located. The measurements are hourly averaged for 48-h periods at the beginning and end of each month, with a total measurement time of four months. The line is for one set of 48 h measurements at the beginning of the trial. Source: Reproduced with permission from Ref. [54].

error. Then for some subsequent measurement of an ozone concentration, actual value P_{O_3}, using the measured resistances, the difference between the actual and estimated concentrations can be expressed as a Taylor series in a small variation of the parameters:

$$\delta P = P_{O_3} - P_{est} \approx \frac{\partial f}{\partial a}\delta a + \frac{\partial f}{\partial b}\delta b \tag{9.25}$$

Typically, the leading error term is linear in the ozone concentration, so a linear or at worst a quadratic correction can be derived to convert the estimated ozone concentration to the actual value. The linear dependence of reference analyzer measurement on P_{est} confirms this analysis, which is also supported by simulations with different assumptions concerning the errors in measurement of the resistances and the number of different calibration concentrations and their values [55]. The errors are acceptable for many applications, but for extended campaigns of environmental measurement, therefore, we first mounted instruments next to a reference analyzer for at least 48 h in order to align the calibration with a local reference.

9.3.3 Signal Drift Detection

The instrument described in the previous section has a very simple maintenance philosophy: the sensor head is simply exchanged for another if there is any reason to doubt the measurement reliability. There are some clearly understood failure modes. Sublimation of WO_3 at the reset temperature, particularly in a humid atmosphere, causes a change of microstructure and an increase in sensitivity. Cracking and spallation of the oxide layer can result from the repeated thermal stress of the temperature switching. The air flow rate can fluctuate or drop as a consequence of a failure of the air pump or fan. Multielectrode devices can detect microstructure change, as noted above. There are also some simple diagnostics of sensor stability and measurement reliability available from straightforward measurements on the device [55, 58, 116, 117]. First, the "zero" ozone measurement (low air flow rate) should be reasonably stable. If the zero resistance rises significantly, this is an indication of a change in the microstructure of the sensor [54]. Second, the accuracy is dependent on the high air flow rate being sufficiently high and stable. The sensor itself can function as a hot wire anemometer: the power required to maintain the sensor at a fixed temperature is sensitively dependent on the air flow rate, particularly during the high-temperature ("reset") part of the cycle. The heater power requirement is easy to measure and a good indicator that the air pump is operating correctly and that the inlet is not obstructed [56]. Another failure mode is that dirt can deposit on the inlet filters, in the inlet line and upon the sensor surface, resulting in catalyzed decomposition of ozone and hence a measurement lower than the actual ozone concentration. This is more tricky to detect from measurements on the sensor alone, since that may be working reliably, telling what it actually sees. Additional information is needed: that problem is addressed in Section 9.3.5.

These methods of verifying data reliability come into their own when instruments are deployed in a network for a significant time, without calibration, and a decision needs to be made as to whether the sensor head should be changed.

9.3.4 A Low-Cost Instrument for Measurement of Atmospheric Nitrogen Dioxide

Previous literature reports have noted the high sensitivity of WO_3 elements to NO_2 and relatively very low sensitivity to NO [90, 121]. We discovered that almost arbitrarily large sensitivity could be achieved by manipulation of the microstructure to yield an extremely porous device with a very large internal surface area. The key problem to solve was the overwhelmingly higher response to ozone. The key development has been that of a simple system to scrub ozone from the gas stream presented to the device [56]. Because ozone decomposes extremely rapidly at elevated temperature, both by gas-phase reaction and by catalyzed decomposition on metal surfaces, we used a simple thermal scrubber, decomposing the trace ozone in air by passage through a heated stainless steel tube. With a residence time of 3 s, at tube temperatures above 180 °C ozone was completely absent from the exit gas. Figure 9.27 shows that the NO_2 measurement system effectively rejected interference from NO and ozone. It also illustrates that the sensor resistance response to change of NO_2 concentration at the level required for atmosphere monitoring was more than adequate, but that the response transient was slow. This is a current limitation of the device. Even so, the instrument could be calibrated in the atmosphere by assuming that the mean and variance

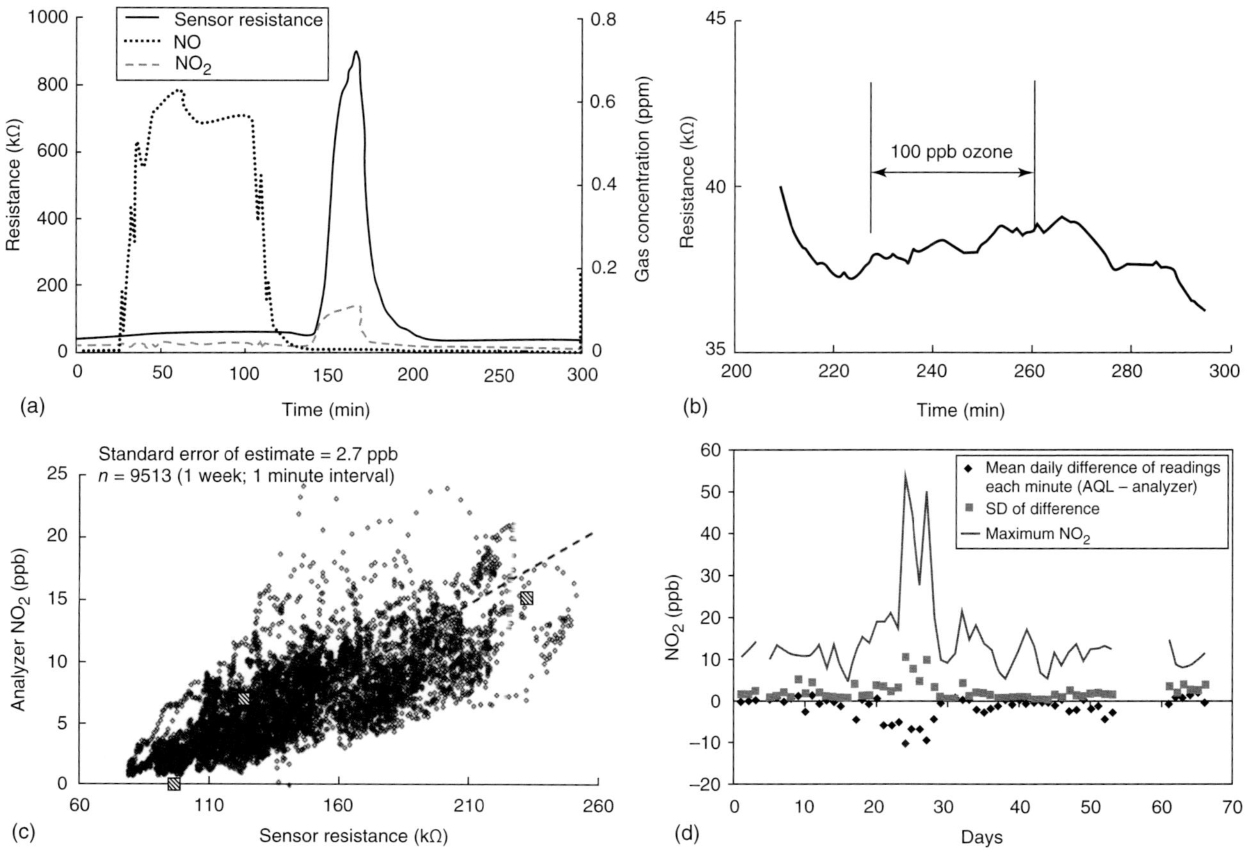

Figure 9.27 Performance of WO_3 sensor for NO_2 in the atmosphere, using a simple thermal scrubber to remove ozone interference. (a) High sensitivity to ppb levels of NO_2 in air and lack of sensitivity to NO. (b) Illustration of the efficiency of the ozone scrubber (a heated length of stainless steel tube). (c) Sensor resistance against signal from co-located NO_2 reference analyzer, 1 week of measurements every minute. The square symbols and dashed line are the laboratory calibration. (d) Mean daily difference of NO_2 concentration derived from the sensor resistance calibration in (c) from the analyzer measurement, for measurements every minute, for 1500 h. Source: Adapted from Ref. [56].

of the signal delivered by a sensor co-located with a reference analyzer should match the mean and variance of the concentration as measured by the analyzer and then the sensor tracked a reference analyzer reliably for >1000 h [56]. The major failure mode of this device was performance degradation of the ozone scrubber, as if ozone leaked through the interference signal was extremely large.

9.3.5 Networks of Instruments in the Atmosphere

Developments as have been described above, together with advances in data management, processing, and communications, have made it financially and logistically conceivable to operate a spatially dense network of monitors with high temporal resolution. It has been termed a "changing paradigm" for air quality measurement [122]. Such networks would have the potential to resolve the complex spatial and temporal heterogeneity of air pollution concentrations in urban centers in near-real time and would make it possible to answer new questions about the underlying causes of poor air quality, ensure more accurate modeling and prediction at local scales, improve the ability to identify the links between air quality and human health or environmental degradation, identify potential air pollution "hot spots," and enhance the ability to quantify the impacts of pollutant mitigation techniques. Experimental networks have certainly demonstrated feasibility and explored issues [55, 116, 117, 123]. However, these possibilities can only be fully realized if the maintenance requirement to provide reliable data is low. The data have to be believable without the need for regular on-site calibration because the necessary network scale may be very large – hundreds or thousands of instruments. This is an extremely challenging problem.

The discussion emphasizes instrument "reliability," in order to describe instrument performance [8, 17, 18]. This measure is less restrictive than compliance. However, its definition is not necessarily obvious and does require clarity such that users have confidence in the data within known and defined constraints. A first requirement is a specification, so we started by reflecting on the stated purpose of a dense low-cost network. Following Snyder et al. [122], we specified this purpose as supplementing a compliant ambient air monitoring network, extending coverage and providing reliable information for communities, including improved local coverage for exposure assessment and enhancing source compliance monitoring. We framed this specification in terms of reliability of determination of the statistics of variation of the pollutant. We also interpreted reliability in terms of an instrument specification: slope 1 ± 0.3 and offset 0 ± 5 ppb based on the maximum variability for transfer standard accuracy (slope) and for indicative measurement (offset) suggested by the US EPA [124, 125]. The specification can alternatively be framed in terms of reliability of detection of exceedances: this is equivalent to reliability in determining the statistics of variation.

Reliability can be assessed using temporary or permanent co-location of one or more instruments. There have been studies using low-cost devices where the random co-location of devices allows spot checking of one against another [126]. However, there remains the possibility that unusual trends in a monitor that is not currently co-located against another may be misinterpreted erroneously as either fluctuations in environmental processes or as instrument error. Reliability can also be assessed using computational techniques to detect and compensate for changes from an expected pattern [127], and for specifically defined instrument conditions [128, 129]. These methods generally assume either a model for the phenomenon being sensed, typically exploiting correlations across the network, or a model for the behavior of the sensor within the instrument. Simple multivariate time series, principal component analysis, or "soft sensor" methods [130–134] require a longtime run of data to establish the model from which drifts and malfunctions can be detected. General limitations of such approaches include the accuracy and reliability of models and the data upon which they are built, and the model stability over time. All methods suffer from to some degree from these limitations.

In our work, we have aimed to develop methodologies for assessing the reliability of the datasets provided from these instruments, in real time during deployment, at minimal cost [58, 119]. We have developed a framework such that general qualitative knowledge can be used, both about the spatio-temporal behavior of the pollutant as can be derived from land-use regression (LUR) models, and about the performance characteristics of the measurement system, which encompasses the instrument and its sampling system and the location characteristics.

If $Y_k(t)$ denotes the signal at time t from an instrument at position k and $X_k(t)$ denotes the "true" value of the target analyte at position k, then Y_k is a predictor of the unknown X_k and if Y_k is the linearized instrument output, rather than the raw data from the sensor from which the output is derived, then

$$X_k(t) = a_1 Y_k(t) + a_0 + \varepsilon_X \tag{9.26}$$

The initial calibration (both linearization and calibration alignment) establishes $a_1 = 1 \pm \varepsilon_\alpha$ and $a_0 = 0 \pm \varepsilon_\beta$ where the ε are normally distributed random errors that satisfy some specification for the instrument: $\varepsilon < \varepsilon^*$. The question about calibration stability might then be framed as: what is the probability that the instrument result, Y_k, correctly predicts the true X_k: $\boldsymbol{P}_t(Y_k = X_k \pm \varepsilon_X)$ or what is $\boldsymbol{P}_t(a_1 = 1 \pm \varepsilon_\alpha)$ and $\boldsymbol{P}_t(a_0 = 0 \pm \varepsilon_\beta)$ where $\boldsymbol{P}_t$ denotes probability and is a function of the time, t, since calibration and installation? In principle, Bayes' theorem provides a way for assessing $\boldsymbol{P}_t$. We use it in the framework of "a probability of an event based on prior knowledge of conditions," and as "a method that allows new evidence to update beliefs." However, the formal application of Bayes theorem is fraught with practical difficulties. Thus, if $Z_k(t)$ denotes the time series of all other data ("evidence") that might be pertinent to our question, and we wish to establish that $\boldsymbol{P}_t(Y_k = X_k \pm \varepsilon_X) > \boldsymbol{P}^*$, where $\boldsymbol{P}^*$ is a critical level below which some action such as maintenance or investigation would be triggered ("hypothesis"), then tracking back through the time series to the initial installation of a calibrated instrument where each estimate constitutes the prior for the next, Bayes' theorem states

$$\boldsymbol{P}_t(Y_k = X_k \pm \varepsilon_Y | Z_k) = \left(\prod_0^t \frac{\boldsymbol{P}_t(Z | Y_k = X_k \pm \varepsilon_Y)}{\boldsymbol{P}_t(Z)} \right) \tag{9.27}$$

where it is assumed that, at installation, $\boldsymbol{P}_t(Y_k = X_k \pm \varepsilon_X) = 1$. The problem is first the reliable determination of the conditional probability distributions and second that $\boldsymbol{P}_t$ determined in this way will tend to zero very quickly unless the conditional probabilities $\boldsymbol{P}(Z|Y)$ are extremely high, that is the evidence or model needs to be extremely good. An alternative is a scheme whereby if $\boldsymbol{P}_t(Y|Z) < \boldsymbol{P}^*$ for some time $t > t_f$, then the data are accepted as "ground truth" and t is reset to $t = 0$. The choice of t_f becomes critical. The "evidence" Z can be a model for the sensor (for example, using the movement of an operating point off an operating line as discussed above, or some "soft sensor" model) or it could be a model for the atmosphere and for pollutant dispersion, or it could be a correlation, of the local instrument reading with those from other instruments in the network. A large number of schemes have been given. Most require large training sets of data [119] or involve empirical models for sensor drift. We sought a method that did not require a large training set and which could indicate issues within a reasonable time (e.g. 1 week). We started from the position that, if there was no evidence to the contrary, the instrument result was the best estimate available of "ground truth": that is, we do not attempt to "re-calibrate." This is philosophically different from much of the literature in this area, which is concerned with "blind" or "semi-blind" calibration: that is, with methods to adjust the instrument parameters α and β to obtain some sort of consistency with an interpolated field and which is therefore strongly dependent on model assumptions.

The state of the local and regional atmosphere is driven by several processes with different spatial and temporal characteristics, such as traffic, industrial emissions, sunlight, and wind. Correlations across the region are to be expected, as are patterns that may remain stable on average over time. These patterns are revealed, for example, by LUR, a method that is widely used as a predictive tool [135]. Naturally, with increasing density of instruments, the cross-correlations between different locations tend to increase. The construction of a network is subject to the dual problem: (i) the justification for a particular instrument is to observe and detect local, spatial, and temporal variations, that is, the deviation from the state that can be predicted (interpolated) by the other sensors. This means that it is not too correlated with other instruments, otherwise it would be superfluous; (ii) that there is a sufficient degree of predictive redundancy that methods to detect instrument problems can work reliably. Another way of expressing this is to say that the spatial field of the pollutant is sufficiently over-sampled [136]. If the instruments are sufficiently inexpensive and reliable, then the spatial density can be large, but here, reliability needs to be emphasized because the cost of maintenance, even such a simple step as changing a sensor head, can rapidly swamp all other costs as the network scale increases.

We described methods using "proxies" and demonstrated these using the results from an extensive and long-term field deployment of low-cost ozone instruments in the Lower Fraser Valley, British Columbia, which included the city and suburbs of Vancouver [58]. The proxies were chosen based on LUR, that is, either a reference station that shared a land-use category with the instrument under evaluation or the median of all other instruments in the same land-use category. The methods were based on the statistics of the variation of ozone concentration. Formally, as above, we can consider that $Y_k(t)$ is a predictor of

$X_k(t)$, or we can consider that the conditional probability distribution of Y_k given X_k is stable and time invariant; in other words, although the ozone concentration (actual or measured) is highly time varying, the measured process conditional on the underlying actual process is stationary, provided that the data Y_k are measuring reliably. Of course, the reliability of $Y_k(t)$ cannot be assessed directly, since $X_k(t)$ is by definition unknown. Hence, we seek some proxy, $Z_k(t)$ against which $Y_k(t)$ can be assessed.

Let $F_{X_k}(x; t_1, \ldots, t_1 + t_d)$ denote the empirical cumulative distribution of $X_k(t)$ obtained over a time index t_d (an integer that counts the time steps of measurement) running over the data stepping by one time step. As noted, we defined the purpose of the network as supplementing a compliant ambient air monitoring network, specifically to provide reliable local determination of the short-term statistics of the ozone concentration. Hence, we defined the purpose of the network to be to deliver a reliable local estimate of $F_{X_k}(x; t_1, \ldots, t_1 + t_d)$. The data, $Y_k(t)$ and the proxy, $Z_k(t)$, if the proxy is appropriately chosen, provide two different estimates, $F_{Y_k}(x; t_1, \ldots, t_1 + t_d)$ and $F_{Z_k}(x; t_1, \ldots, t_1 + t_d)$, of this distribution. Hence, a straightforward procedure is to ask with what probability these estimates are the same. The two-sample Kolmogorov–Smirnov (K–S) test gives this probability. The number of time steps, t_d, used for the determination of the distribution is empirically chosen to obtain a reasonably representative estimate of the variation of the ozone concentration that is sufficiently large that the distribution is well estimated and missing values have a small effect, but sufficiently small that action can be taken in a reasonable time depending on the result of the test. We have used hourly averaged data for $Y_k(t)$ and $Z_k(t)$, and so the time step is one hour.

Let $\boldsymbol{P}^*_{\text{KS}}(Z_k, Y_k)$ denote the critical value for the K–S statistic. If $\boldsymbol{P}_{\text{KS}}(Z_k, Y_k) > \boldsymbol{P}^*_{\text{KS}}(Z_k, Y_k)$ then $F_{Y_k}(x; t_1, \ldots, t_1 + t_d)$ and $F_{Z_k}(x; t_1, \ldots, t_1 + t_d)$ can be considered to be estimates of the same distribution. In this case, the instrument is then considered to be functioning as expected (termed "intact") and we take $X_k(t) = Y_k(t)$ over the interval $(t_1, \ldots, t_1 + t_d)$. However, if $\boldsymbol{P}_{\text{KS}} \leq \boldsymbol{P}^*_{\text{KS}}$, then an alarm is signaled. Either the instrument is not intact (a "true alarm") or the local environment has changed either with respect to the proxy or to the measurement location, or the proxy is not suitable for ozone concentration variation at the measurement location (a "false alarm"). A criterion has to be defined to discriminate in a practically acceptable way between these two alternatives. Site-specific, weather-specific, and large-scale event-specific phenomena can be expected when analyzing air quality time series, and so periodic alarms may occur even for intact measurements, where the proxy signal is not a suitable estimator for $F_{X_k}(x; t_1, \ldots, t_1 + t_d)$. For such variations, if the low-cost instrument were intact, then while an alarm might be signaled, after some time $\boldsymbol{P}_{KS}$ could be expected to once more increase above the threshold. However, for instrument failure – a drift or sustained change in calibration parameters – then a clear and sustained pattern of change in $F_{Y_k}(x; t_1, \ldots, t_1 + t_d)$ with respect to $F_{Z_k}(x; t_1, \ldots, t_1 + t_d)$ would be observed and hence a sustained pattern of change in $\boldsymbol{P}_{\text{KS}}$. We therefore defined

$$X_k(t) \neq Y_k(t) \text{ if } \boldsymbol{P}_{\text{KS}} < \boldsymbol{P}^*_{\text{KS}} \text{ for } t > t_f \tag{9.28}$$

where the time index, t_f, is considered sufficient for confidence that the two distributions remain different and is empirically determined. The threshold probability, $\boldsymbol{P}^*_{\mathrm{KS}}(Z_k, Y_k)$, and the averaging time, t_d, are also to be specified. The tests run over the data, one time step at a time, and so the time series of $\boldsymbol{P}_{\mathrm{KS}}$ evolves as the data increment.

A second approach is to ask whether any drift in the parameters a_0, a_1, or ε_X remains within bounds defined by the network specification. In this case, we assumed that the mean and variance of the data, $Y_k(t)$ and the proxy, $Z_k(t)$ are both estimates of the mean $\mu(X_k)$ and variance $\mathrm{var}(X_k)$ of $X_k(t)$ evaluated over the time t_d. Thus,

$$\mu(X_k) = a_0 + a_1\mu(Y_k) \text{ and } \mathrm{var}(X_k) = a_1^2\mathrm{var}(Y_k) + \mathrm{var}(\varepsilon_X) \tag{9.29}$$

Indeed, if the instrument remains in calibration, $a_1 = 1 \pm \varepsilon_\alpha$ and $a_0 = 0 \pm \varepsilon_\beta$.

The proxy should be chosen such that mean $\mu(Z_k(t_1, \ldots, t_1 + t_d))$ and variance $\mathrm{var}(Z_k(t_1, \ldots, t_1 + t_d))$ evaluated over the interval $(t_1, \ldots, t_1 + t_d)$ satisfy

$$\mu(X_k(t_1, \ldots, t_1 + t_d)) \cong b_0 + b_1\mu(Z_k(t_1, \ldots, t_1 + t_d)) \tag{9.30}$$

$$\mathrm{var}(X_k(t_1, \ldots, t_1 + t_d)) \cong b_1^2\mathrm{var}(Z_k(t_1, \ldots, t_1 + t_d)) + \mathrm{var}(\varepsilon_{Z,k}(t_1, \ldots, t_1 + t_d)) \tag{9.31}$$

where a good proxy implies $\mathrm{var}(\varepsilon_{Z,k}) \ll \mathrm{var}(Z_k)$ and the parameters b_0, b_1 and $\mathrm{var}(\varepsilon_{Z,k})$ at most fluctuate within defined bounds over the observation period. Given the definition of the proxy (Eqs. (9.29) and (9.30)), we defined estimators for the slope $\hat{a}_1$ and offset $\hat{a}_0$ as

$$\hat{a}_1 = \left(\frac{\mathrm{var}(Z_k(t_1, \ldots, t_1 + t_d))}{\mathrm{var}(Y_k(t_1, \ldots, t_1 + t_d))}\right)^{1/2} \tag{9.32}$$

$$\hat{a}_0 = \mu(Y_k(t_1, \ldots, t_1 + t_d)) - \mu(Z_k(t_1, \ldots, t_1 + t_d))\left(\frac{\mathrm{var}(Z_k(t_1, \ldots, t_1 + t_d))}{\mathrm{var}(Y_k(t_1, \ldots, t_1 + t_d))}\right)^{1/2} \tag{9.33}$$

The variation of $\hat{a}_0$ and $\hat{a}_1$, determined over time, t_d, is tracked and can be evaluated in real time. An alarm is signaled if these quantities move out of defined bounds, which are now specified based on an acceptable instrument specification for the error in a_0 and a_1. Again, the idea is to distinguish between instrument drift and normal atmosphere variations: periodic alarms may occur even for intact measurements, where the proxy signal is not a suitable estimator, but normally one would expect that the parameters would come back into range after some time. Hence, if $\hat{a}_1$ and $\hat{a}_0$ remain outside the bounds for time $t > t_f$, then either $X_k(t) \neq Y_k(t)$ (i.e. $a_0 \neq 0$ and/or $a_1 \neq 1$ within the bounds of the instrument specification) or $Z_k(t)$ has ceased to be a good proxy, i.e. Eqs. (9.30) and (9.31) no longer apply.

The choice of proxies is clearly critical to the success of these procedures. We used the ideas of LUR, where it had been shown that long-term mean ozone concentrations are correlated with site descriptions, such as land-use and location [135]. We evaluate the choice of $Z_k(t)$ as a reference station signal or a network median having similar land use to the location, k. We showed that simple general land-use descriptors appeared to be sufficient: ozone measurements from different instruments in the same general land use appeared to be highly correlated and those in different land uses to be distinct. In effect, we are, in a rather simple way, including LUR in a Bayesian framework to assess the reliability of a result. We compared the results of the two assessments with results from manual data scanning, and evaluation of the raw resistance data from the sensors (as described in Section 9.3.3). The methods reliably detected instrument failures within a few days, especially when used in conjunction with conditions imposed on the raw resistance data from the sensors [58]. Examples are shown in Figure 9.28. The results using a reference station proxy remote but in the same land use were essentially the same as those using a reference station with which the instruments were co-located. The K–S test (Eq. (9.28) and Eqs. (9.32) and (9.33)) detected a drift in the result from the Maple Ridge instrument essentially simultaneously, and a few days before manual inspection of the data. The issue was identified as the effect of a drifting baseline (low air flow resistance signal). The tests also detected a subtle issue with the Abbottsford instrument toward the end of the deployment: this was identified as an effect when the ozone concentration was low; it was probably weather related and due to a subtle effect of the siting of the low-cost instrument relative to the reference station intake. Thus, we have demonstrated powerful automated methods for confirming continued reliability of data from a dense network of low-cost air quality instruments. Given specific knowledge about the sensor and instrument, other information can be extracted about the nature of any failure. For example, detection of lower than expected ozone by one of the statistical proxy comparison tests in the absence of any issue with sensor signals (baseline resistance and operating line stable, heater power stable, heater resistance constant) would imply that there was some very local effect that was consuming ozone; if the effect persisted, then the most likely explanation would be ozone decomposing in the instrument inlet system – pointing at dirt in the inlet lines [55].

9.4 Conclusion and Prospects

The work has shown how a sensor principle that is well grounded in good science can be turned into instruments that are extremely well adapted to their market niche. The key has been to control the device, in the instrument, in totality: a heated semiconducting oxide is a very robust device, and the signals that it delivers tell its temperature and its immediate environment, to which it is exquisitely sensitive. Many of the issues fell away when the sensor element was correctly controlled and supplied with a well-controlled and conditioned sample, no different in many ways to any other sensitive analytical device.

The major costs associated with any instrument are calibration, maintenance, and the cost of an erroneous reading. Thus, low-cost instruments based on simple sensors are only truly low-cost if these other costs are also low. Sensors and instruments of the type I have described in this chapter come into their own when

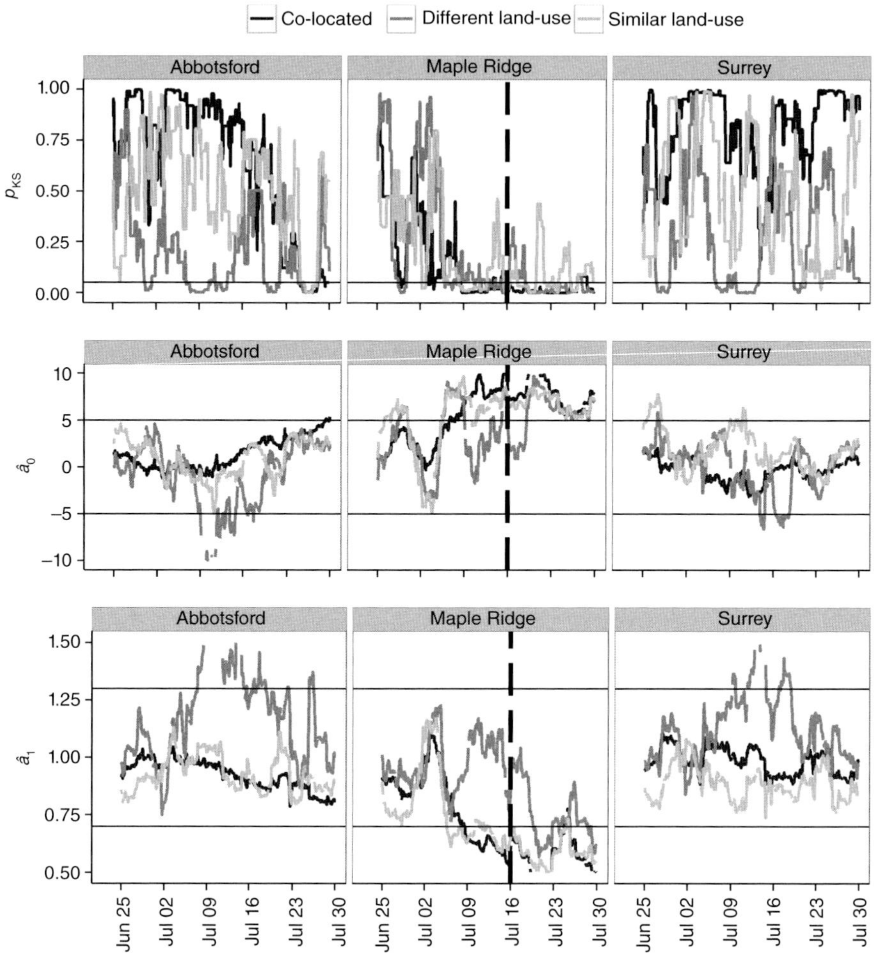

Figure 9.28 Demonstration of application of Eqs. (9.28), (9.32), and (9.33) to detect drift of low-cost instruments at three different locations with respect to reference stations as proxies. Three different proxies are shown: reference station with which the low-cost instrument was co-located, remote reference station in the same land use in the same air-shed, and remote reference station in the same air shed but different land use. The vertical line shows when drift of the instrument at Maple Ridge was detected by manual inspection of the data. Source: Miskell et al. 2016 [58]. Reproduced with permission of American Chemical Society.

there is an application that requires a lot of instruments or one where small size and portability are important. However, in these cases, one immediately comes up against the reliability of the result and the cost of confirming reliability. Here is where science and the market meet in interesting ways that demand deep study and innovative solutions.

As has been explored earlier in the chapter, semiconducting oxides are very sensitive to traces of volatile organic compounds (VOC) in air. The selectivity is generally poor, although it can be modified somewhat by adjustment of the sensor-operating temperature. Signals can also be modified by deposition of catalysts on the surface. These effects have been exploited in "electronic noses" using arrays of sensors with different composition and temperature to give some variance in response to an unknown gas mixture. This remains an active area of research. We have shown that being more sophisticated in the choice of catalyst and also manipulating the way in which the catalyst is either deposited on top of the sensor or dispersed within the sensing layer has a very significant effect on the response and greatly increases the variance of signals in response to different VOC. The idea that the sensing layer can be treated as a chemical reactor that also detects its reactants and products, and the distribution of these throughout the layer, is a powerful one and there is much development that can be foreseen to produce devices that are tuned to different applications [62, 64, 65, 85, 86].

There is certainly a need for speciation and measurement of volatile organics in the atmosphere because present methods involving automated gas chromatographs and mass spectrometers are too expensive for widespread deployment. The variation of baseline signal with relative humidity is a problem. One way forward is to combine a semiconductor sensor array with a simple temperature-programmed desorption bed. This system proved very effective [137]. It provided a defined baseline and concentrated the target gases then desorbed them at characteristic temperatures. It provided a second axis of discrimination; the sensors discriminate on surface reactivity and the desorption bed on interaction energy with the adsorbent. This method proved powerful for measurement at concentrations relevant to atmosphere studies with a general discrimination. Certainly, such a device could be further developed using miniature heaters and adsorbent beds. Whether devices like this get further developed and commercialized will depend on the "pull" from the market and clarification of the measurement problem that is to be addressed.

This chapter has described a personal journey that illustrates how basic scientific threads unite along the path from discovery to product, from the surface reactions of a semiconducting oxide, to sensor development, to instrument development, to application, and finally to big data. It has shown how a quest for thorough understanding at a deep scientific level has led to successful products, and how the development of these products and the development of markets for them have thrown up questions that have stimulated further quest for knowledge and understanding, in fields that were never in view at the start.

Acknowledgment

Many wonderful people have contributed to the work that I have described here. I have been exceptionally fortunate to have been able to work with them. The work would not have come to fruition in practice without the energy and enthusiasm of those who drove the success of the companies formed to exploit the work and who equally contributed to the science that underpinned the success of these endeavors: Peter McGeehin and Pat Moseley with Capteur Sensors and Analysers Ltd. (UK) and Geoff Henshaw and Brett Wells with Aeroqual Ltd. and Mote Ltd. (NZ). Rod Jones and Tony Cox at Cambridge University triggered the work on ozone measurement in the atmosphere that has been so productive. I acknowledge particularly the efforts of the postdoctoral researchers and students whose names appear on the papers I have cited, with special thanks to Keith Pratt, who also provided the information for Figures 9.19 and 9.20. Simon Naisbitt provided the previously unpublished data for Figure 9.11.

References

1 Heiland, G. (1982). Homogeneous semiconducting gas sensors. *Sens. Actuators* 2: 343–361.

2 Williams, D.E. (1999). Semiconducting oxides as gas-sensitive resistors. *Sens. Actuators B Chem.* 57 (1–3): 1–16.

3 Brattain, W.H. and Bardeen, J. (1953). Surface properties of germanium. *Bell Syst. Tech. J.* 32 (1): 1–41.

4 Morrison, S.R. (1953). Changes of surface conductivity of germanium with ambient. *J. Phys. Chem.* 57 (8): 860–863.

5 Morrison, S.R. (1955). Surface barrier effects in adsorption, illustrated by zinc oxide. *Adv. Catal.* 7: 259–301.

6 Seiyama, T., Kato, A., Fujiishi, K. et al. (1962). A new detector for gaseous components using semiconductive thin films. *Anal. Chem.* 34 (11): 1502–1503.

7 Chiba, A. (2012). The development of the TGS gas sensor. In: *Chemical Sensor Technology* (ed. S. Yamauchi), 1–18. Tokyo, Amsterdam: Kodansha and Elsevier.

8 Windischmann, H. and Mark, P. (1979). Model for the operation of a thin-film SnO_x conductance-modulation carbon monoxide sensor. *J. Electrochem. Soc.* 126 (4): 627–633.

9 McAleer, J.F., Mosely, P.T., Norris, J.O.W. et al. (1987). Tin dioxide gas sensors. 1. Aspects of the surface-chemistry revealed by electrical conductance variations. *J. Chem. Soc. Faraday Trans. I* 83: 1323–1346.

10 Williams, D.E. (1987). Defect chemistry and gas sensing properties of semiconducting ceramic oxides. *Cryst. Lattice Defects Amorphous Mater.* 14 (3–4): 341–349.

11 Williams, D.E., Tofield, B.C., and McGeehin, P. (1985). Sensors for use in gases or gas mixtures, GB patent 2149121A. United Kingdom Atomic Energy Authority, UK. p. 11.

12 Williams, D.E., Tofield, B.C., and McGeehin, P. (1985). Sensors for use in gases or gas mixtures, GB patent 2149122A. 1985, United Kingdom Atomic Energy Authority, UK. p. 18.

13 Williams, D.E., Tofield, B.C., and McGeehin, P. (1985). Sensors for use in gases and gaseous mixtures, GB patent 2149123A. United Kingdom Atomic Energy Authority, UK. p. 12.

14 Williams, D.E., Tofield, B.C., and McGeehin, P. (1985). Sensors for use in gases or gas mixtures, GB patent 2149120A. United Kingdom Atomic Energy Authority, UK. p. 10.

15 Moseley, P.T., Tofield, B.C., and Williams, D.E. (1986). Metal oxide gas sensors, GB patent 2166244A. United Kingdom Atomic Energy Authority, UK. p. 11.
16 Moseley, P.T., Tofield, B.C., and Williams, D.E. (1986). Gas sensors, GB patent 2165054A. United Kingdom Atomic Energy Authority, UK. p. 9.
17 Moseley, P.T., Williams, D.E., and Norris, J.O.W. (1988). Gas sensor for combustible gases in air, GB patent 2192710A. United Kingdom Atomic Energy Authority, UK. p. 4.
18 Williams, D.E., Moseley, P.T., and McGeehin, P. (1988). Gas sensor, GB patent 2198844A. United Kingdom Atomic Energy Authority, UK. p. 9.
19 Moseley, P.T., Peat, R., and Williams, D.E. (1992). Carbon dioxide sensor from semiconducting oxide material, GB patent 2249635A. United Kingdom Atomic Energy Authority, UK. p. 11.
20 McGeehin, P., Moseley, P.T., and Williams, D.E. (1995). Gas sensing apparatus and sensors therefor, patent WO9504926A1. Capteur Sensors and Analysers ltd., UK. p. 43.
21 McGeehin, P., Moseley, P.T., Williams, D.E. et al. (1995). Semiconducting oxide gas sensors and materials for them, patent WO9500836A1. Capteur Sensors and Analysers ltd., UK. p. 64.
22 Williams, D.E., Moseley, P.T., and McGeehin, P. (1995). Resistive gas sensing, especially for detection of ozone, patent WO9535495A1. Capteur Sensors and Analysers Ltd., UK. p. 44.
23 Williams, D.E., Ridley, R., and Sizeland, E. (1995). Resistive gas sensing of carbon monoxide with tin dioxide based sensor, patent WO9530143A1. Capteur Sensors and Analysers ltd., UK. p. 55.
24 McGeehin, P., Moseley, P.T., Williams, D.E. et al. (1997). Carbon dioxide sensor based on antimony barium stannate, patent WO9728441A1. Capteur Sensors and Analysers Ltd., UK.
25 Williams, D.E. (1999). Gas sensor, US patent 5858739A. Capteur Sensors & Analysers, Ltd., UK. p. 10, Cont.-in-part of U.S. Ser. No. 621,901.
26 McGeehin, P., Moseley, P.T., Williams, D.E. et al. (2000). Semiconducting oxide gas sensors, US patent 6046054A. Capteur Sensors & Analysers, Ltd., UK. p. 15, Cont.-in-part of U.S. Ser. No. 387,794.
27 Williams, D.E. (2000). Materials for solid-state gas sensors for detecting carbon monoxide in air, patent WO2000024677A1. 2000, Capteur Sensors and Analysers Limited, UK. p. 19.
28 Moseley, P.T., Williams, D.E., Norris, J.O.W. et al. (1988). Electrical-conductivity and gas sensitivity of some transition-metal tantalates. *Sens. Actuators* 14 (1): 79–91.
29 Moseley, P.T. and Williams, D.E. (1989). Gas sensors based on oxides of early transition-metals. *Polyhedron* 8 (13–14): 1615–1618.
30 Moseley, P.T., Williams, D.E., and Stoneham, A.M. (1991). Oxide semiconductors: patterns of gas response behaviour according to material type. In: *Techniques and Mechanisms in Gas Sensing* (ed. P.T. Moseley, J.O.W. Norris and D.E. Williams), 108. Bristol: Adam Hilger.

31 Williams, D.E. and Moseley, P.T. (1991). Dopant effects on the response of gas-sensitive resistors utilizing semiconducting oxides. *J. Mater. Chem.* 1 (5): 809–814.
32 Moseley, P.T., Stoneham, A.M., and Williams, D.E. (1991). Oxide semiconductors: patterns of gas response behaviour according to material type. In: *Techniques and Mechanisms in Gas Sensing* (ed. P.T. Moseley, J.O.W. Norris and D.E. Williams), 108–138. Bristol: Adam Hilger.
33 Degler, D., Wicker, S., Weimar, U. et al. (2015). Identifying the active oxygen species in SnO_2 based gas sensing materials: an operando IR spectrsocopy study. *J. Phys. Chem. C* 119 (21): 11792–11799.
34 McAleer, J.F., Mosely, P.T., Norris, J.O.W. et al. (1987). Tin oxide based gas sensors. *Mater. Chem. Phys.* 17 (6): 577–583.
35 Williams, D. (1987). Conduction and gas response of semiconductor gas sensors. In: *Solid State Gas Sensors* (ed. P.T. Moseley and B.C. Tofield), 71. Bristol: Adam Hilger.
36 McAleer, J.F., Mosely, P.T., Norris, J.O.W. et al. (1988). Tin dioxide gas sensors. 2. The role of surface additives. *J. Chem. Soc. Faraday Trans. I* 84: 441–457.
37 McAleer, J.F., Maignan, A., Mosely, P.T. et al. (1989). Tin dioxide gas sensors. 3. Solid-state electrochemical investigations of reactions taking place at the oxide surface. *J. Chem. Soc. Faraday Trans. I* 85: 783–799.
38 McAleer, J.F., Mosely, P.T., Williams, D.E. et al. (1989). The use of an oxygen-ion conducting ceramic to probe reactions taking place on the surface of a semiconducting oxide. *Sens. Actuators* 17 (1–2): 313–318.
39 Naik, A.J.T., Parkin, I.P., and Binions, R. (2014). Gas sensing studies of a n-n hetero-junction array based on WO_3 and ZnO composites. *IEEE Sens. J.* 14 (9): 3137–3147.
40 Sun, C.H. and Dutta, P.K. (2016). Selective detection of part per billion concentrations of ammonia using a p-n semiconducting oxide heterostructure. *Sens. Actuators B Chem.* 226: 156–169.
41 Williams, D.E. (1987). Conduction and gas response of semiconductor gas sensors. In: *Solid State Gas Sensors* (ed. P.T. Moseley and B.C. Tofield), 71–123. Adam Hilger.
42 Gopel, W. (1985). Chemisorption and charge transfer at ionic semiconductor surfaces – implications in designing gas sensors. *Prog. Surf. Sci.* 20 (1): 9–103.
43 Batzill, M., Katsiev, K., Burst, J.M. et al. (2005). Gas-phase-dependent properties of SnO_2 (110), (100), and (101) single-crystal surfaces: structure, composition, and electronic properties. *Phys. Rev. B* 72 (16): 20.
44 Dusastre, V. and Williams, D.E. (1999). Selectivity and composition dependence of response of Wolframite-based gas sensitive resistors $(MWO_4)_x([Sn\text{-}Ti]O_2)_{(1-x)}$, $(0 < x < 1$, M = Mn, Fe, Co, Ni, Cu, Zn). *J. Mater. Chem.* 9 (4): 965–971.
45 Henshaw, G.S., Dusastre, V., and Williams, D.E. (1996). Selectivity and composition dependence of response of gas-sensitive resistors. 3. Properties of the solid solution series $(CrNbO_4)_{(x)}(Sn_{1-y}Sb_yO_2)_{(1-x)}(0 \leq x \leq 1, y = 0, 0.01, 0.05)$. *J. Mater. Chem.* 6 (8): 1351–1354.

46 Williams, D.E. and Pratt, K.F.E. (1998). Classification of reactive sites on the surface of polycrystalline tin dioxide. *J. Chem. Soc. Faraday Trans.* 94 (23): 3493–3500.

47 Dusastre, V. and Williams, D.E. (1998). Sb(III) as a surface site for water adsorption on $Sn(Sb)O_2$, and its effect on catalytic activity and sensor behavior. *J. Phys. Chem. B* 102 (35): 6732–6737.

48 Williams, D.E., Pratt, K.F.E., Smith, P.J. et al. (2001). Gas sensors with improved resistance to humidity interference, US Patent 7010957 published as WO 01/88517 20 November 2001 and issued March 14 2006.

49 Niemeyer, D., Williams, D.E., Smith, P. et al. (2002). Experimental and computational study of the gas-sensor behaviour and surface chemistry of the solid-solution $Cr_{2-x}Ti_xO_3$ ($x \leq 0.5$). *J. Mater. Chem.* 12 (3): 667–675.

50 Aliwell, S.R., Halsall, J.F., Pratt, K.F.E. et al. (2001). Ozone sensors based on WO_3: a model for sensor drift and a measurement correction method. *Meas. Sci. Technol.* 12 (6): 684–690.

51 Williams, D.E., Aliwell, S.R., Pratt, K.F.E. et al. (2002). Modelling the response of a tungsten oxide semiconductor as a gas sensor for the measurement of ozone. *Meas. Sci. Technol.* 13 (6): 923–931.

52 Hansford, G.M., Freshwater, R.A., Bosch, R.A. et al. (2005). A low cost instrument based on a solid state sensor for balloon-borne atmospheric O_3 profile sounding. *J. Environ. Monit.* 7 (2): 158–162.

53 Chabanis, G., Parkin, I.P., and Williams, D.E. (2003). A simple equivalent circuit model to represent microstructure effects on the response of semiconducting oxide-based gas sensors. *Meas. Sci. Technol.* 14 (1): 76–86.

54 Williams, D.E., Henshaw, G.S., Bart, M. et al. (2013). Validation of low-cost ozone measurement instruments suitable for use in an air-quality monitoring network. *Meas. Sci. Technol.* 24 (6): 065803.

55 Bart, M., Williams, D.E., Ainslie, B. et al. (2014). High density ozone monitoring using gas sensitive semi-conductor sensors in the lower Fraser Valley, British Columbia. *Environ. Sci. Technol.* 48 (7): 3970–3977.

56 Williams, D.E., Salmond, J., Yung, Y.F. et al. (2009). Development of Low-Cost Ozone and Nitrogen Dioxide Measurement Instruments Suitable for Use in an Air Quality Monitoring Network. *2009 IEEE Sensors, Vols 1–3*. pp. 1053–1058.

57 Aeroqual Ltd. http://www.aeroqual.com/company (accessed 3 July 2017).

58 Miskell, G., Salmond, J., Alavi-Shoshtari, M. et al. (2016). Data verification tools for minimizing management costs of dense air-quality monitoring networks. *Environ. Sci. Technol.* 50 (2): 835–846.

59 Mote Ltd. http://airqualityltd.com/; http://mote.io (accessed 3 July 2017).

60 Williams, D.E., Henshaw, G.S., and Pratt, K.F.E. (1995). Detection of sensor poisoning using self-diagnostic gas sensors. *J. Chem. Soc.Faraday Trans.* 91 (18): 3307–3308.

61 Pratt, K.F.E. and Williams, D.E. (1997). Self diagnostic gas sensitive resistors in sour gas applications. *Sens. Actuators B Chem.* 45 (2): 147–153.

62 Mann, D.P., Pratt, K.F.E., Paraskeva, T. et al. (2007). Transition metal exchanged zeolite layers for selectivity enhancement of metal-oxide semiconductor gas sensors. *IEEE Sens. J.* 7 (3–4): 551–556.

63 Hernandez, P.T., Hailes, S.M.V., and Parkin, I.P. (2017). Hydrocarbon detection with metal oxide semiconducting gas sensors modified by overlayer or admixture of zeolites Na-A, H-Y and H-ZSM-5. *Sens. Actuators B Chem.* 242: 1281–1295.

64 Binions, R., Afonja, A., Dungey, S. et al. (2011). Discrimination effects in zeolite modified metal oxide semiconductor gas sensors. *IEEE Sens. J.* 11 (5): 1145–1151.

65 Binions, R., Davies, H., Afonja, A. et al. (2009). Zeolite-modified discriminating gas sensors. *J. Electrochem. Soc.* 156 (3): J46–J51.

66 Pugh, D.C., Hailes, S.M.V., and Parkin, I.P. (2015). A gas-sensing array produced from screen-printed, zeolite-modified chromium titanate. *Meas. Sci. Technol.* 26 (8): 9.

67 Williams, D.E. and Pratt, K.F.E. (2000). Microstructure effects on the response of gas-sensitive resistors based on semiconducting oxides. *Sens. Actuators B-Chem.* 70 (1–3): 214–221.

68 Naisbitt, S.C., Pratt, K.F.E., Williams, D.E. et al. (2006). A microstructural model of semiconducting gas sensor response: the effects of sintering temperature on the response of chromium titanate (CTO) to carbon monoxide. *Sens. Actuators B Chem.* 114 (2): 969–977.

69 Dusastre, V. and Williams, D.E. (1999). Gas-sensitive resistor properties of the solid solution series $Ti_x(Sn_{1-y}Sb_y)_{(1-x)}O_2$ ($0 < x < 1$, $y = 0$, 0.01, 0.05). *J. Mater. Chem.* 9 (2): 445–450.

70 Slater, B., Catlow, C.R.A., Gay, D.H. et al. (1999). Study of surface segregation of antimony on SnO_2 surfaces by computer simulation techniques. *J. Phys. Chem. B* 103 (48): 10644–10650.

71 Ducere, J.M., Hemeryck, A., Esteve, A. et al. (2012). A computational chemist approach to gas sensors: modeling the response of SnO_2 to CO, O_2, and H_2O gases. *J. Comput. Chem.* 33 (3): 247–258.

72 Habgood, M. and Harrison, N. (2008). An ab initio study of oxygen adsorption on tin dioxide. *Surf. Sci.* 602 (5): 1072–1079.

73 Ponce, C.A., Caravaca, M.A., and Casali, R.A. (2015). Ab initio studies of structure, electronic properties, and relative stability of SnO_2 nanoparticles as a function of stoichiometry, temperature, and oxygen partial pressure. *J. Phys. Chem. C* 119 (27): 15604–15617.

74 Goniakowski, J. and Gillan, M.J. (1996). The adsorption of H_2O on TiO_2 and SnO_2(110) studied by first-principles calculations. *Surf. Sci.* 350 (1): 145–158.

75 Manassidis, I., Goniakowski, J., Kantorovich, L.N. et al. (1995). The structure of the stoichiometric and reduced SnO_2(110) surface. *Surf. Sci.* 339 (3): 258–271.

76 Oviedo, J. and Gillan, M.J. (2000). The energetics and structure of oxygen vacancies on the SnO_2(110) surface. *Surf. Sci.* 467 (1–3): 35–48.

77 Oviedo, J. and Gillan, M.J. (2001). First-principles study of the interaction of oxygen with the SnO(2)(110) surface. *Surf. Sci.* 490 (3): 221–236.

78 Sensato, F.R., Custodio, R., Calatayud, M. et al. (2002). Periodic study on the structural and electronic properties of bulk, oxidized and reduced

SnO_2(110) surfaces and the interaction with O_2. *Surf. Sci.* 511 (1–3): 408–420.
79 Wang, X.F., Qin, H.W., Chen, Y.P. et al. (2014). Sensing mechanism of SnO_2 (110) surface to CO: density functional theory calculations. *J. Phys. Chem. C* 118 (49): 28548–28561.
80 Henshaw, G.S., Morris, L., Gellman, L.J. et al. (1996). Selectivity and compositional dependence of response of gas-sensitive resistors. 4. Properties of some rutile solid solution compounds. *J. Mater. Chem.* 6 (12): 1883–1887.
81 Williams, D.E. and Pratt, K.F.E. (1995). Theory of self-diagnostic sensor array devices using gas-sensitive resistors. *J. Chem. Soc. Faraday Trans.* 91 (13): 1961–1966.
82 Gardner, J.W. (1989). Electrical conduction in solid-state gas sensors. *Sens. Actuators* 18 (3–4): 373–387.
83 Williams, D.E. and Pratt, K.F.E. (1999). Multi-electrode gas sensors and methods of making and using them, US Patent 5,918,261. Google Patents.
84 Williams, D.E. and Pratt, K.F.E. (1996). Resolving combustible gas mixtures using gas sensitive resistors with arrays of electrodes. *J. Chem. Soc. Faraday Trans.* 92 (22): 4497–4504.
85 Afonja, A., Binions, R., Dungey, S. et al. (2010). Zeolites as transformation elements in discriminating semiconductor metal oxide sensors. In: *Eurosensors XXIV Conference* (ed. B. Jakoby and M.J. Vellekoop). pp. 103–106.
86 Mann, D.P., Paraskeva, T., Pratt, K.F.E. et al. (2005). Metal oxide semiconductor gas sensors utilizing a Cr-zeolite catalytic layer for improved selectivity. *Meas. Sci. Technol.* 16 (5): 1193–1200.
87 Williams, D.E., Henshaw, G.S., Pratt, K.F.E. et al. (1995). Reaction-diffusion effects and systematic design of gas-sensitive resistors based on semiconducting oxides. *J. Chem. Soc. Faraday Trans* 91 (23): 4299–4307.
88 Xu, C.N., Tamaki, J., Miura, N. et al. (1991). Grain-size effects on gas sensitivity of porous SnO_2-based elements. *Sens. Actuators B Chem.* 3 (2): 147–155.
89 Dusastre, V. (1998). Semiconducting oxide gas-sensitive resistors. PhD thesis. University College London.
90 Tamaki, J., Zhang, Z., Fujimori, K. et al. (1994). Grain-size effects in tungsten oxide-based sensor for nitrogen oxides. *J. Electrochem. Soc.* 141 (8): 2207–2210.
91 Yamazoe, N., Tamaki, J., and Miura, N. (1996). Role of hetero-junctions in oxide semiconductor gas sensors. *Mater. Sci. Eng. B Solid State Mater. Adv. Technol.* 41 (1): 178–181.
92 Scott, R.W.J., Yang, S.M., Chabanis, G. et al. (2001). Tin dioxide opals and inverted opals: near-ideal microstructures for gas sensors. *Adv. Mater.* 13 (19): 1468–1472.
93 Kirkpatrick, S. (1973). Percolation and conduction. *Rev. Mod. Phys.* 45 (4): 574–588.
94 Chabanis, G., Parkin, I.P., and Williams, D.E. (2001). Microspheres of the gas sensor material $Cr_{2-x}Ti_xO_3$ prepared by the sol-emulsion-gel route. *J. Mater. Chem.* 11 (6): 1651–1655.

95 Shaw, G.A., Parkin, I.P., and Williams, D.E. (2005). Gas sensing properties of thin film ($\leq$3 μm) $Cr_{2-x}Ti_xO_3$ (CTO) prepared by atmospheric pressure chemical vapour deposition (APCVD), compared with that prepared by thick film screen-printing. *Sens. Actuators B Chem.* 104 (1): 151–162.

96 Henshaw, G.S., Dawson, D.H., and Williams, D.E. (1995). Selectivity and composition dependence of response of gas-sensitive resistors. 2. Hydrogen-sulfide response of $Cr_{2-x}Ti_xO_{3+y}$. *J. Mater. Chem.* 5 (11): 1791–1800.

97 Blacklocks, A.N., Atkinson, A., Packer, R.J. et al. (2006). An XAS study of the defect structure of Ti-doped α-Cr_2O_3. *Solid State Ionics* 177 (33): 2939–2944.

98 Kaspar, T.C., Sushko, P.V., Bowden, M.E. et al. (2016). Defect compensation by Cr vacancies and oxygen interstitials in Ti^{4+}-doped Cr_2O_3 epitaxial thin films. *Phys. Rev. B* 94 (15): 155409.

99 Williams, D.E., Pratt, K.F.E., Smith, P. et al. (2001). Gas sensors with improved resistance to humidity interference, patent WO2001088517A2. Capteur Sensors and Analysers Limited, UK. p. 11.

100 Henshaw, G.S., Ridley, R., and Williams, D.E. (1996). Room-temperature response of platinised tin dioxide gas-sensitive resistors. *J. Chem. Soc. Faraday Trans.* 92 (18): 3411–3417.

101 Morris, L. and Williams, D.E. (2001). Pt(11) as an electronically active surface site in the room temperature CO response of Pt modified gas sensitive resistors. *J. Phys. Chem. B* 105 (30): 7272–7279.

102 Morris, L., Williams, D.E., Kaltsoyannis, N. et al. (2001). Surface grafting as a route to modifying the gas-sensitive resistor properties of semiconducting oxides: studies of Ru-grafted SnO_2. *Phys. Chem. Chem. Phys.* 3 (1): 132–145.

103 Slater, B., Catlow, C.R.A., Williams, D.E. et al. (2000). Dissociation of O_2 on the reduced SnO_2 (110) surface. *Chem. Commun.* 14: 1235–1236.

104 Slater, B., Catlow, C.R.A., Williams, D.E. et al. (2000). Competitive Adsorption of O_2 and H_2O at the Neutral and Defective SnO_2 (110) surface. *MRS Proceedings*. Cambridge University Press.

105 Williams, D. and McGeehin, P. (1984). Solid-state gas sensors and monitors. In: *Electrochemistry*, vol. 9 (ed. D. Pletcher), 246–290. Royal Society of Chemistry.

106 Moseley, P.T. and Williams, D.E. (1990). A selective ammonia sensor. *Sens. Actuators B Chem.* 1 (1–6): 113–115.

107 Dawson, D.H., Henshaw, G.S., and Williams, D.E. (1995). Description and characterization of a hydrogen-sulfide gas sensor-based on $Cr_{2-y}Ti_yO_{3+x}$. *Sens. Actuators B Chem.* 26 (1–3): 76–80.

108 Dawson, D.H. and Williams, D.E. (1996). Gas-sensitive resistors: surface interaction of chlorine with semiconducting oxides. *J. Mater. Chem.* 6 (3): 409–414.

109 Williams, D.E., Jones, C.P., and El Murr, N. 1989. Microelectrode assembly, patent application GB2217461A. United Kingdom Atomic Energy Authority, UK. p. 13.

110 Craston, D.H., Jones, C.P., Williams, D.E. et al. (1991). Microband electrodes fabricated by screen printing processes – applications in electroanalysis. *Talanta* 38 (1): 17–26.

111 Johnston, J.S. (1979). Solid state sensor element, US Patent 4 307 373. 1979, Rosemount Engineering Co. Ltd., UK. p. 15.

112 Williams, D.E., Pratt, K.F.E., Jones, R.L. et al. (2000). Semiconductor gas sensing, patent WO2000014531A1. Capteur Sensors and Analysers Limited, UK. p. 27.

113 Utembe, S.R., Hansford, G.M., Sanderson, M.G. et al. (2006). An ozone monitoring instrument based on the tungsten trioxide (WO_3) semiconductor. *Sens. Actuators B Chem.* 114 (1): 507–512.

114 Thompson, T.M. and Selin, N.E. (2012). Influence of air quality model resolution on uncertainty associated with health impacts. *Atmos. Chem. Phys.* 12 (20): 9753–9762.

115 Williams, D.E., Henshaw, G.S., Wells, D.B. et al. (2009). Development of Low-Cost Ozone and Nitrogen Dioxide Measurement Instruments Suitable For Use in an Air Quality Monitoring Network. 35 Years of Chemical Sensors – an Honorary Symposium for Professor Jiri Janata's 70th Birthday Celebration (ed. Li, J., Vanýsek, P., Brown, R. et al.). pp. 251–254.

116 Miskell, G., Salmond, J., and Williams, D.E. (2017). Low-cost sensors and crowd-sourced data: observations of siting impacts on a network of air-quality instruments. *Sci. Total Environ.* 575: 1119–1129.

117 Weissert, L.F., Salmond, J.A., Miskell, G. et al. (2017). Use of a dense monitoring network of low-cost instruments to observe local changes in the diurnal ozone cycles as marine air passes over a geographically isolated urban Centre. *Sci. Total Environ.* 575: 67–78.

118 Salmond, J.A., Williams, D.E., Laing, G. et al. (2013). The influence of vegetation on the horizontal and vertical distribution of pollutants in a street canyon. *Sci. Total Environ.* 443: 287–298.

119 Alavi-Shoshtari, M., Williams, D.E., Salmond, J.A. et al. (2013). Detection of malfunctions in sensor networks. *Environmetrics* 24 (4): 227–236.

120 Tzortzakis, N., Singleton, I., and Barnes, J. (2007). Deployment of low-level ozone-enrichment for the preservation of chilled fresh produce. *Postharvest Biol. Technol.* 43 (2): 261–270.

121 Kida, T., Nishiyama, A., Yuasa, M. et al. (2009). Highly sensitive NO_2 sensors using lamellar-structured WO_3 particles prepared by an acidification method. *Sens. Actuators B Chem.* 135 (2): 568–574.

122 Snyder, E.G., Watkins, T.H., Solomon, P.A. et al. (2013). The changing paradigm of air pollution monitoring. *Environ. Sci. Technol.* 47 (20): 11369–11377.

123 Mead, M.I., Popoola, O.A.M., Stewart, G.B. et al. (2013). The use of electrochemical sensors for monitoring urban air quality in low-cost, high-density networks. *Atmos. Environ.* 70: 186–203.

124 Environmental Protection Agency (2017). Quality Assurance Handbook for Air Pollution Measurement Systems Volume II Ambient Air Quality Monitoring Program. United States Environmental Protection Agency, report

EPA-454/B-17-001. Available at: https://www3.epa.gov/ttn/amtic/qalist.html (accessed 13 April 2018).
125 Williams, R., Kilaru, V., Snyder, E. et al. (2014). Air Sensor Guidebook. United States Environmental Protection Agency, Washington, DC, EPA/600/R-14/159 (NTIS PB2015-100610). Available at: https://cfpub.epa.gov/si/si_public_record_report.cfm?dirEntryId=277996 (accessed 23 April 2018).
126 Hasenfratz, D., Saukh, O., and Thiele, L. (2012). On-the-fly calibration of low-cost gas sensors. Lecture Notes in Computer Science (including subseries Lecture Notes in Artificial Intelligence and Lecture Notes in Bioinformatics). pp. 228–244.
127 Liu, J., Chen, D.S., and Shen, J.F. (2010). Development of self-validating soft sensors using fast moving window partial least squares. *Ind. Eng. Chem. Res.* 49 (22): 11530–11546.
128 Talampas, M.C.R. and Low, K.S. (2013). Maximum likelihood estimation of ground truth for air quality monitoring using vehicular sensor networks. *IEEE Region 10 Annual International Conference, Proceedings/TENCON*. IEEE.
129 Tsujita, W., Yoshino, A., Ishida, H. et al. (2005). Gas sensor network for air-pollution monitoring. *Sens. Actuators B Chem.* 110 (2): 304–311.
130 De Vito, S., Piga, M., Martinotto, L. et al. (2009). CO, NO_2 and NO_x urban pollution monitoring with on-field calibrated electronic nose by automatic Bayesian regularization. *Sens. Actuators B Chem.* 143 (1): 182–191.
131 Harkat, M.F., Mourot, G., and Ragot, J. (2006). An improved PCA scheme for sensor FDI: application to an air quality monitoring network. *J. Process Control* 16 (6): 625–634.
132 Joseph, J., Sharif, H.O., Sunil, T. et al. (2013). Application of validation data for assessing spatial interpolation methods for 8-h ozone or other sparsely monitored constituents. *Environ. Pollut.* 178: 411–418.
133 Zhang, L., Tian, F.C., Peng, X.W. et al. (2014). A rapid discreteness correction scheme for reproducibility enhancement among a batch of MOS gas sensors. *Sens. Actuators A* 205: 170–176.
134 Hasenfratz, D., Saukh, O., and Thiele, L. (2013). Model-driven accuracy bounds for noisy sensor readings. *2013 9th IEEE International Conference on Distributed Computing in Sensor Systems (IEEE Dcoss 2013).* pp. 165–174.
135 Hoek, G., Beelen, R., de Hoogh, K. et al. (2008). A review of land-use regression models to assess spatial variation of outdoor air pollution. *Atmos. Environ.* 42 (33): 7561–7578.
136 Balzano, L. and Nowak, R. (2007). Blind calibration of sensor networks. Proceedings of the Sixth International Symposium on Information Processing in Sensor Networks. 2007, New York: Assoc Computing Machinery. pp. 79–88.
137 Morris, L., Caruana, D.J., and Williams, D.E. (2002). Simple system for part-per-billion-level volatile organic compound analysis in groundwater and urban air. *Meas. Sci. Technol.* 13 (4): 603–612.

Index

Electrochemical Engineering: The Path from Discovery to Product,
First Edition. Edited by Richard C. Alkire, Philip N. Bartlett, and Marc T. Koper.